Physics of Tsunamis

Boris Levin · Mikhail Nosov

Physics of Tsunamis

Dr. Boris Levin
Institute of Marine Geology and Geophysics
Yuzhno-Sakhalinsk
Russia
levinbw@mail.ru

Dr. Mikhail Nosov
Faculty of Physics
Moscow State University
Russia
nosov@phys.msu.ru

Translated by Gil Pontecorvo

ISBN 978-1-4020-8855-1   e-ISBN 978-1-4020-8856-8

Library of Congress Control Number: 2008933596

All Rights Reserved
© 2009 Springer Science + Business Media B.V.
No part of this work may be reproduced, stored in a retrieval system, or transmitted in any form or by any means, electronic, mechanical, photocopying, microfilming, recording or otherwise, without written permission from the Publisher, with the exception of any material supplied specifically for the purpose of being entered and executed on a computer system, for exclusive use by the purchaser of the work.

Printed on acid-free paper

9 8 7 6 5 4 3 2 1

springer.com

Preface

Till the very end of the twentieth century tsunami waves (or 'waves in a harbour', translated from Japanese) were considered an extremely rare and exotic natural phenomenon, originating in the ocean and unexpectedly falling upon the seaside as gigantic waves. The 26th of December 2004, when tsunami waves wiped out, in a single day, more than 250,000 human lives, mourned in many countries, turned out to be a tragic date for all mankind.

The authors of this book, who have studied tsunami waves for many years, intended it to be a systematic exposition of modern ideas concerning

- The mechanisms of tsunami wave generation
- The peculiarities of tsunami wave propagation in the open ocean and of how waves run-up beaches
- Methods for tsunami wave registration and the operation of a tsunami warning system
- The mechanisms of other catastrophic processes in the ocean related to the seismic activity of our planet

The authors considered their main goal to be the creation of book presenting modern knowledge of tsunami waves and of other catastrophes in the ocean to scientific researchers and specialists in geophysics, oceanography, seismology, hydroacoustics, geology, geomorphology, civil and seaside engineering, postgraduate students and students of relevant professions. At present, it has become clear that the demand for the information and scientific results presented in the book may be significantly broader and that they may be of interest to a large part of the population. Politicians, administrators, mass media, insurance companies, owners of seaside resorts and hotels, the civil fleet and the navy, oil-extracting companies, security services, space agencies, publishing houses, public education systems, etc., is a short list of possible users interested today in assimilating and spreading knowledge of the nature and manifestations of tsunami waves.

Waves that regularly devastate the coasts of oceanic islands and are called tsunami in Japan have been known for several centuries. The European civilization first encountered such catastrophic waves in 1755, when an exceptionally strong earthquake took place in the Atlantic Ocean near the coast of Portugal and gave rise to a tsunami wave that immediately killed over 50,000 people in the blooming city of Lisbon, which was about a quarter of the city's population. In the USSR, the Kamchatka tsunami of 1952 (2,336 victims) resulted in creation of a State tsunami warning system. During the past 10 years (not counting the tragedy caused by the Indonesian tsunami in 2004) tsunami waves in the Pacific Ocean took the lives of more than 10,000 people.

According to UNESCO information, by the year 2010 residents of the coasts of oceans and seas will represent about 70% of the total population of our planet. One should add persons visiting numerous seaside resorts, those who like to celebrate the New Year on exotic oceanic islands and, also, individuals seeking maritime adventures. All these people may happen to be within the reach of one of the oceanic catastrophes, of which tsunami waves are the most dangerous.

Today, many states of the Pacific region, Russia, Japan, the USA, and Chile, operate tsunami warning systems. The Russian system includes two tsunami Centers, situated in Yuzhno-Sakhalinsk and Petropavlovsk-Kamchatskii that are managed by the respective Board of the State Committee (Goskomitet) for hydrometeorology of Russia. The tsunami centres receive online information from seismic stations that carry out round-the-clock observation within the framework of the Geophysical Service of the Russian Academy of Sciences (RAS). In former times there were six such specialized seismic stations functioning along the Far East coast of the USSR. At present only three stations (Yuzhno-Sakhalinsk, Petropavlovsk-Kamchatskii, Severo-Kuril'sk) are in operation, and they all long need to be modernized and re-equipped.

The International Tsunami Information Center, the Pacific Tsunami Warning Center, and the Alaska Tsunami Warning Center function successfully within the framework of the USA National Oceanic and Atmospheric Administration with participation of the UNESCO Intergovernmental Oceanic Commission (IOC/UNESCO). In Japan, the duties of tsunami warning are performed by several hundred seismic and sea-level stations united in a common information system managed by national agencies (JMA, JAMSTEC).

All national Tsunami warning services exchange online information via the Internet, electronic mail and the specialized Tsunami Board Bulletin. Scientific studies of tsunami waves are coordinated by the International Tsunami Commission within the International Union for Geodesy and Geophysics (IUGG). During the period between 1977 and 1979 this commission was led by Academician S. L. Soloviev, who founded the Soviet Tsunami School. Another Russian scientist, Dr. V. K. Gusyakov (Novosibirsk) occupied this position from 1995 up to 2003. In 2003, Professor K. Satake (Japan) was elected Chairman of the Commission. The Tsunami Commission and the International Group of the UNESCO Intergovernmental Oceanographic Commission (IOC/UNESCO) organize regular international scientific and practical conferences, devoted to the problem of tsunami waves, in situ inspections of coasts that were victims of tsunami waves; they publish reviews, information bulletins, national reports, general-education literature; and support the creation of databases.

In 1996, The European Geophysical Society (EGS) established the Sergei Soloviev medal to mark the recognition of S. L. Soloviev's scientific achievements. This medal is presented to scientists who have made essential contributions to the investigation of natural catastrophes.

The Russian school of tsunami researchers organized and led for many years by Academician S. L. Soloviev is still considered a leading team in this scientific sector. A large contribution to the development of tsunami studies has been made by RAS Corresponding members S. S. Lappo and L. N. Rykunov; the Doctors of Sciences, who grew up in the Russian Tsunami School, A. V. Nekrasov, A. A. Dorfman (Leningrad), B. W. Levin, M. A. Nosov, A. B. Rabinovich, E. A. Kulikov, L. I. Lobkovsky (Moscow), E. N. Pelinovsky, V. E. Friedman, T. K. Talipova (Nizhny Novgorod), V. K. Gusyakov, L. B. Chubarov, An. G. Marchuk (Novosibirsk), P. D. Kovalev, V. V. Ivanov (Yuzhno-Sakhalinsk), and their pupils have done much for successful development of the science of tsunami waves. Specialized tsunami laboratories and several scientific groups work in the M. V. Lomonosov Moscow State University (MSU) and in various RAS institutes: the Institute of Oceanology (Moscow), the Institute of Applied Physics (Nizhny Novgorod), the Institute of Computational Mathematics and Mathematical Geophysics of the RAS Siberian Branch (RAS SB) (Novosibirsk), the Institute of Marine Geology and Geophysics of the RAS Far-East Branch (RAS FEB) (Yuzhno-Sakhalinsk), the Institute of Vulcanology and Seismology of RAS FEB (Petropavlovsk-Kamchatskii).

Many Russian specialists in tsunami waves, including the authors and the editor of this book, have acquired significant teaching experience not only in the universities of Russia (MSU, MSGU, NSU, NNSU, NSTU, SakhSU), but also in universities of the USA, France, Guadelupa, Australia, and Columbia. Recently, owing to the development of new computer technologies and software, original models have appeared of rare phenomena in the ocean, that were hitherto beyond the reach of scientific analysis. The experience of elaborating original ideas accumulated by Russian scientists in the research of seaquakes, killer waves, temperature anomalies above underwater earthquakes, the formation of cavitation zones, plumes and surges of water require detailed exposition and physical analysis. The experience of collaboration with foreign colleagues, regular participation in international meetings, as well as experience in organizing international conferences in Russia (the Tsunami conferences of 1996, 2000, 2002) have revealed an increased demand in tsunami wave specialists and in systematization of the knowledge accumulated in this field.

At present, no proof is needed of the fact that the influence of tsunami waves on the coasts of continents and islands is of a global nature. This catastrophic phenomenon cares nothing about the borders of states and of the nationalities of individuals, who happen to be in the zone within the reach of the catastrophe. In the nearest future the politicians of civilized countries will be compelled to start resolving the issue of creating a global tsunami warning system, something similar to the World Meteorological Organization. This task will require scientists from all countries to make enormous efforts for systematization of the knowledge on tsunami waves, for the preparation of national experts, specialists and teachers in the problem of tsunami waves, for developing new methods and means of monitoring, for publishing series of textbooks, scientific and general-education literature.

The authors hope that this book will contribute to the formation of a general collection of knowledge on tsunami waves. The necessity of such a book has ultimately become evident.

Many of our colleagues have taken part in completing the book and preparing it for publication. Section 6.1 was in part prepared by the Director of the SakhUSMS Tsunami Center T. N. Ivelskaya (Yuzhno-Sakhalinsk), Sect. 6.2 was written by Dr. T. K. Pinegina (Petropavlovsk-Kamchatskii), a well-known specialist in palaeotsunami. The illustrations, used in the book and based on computer graphics, were prepared by the leading scientific researcher of the RAS Institute of Oceanology Dr. E. V. Sasorova (Moscow). The image of the word 'tsunami' in the form of Japanese hieroglyphs was prepared for the book by Dr. H. Matsumoto (Japan, Tokyo). Certain material, put at our disposal by Dr. E. A. Kulikov (Moscow), Dr. V. K. Gusyakov (Novosibirsk), Dr. V. V. Titov (Seattle, USA) and other colleagues of ours, has been included in the book. The authors express their sincere gratitude to all of them.

We are grateful to our teachers S. L. Soloviev and L. N. Rykunov for the good school, and we revere their memory. We are grateful to our pupils and colleagues, whose friendly participation and help promoted the appearance of this book. We wish to express particular gratitude to Professor E. N. Pelinovsky, referee of the Russian issue of this book. The support of the Russian Foundation for Basic Research and of the Russian Academy of Sciences was an enormous stimulus for the preparation of this book. The authors are especially grateful to G. Pontecorvo, who translated the original Russian text into English and to V. E. Rokolyan, who prepared the text of the book for typesetting.

B. W. Levin
M. A. Nosov

Contents

1 General Information on Tsunami Waves, Seaquakes and Other Catastrophic Phenomena in the Ocean 1
  1.1 Tsunami: Definition of Concepts 2
  1.2 Manifestations of Tsunami Waves on Coasts 5
  1.3 Tsunami Magnitude and Intensity 10
  1.4 Tsunami Warning Service: Principles and Methods 15
  1.5 Databases and Tsunami Statistics 17
  1.6 Seaquakes: General Ideas 20
  1.7 Hydroacoustic Signals in the Case of Underwater Earthquakes 22
  1.8 Killer Waves in the Ocean 24
  References 27

2 Physical Processes at the Source of a Tsunami of Seismotectonic Origin 31
  2.1 Seismotectonic Source of a Tsunami: The Main Parameters and Secondary Effects 32
    2.1.1 The Main Parameters 32
    2.1.2 Secondary Effects 39
    2.1.3 Calculation of Deformations of the Ocean Bottom 44
  2.2 General Solution of the Problem of Excitation of Gravitational Waves in a Layer of Incompressible Liquid by Deformations of the Basin Bottom 49
    2.2.1 Cartesian Coordinates 49
    2.2.2 Cylindrical Coordinates 53
  2.3 Plane Problems of Tsunami Excitation by Deformations of the Basin Bottom 56
    2.3.1 Construction of the General Solution 57
    2.3.2 Piston and Membrane Displacements 61
    2.3.3 Running and Piston-like Displacements 71
    2.3.4 The Oscillating Bottom 77

		2.4	Generation of Tsunami Waves and Peculiarities of the Motion of Ocean Bottom at the Source 82
		References 94

3 Role of the Compressibility of Water and of Non-linear Effects in the Formation of Tsunami Waves 99
    3.1 Excitation of Tsunami Waves with Account of the Compressibility of Water 100
        3.1.1 Preliminary Estimates 100
        3.1.2 General Solution of the Problem of Small Deformations of the Ocean Bottom Exciting Waves in a Liquid 103
        3.1.3 Piston and Membrane Displacements 107
        3.1.4 The Running Displacement 113
        3.1.5 Peculiarities of Wave Excitation in a Basin of Variable Depth 116
        3.1.6 Elastic Oscillations of the Water Column at the Source of the Tokachi-Oki Tsunami, 2003 124
    3.2 Non-linear Mechanism of Tsunami Generation 132
        3.2.1 Base Mathematical Model 132
        3.2.2 Non-linear Mechanism of Tsunami Generation by Bottom Oscillations in an Incompressible Ocean 136
        3.2.3 Non-linear Tsunami Generation Mechanism with Account of the Compressibility of Water 144
    References 150

4 The Physics of Tsunami Formation by Sources of Nonseismic Origin 153
    4.1 Tsunami Generation by Landslides 154
    4.2 Tsunami Excitation Related to Volcanic Eruptions 165
    4.3 Meteotsunamis 171
    4.4 Cosmogenic Tsunamis 183
    References 193

5 Propagation of a Tsunami in the Ocean and Its Interaction with the Coast 197
    5.1 Traditional Ideas Concerning the Problem of Tsunami Propagation 198
    5.2 Numerical Models of Tsunami Propagation 213
    5.3 Tsunami Run-up on the Coast 223
    References 229

6 Methods of Tsunami Wave Registration 233
    6.1 Coastal and Deep-water Measurements of Sea Level 234
    6.2 Geomorphological Consequences of Tsunami: Deposits of Paleotsunamis 241
    6.3 Tsunami Detection in the Open Ocean by Satellite Altimetry 247
    References 254

7	Seaquakes: Analysis of Phenomena and Modelling 257
	7.1 Manifestations of Seaquakes: Descriptions by Witnesses and Instrumental Observations 259
	7.1.1 Historical Evidence 259
	7.1.2 Analysis of Historical Testimonies and the Physical Mechanisms of Vertical Exchange 274
	7.1.3 Instrumental Observations of Variations of the Ocean's Temperature Field After an Earthquake 278
	7.2 Estimation of the Possibility of Stable Stratification Disruption in the Ocean Due to an Underwater Earthquake 283
	7.3 Parametric Generation of Surface Waves in the Case of an Underwater Earthquake 293
	7.4 Experimental Study of Wave Structures and of Stable Stratification Transformation in a Liquid in the Case of Bottom Oscillations 297
	References 305

Colour Plate Section 309

Index 325

Chapter 1
General Information on Tsunami Waves, Seaquakes and Other Catastrophic Phenomena in the Ocean

Abstract Fundamental information on the physics and geography of tsunami waves is presented. Examples are given of known historical events, illustrating the character of tsunami manifestation on coasts. Quantitative characteristics are introduced that describe tsunami strength: magnitude and intensity. Physical principles of the operation of tsunami warning systems are described. Information is provided on tsunami catalogues and electronic databases. The seaquake phenomenon is defined, and a synthesized description is given. Information is presented on the main hydroacoustic effects, related to underwater earthquakes: the T-phase, low-frequency elastic oscillations, and cavitation. Basic information on killer waves is given.

Keywords Tsunami · seaquake · surface gravitational waves · long waves · run-up · sudden inundation · impact of waves · erosion · damage · fires · environment pollution · epidemics · human casualties · local tsunami · regional tsunami · teletsunami · tsunami catalogue · historical tsunami database · tsunami magnitude · tsunami intensity · tsunami warning · hydroacoustic signals · T-phase · cavitation · freak waves

Catastrophic oceanic waves, termed 'tsunami' back in the 1960s, were considered a mysterious and inexplicable phenomenon of the life of the ocean. The sudden onslaught on the coast by a rabid giant wave would take the lives of tens of thousands of people and leave memories engraved for a long time on the minds of those who remained alive. Scientists of many countries have united their efforts to understand the secret of this awe-inspiring phenomenon and to bring nearer resolution of the problem of tsunami waves. At present, scientists have at their disposal the information on 1,500 events in oceans and seas that have given rise to tsunami waves.

The Pacific is considered the most tsunami-dangerous region, in which approximately 1,300 events are known [Soloviev et al. (1974), (1975), (1986)]. About 300 tsunami events are known to have taken place in the Mediterranean Sea. There exists information on tsunamis in the Atlantic Ocean and the Caribbean sea, in the Black and Caspian Seas [Nikonov(1997), Dotsenko et al. (2000), Lander et al. (2002)]. Insignificant tsunamis also occurred on lake Baikal [Soloviev, Ferchev (1961)]. Europe was exposed to the action of the catastrophic tsunami of 1755, during

Fig. 1.1 The 1775 Lisbon earthquake and tsunami. Old engraving by unknown author

which the city of Lisbon was destroyed. This event was reflected in an old engraving (Fig. 1.1). At present, researchers are paying particular attention to the Indian Ocean, although in the past, also, its coasts were repeatedly attacked by catastrophic tsunami waves.

The seaquake phenomenon caused by seismic oscillations of the sea-floor is only known to specialists and to experienced seafarers. Even the edition of the Grand Soviet Encyclopedia had no place for this term, although the amount of registered natural events of this type already exceeded 250. Other transitory, but violent, phenomena in the ocean (killer waves, temperature anomalies and acoustic effects) have only recently attracted the interest of scientists owing to the rapid development of methods for remote observation, the improvement of methods of data handling and the accessibility of electronic databases and catalogues. Investigation of the entire complex of mentioned phenomena in the Ocean sheds light on the interaction mechanisms of various media in the communicating and interpenetrative lithosphere–hydrosphere–atmosphere system.

1.1 Tsunami: Definition of Concepts

The word tsunami originates from a combination of two Japanese hierogliphs (Fig. 1.2), translated together as a 'wave in the harbour'. This term has already been conventionally adopted in scientific literature, although in mass media one may still encounter terms that prevailed some time ago, such as 'high-tide wave', and 'seismic sea wave' and 'seaquake'. Sometimes the antique European terms 'zeebeben' and 'maremoto' are also used.

Usually, tsunami waves are understood to be surface gravitational waves exhibiting periods within the range of $T \sim 10^2$–10^4 s. Tsunamis pertain to long waves;

1.1 Tsunami: Definition of Concepts

Fig. 1.2 Japanese hieroglifs, pronounced as 'tsu-nami' and literally translated as a 'wave in the harbour'

therefore not only the subsurface layer, but also the entire thickness of water becomes involved in the motion. Here, the term 'surface' signifies that the presence of a free surface is a necessary condition for this kind of waves to exist.

The formation of tsunamis is primarily considered to be related to seismic motions of the sea-floor, slides and collapses (underwater, also) and underwater volcanic eruptions. Waves exhibiting similar characteristics may be due to sharp changes in the atmospheric pressure (meteotsunami) and to powerful underwater explosions. Recently, the issue has been actively discussed of tsunami originating as the result of falling meteorites. One must bear in mind the possibility of combinations of various causes. Thus, for example, underwater slides, provocated by earthquakes, may provide an additional contribution to the energy of the tsunami waves, formed by displacements of the sea-floor. We stress that the main cause of destructive tsunami consists in sharp vertical displacements of parts of the sea-floor due to strong underwater earthquakes. Considering all the causes together, it may be asserted that any coast of a large water reservoir is potentially dangerous from the point of view of tsunamis.

Modern ideas of the sources of tsunami waves are not unambiguous. Usually, the source of tsunami waves is characterized by its horizontal dimension $L \sim 100$ km, which significantly exceeds the typical depth of the World Ocean, $H \sim 4$ km. A certain quite rapid transient process results in gravitational waves originating at the source with a wavelength $\lambda \sim L$. From the point of view of hydrodynamics these waves are long ($\lambda \gg H$). The propagation velocity of long waves in a reservoir of depth H is determined by the formula $c = \sqrt{gH}$, where g is the free-fall acceleration of gravity. In the case of a depth $H \sim 4$ km the tsunami wave propagates with a velocity of the order of magnitude of 200 m/s, or about 720 km/h, which is comparable to the velocity of a modern jet aircraft. From the tsunami wavelength and its propagation velocity one can readily estimate the tsunami wave period $T = \lambda/c \approx 500$ s (actually, it varies within the limits of 10^2–10^4 s). The tsunami wave amplitude in the open ocean, even in the case of catastrophic events, is usually limited to tens of centimeters and, most likely, rarely exceeds 1 m. Nevertheless, the displacement amplitude of the water surface at the tsunami source may amount to 10 m and more. But in this case, also, it is essentially inferior to the depth of the ocean.

The small amplitude together with the large period renders the tsunami wave in open ocean practically imperceptible for an observer on board a ship. The catastrophic tsunami wave that took 28,000 lives in Japan (June 15, 1896), destroyed the port of Sanriku and all the settlements along the 275 km coastline was known not to have even been noticed by fishermen, who were only 40 km from the coast.

Dependence of the tsunami wave propagation velocity on the depth renders these waves sensitive to the shape of the sea-floor. Effects peculiar to tsunamis include the capture of wave energy both by underwater ridges and by the shelf, focusing and defocusing exhibited when waves propagate above underwater elevations and depressions. Irregularities of the sea-floor lead to the scattering of tsunami waves.

Actually, the propagation velocity of gravitational waves does not depend only on the depth, but on the wavelength, also. The formula presented above for the velocity of long waves is the limit case (for $\lambda \gg H$) of the more general expression $c = \sqrt{g \tanh(kH)/k}$, where $k = 2\pi/\lambda$. Wave dispersion results in transformation of the initial perturbation into a wave packet, with the most rapid long waves leading. Note that this effect is manifested in the case of tsunami wave propagation over quite extended routes (1,000 km or more). Dispersion, resonance properties of the coastal relief, phenomena such as reverberation (i.e. when the wave perturbation reaches a certain coastal site via different routes) and the peculiarities of wave formation at the source, all these, as a rule, result in a tsunami being manifested not as a solitary wave, but as a series of waves with a period amounting to tens of minutes. In this case, the first wave is often not the strongest. The absence of knowledge of precisely this property of tsunami waves often leads to human casualties, which could have been avoided.

The tsunami wave amplitude increases as it approaches the coast—which to a great extent is what determines the danger of these waves, is also related to the relief of the sea-floor. A decrease in the water depth leads to a decrease in the wave propagation velocity and, consequently, to compression of the wave packet in space and an increase of its amplitude. In the case of catastrophic tsunamis the run-up height reaches 10–30 m, while the wave is capable of inland inundation (runin) of 3–5 km from the coastline. A scheme of the tsunami onshore run-up, explaining the main parameters of this process, is shown in Fig. 1.3. Note that the maximal wave height can be achieved at the shoreline, at the inundation boundary or at any point in between them.

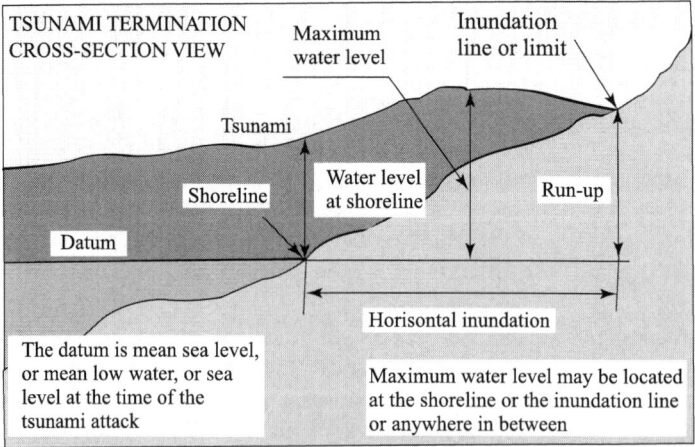

Fig. 1.3 Scheme of tsunami onshore run-up. Adapted from [UNESCO-IOC. Tsunami Glossary (2006)]

The danger carried by tsunami waves is primarily related to the following three factors: the sudden inundation of part of the land, the impact of waves upon buildings and erosion. Strong flows of water, reaching velocities of tens of meters per second, are capable of breaking up houses and of displacing them, washing out substructures of buildings, destroying bridges and buildings in ports. The flows of water often carry pieces broken off buildings and other structures, trees, small and large vessels, which leaves people, picked up by the fast-moving water, no chance of survival. The damage caused by tsunamis may, also, be due to fires, pollution of the environment and epidemics resulting from devastation of the coastal infrastructure.

Depending on the scale of the area, in which the destructive force of tsunamis is manifested, one conventionally distinguishes local, regional and remote (teletsunami) events. The latter are sometimes termed transoceanic tsunamis. Local tsunamis include events, the destructive effect of which is concentrated within distances not exceeding 100 km from the source. If destruction occurs at distances up to 1,000 km from the source, then such an event is classified as regional; when above 1,000 km, as a teletsunami. Most catastrophic events pertain precisely to local or regional tsunamis. At least 18 such events were recorded in the Pacific during the period between 1975 and 1998. The occurrence of transoceanic tsunamis is much less frequent, but they are, naturally, much more dangerous. After having caused significant destruction in the immediate vicinity of the source, these waves are capable of travelling many thousands of kilometers from the source and to continue carrying with them death and devastation. In the past 200 years at least 17 such events took place in the Pacific Ocean.

1.2 Manifestations of Tsunami Waves on Coasts

There exist numerous descriptions of the effect of tsunamis on a coast that are due to eyewitnesses or scientists investigating the consequences of these events. Detailed information on tsunami manifestations can be found in tsunami catalogues and historical databases, e.g. [Soloviev et al. (1974), (1975), (1997)], http://tsun.sscc.ru/, http://www.ngdc.noaa.gov/.

We shall present brief descriptions of some of the outstanding events.

The 1868 tsunami near the city of Arica (Chile) was caused by an underwater earthquake of magnitude $M = 8.8$. In the evening, after it became dark, an enormous 'wall' of phosphorescent foamy water mixed up with sand arrived from the ocean with a thunderous noise. The height of the waves amounted to 15–18 m. Upon hitting the coast with an enormous force, the wave then carried the large US warship 'Wateree' from the harbour 2 miles inland and gently put it down at the rocky foot of the Andes. This event permitted Gabriel Garcia Marquez to depict the fantastic scene of an encounter with a three-mast sailboat amongst trees in the remote jungles (selva) of South America.

The tsunami reduced the site, where the city of Arica with about 5,000 inhabitants had been, to a smooth sandy valley without any signs of buildings. Only individual structures remained here and there on the mountain slopes.

The catastrophic 1908 Messina tsunami was caused by an earthquake of magnitude 7, the source of which was located under the bottom of the Messina Strait (in between continental Italy and Sicily). The tsunami started nearly immediately after the shaking stopped with a withdrawal of the sea water. Part of the sea-floor, adjacent to the coast, happened to be drained, in some places the sea-floor opened up for nearly 200 m. Then, all of a sudden, waves started to advance, the first three being the strongest. The tsunami was preceded by a strong noise, similar to the noise of a tempest or of waves hitting rocks with force. The maximum run-up height on the coast of Sicily amounted to 11.7 m, on the Calabrian coast to 10.6 m. Noticeable waves reached the coasts of Libya and Egypt. Of the mareographs that were not damaged, the one closest to the tsunami source was located on the Malta island. It recorded a tsunami of amplitude 0.9 m.

The number of tsunami waves observed varied from place to place from 3 to 9, and the period of the waves from 5 to 15 min. The waves washed out the structures destroyed by the earthquake and destroyed many buildings that had survived. Of the buildings and structures only the foundations, sliced off at land level, remained.

Many vessels, having been damaged, either sank or were stranded inland. The tsunami stirred up sea-floor sediments; bubbles of gas came up from the sea-floor to the surface of the strait; sea animals and fish, including deep-water inhabitants, unknown to fishermen, were thrown up onto the beach. Sailors on vessels moored several miles from the coast felt a strong seaquake, but couldn't understand why all the lights had gone out in the towns along the coast.

After the tsunami all the strait was full of broken and overturned boats, other vessels, floating debris, bodies of human beings and animals, washed off the coasts of Messina and Calabria.

The 1952 tsunami that occurred near the eastern coasts of Kamchatka and of the Island of Paramushir is considered one of the most destructive tsunamis of the twentieth century. We shall present the description of this event given in the article by S. Soloviev [Soloviev (1968)]. In the night between November 4 and 5 the inhabitants of Severo-Kurilsk were woken up by an earthquake: stoves were destroyed, chimneys and household utensils fell down. Forty minutes after the earthquake stopped a rumble was heard from the ocean, and a water bore moving with a high velocity fell upon the city. In several minutes the water retreated, carrying away what it had destroyed, and the ocean bottom opened up for several hundred meters. In 15–20 min a wall of water 10 m high once again advanced upon the city. It practically washed away everything in its way, at the most leaving only concrete foundations of various structures. Old pillboxes were wrenched out of the ground and thrown around, in the harbour the walls of a bucket were turned upside down, and launches that happened to be there were stranded hundreds of meters inland.

Several minutes later, after this strongest wave, a third relatively weak wave ran up the devastated coast, leaving much debris after it.

The events of 1952 were totally unexpected for most of the population. Thus, for example, some of the vessels moored near the Island of Paramushir, transmitted messages that the island was sinking into the ocean waters.

A. E. Abaev, captain of a detachment of hydrographic vessels sent to Severo-Kurilsk immediately after the catastrophe, witnessed the strait between the islands of Shumshu and Paramushir to be completely crammed with floating wreckage of wooden houses, logs and barrels. The bodies of human beings were seen on the wreckage—it was practically impossible to survive in the ice-cold water.

Another witness of this tsunami, A. Shabanov, who lived in Severo-Kurilsk and at the time was 14 years old, told one of the authors of this book, that soon after the earthquake the water receded from the coast and left the ocean bottom open. When Shabanov's mother saw this sudden ebb tide she ran with her two sons towards the hills, which saved their lives. Their family was the only family, in which no one was killed. On their way they had difficulty in crossing a deep ditch across which the Japanese in former times had thrown several narrow wooden footbridges. By 1952, most of the footbridges had been used as firewood, since it was not clear to the people arriving from the continent what they were for.

The wave that in some parts of the coastline reached a height of 10–15 m ($H_{max} = 18.6$ m) totally destroyed many buildings and port structures of Severo-Kurilsk (Island of Paramushir) and carried them out to sea, taking the lives of 2,336 people. The source of the tsunami wave generated by an underwater earthequake of magnitude $M_w = 9.0$ extended over 800 km and was about 100 km wide.

The fantastic event that gave rise to a tsunami wave of record height took place on July 9, 1958 in Lituya Bay (Alaska) [Soloviev et al. (1975)]. The bay exhibits a T-like shape. Its length amounts to 11 km, its width in the main external part up to 3 km, and its maximum depth about 200 m. The internal part of the bay is part of the Fairweather canyon. Here the bay resembles a fjord, and its steep walls rise up to heights between 650 and 1,800 m. During the earthquake a gigantic slide of snow-and-ice together with local rock of volume about 0.3 km^3 took place. The water ousted by the falling mass splashed out onto the opposite coast and reached the height of 524 m! The displacement of water was so rapid, that all the trees in the flooded wood were wrenched up and the bark and leaves of the trees were rubbed off. Besides this enormous splash, a wave formed that crossed the whole bay right up to the ocean, devastating the bay's shores. Three fishing-launches were caught by the wave in the bay; one of them sank together with two crewmen. The two other crews were lucky to escape. The fishermen spoke of a wave about 30 m high. Signs of the run-up and of trees broken by the wave remained on the slope during decades after the catastrophe. Note that the expedition led in 1786 by G.-F. La Perouse encountered a similar phenomenon in the French Harbour (presently known as Lituya Bay). An enormous wave carried the two-mast vessel of the expedition through the narrow strait and smashed it against the underwater rocks. Of all the 21 crewmen no one was left alive.

The Chilean tsunami of May 22, 1960 was caused by the strongest earthquake of the twentieth century ($M = 9.4$), the source of which was located in the southern part of central Chile [Soloviev et al. (1975)]. The maximum elevation of water amounted to 25 m in Chile, 10.5 m on the Hawaiian islands, 9 m in the Oceania,

6.5 m in Japan and the USSR and 3.5 m in the USA. About 1,000 persons lost their lives in Chile, 60 on the Hawaiian islands and 200 in Japan. It took approximately 15 h for the waves to cover 10,000 km and to reach the Hawaiian islands and nearly a day and night to reach Japan and the Far-East coast of the USSR. Naturally, the earthquake was felt neither on the Hawaiian islands, nor in Japan, nor in the USSR, so the wave turned out to be unexpected.

The 1994 tsunami caused by an earthquake of magnitude $M = 8.3$ near the Island Shikotan, resulted in the destruction of numerous coastal structures. Part of the island sank by 60 cm, which was recorded by the mareograph in the village of Malokurilsk. In the city of Yuzhno-Kurilsk, located at a distance of 120 km from Shikotan, the tsunami wave tore down a single-storeyed block of flats from its foundation and carried it 300 m inland. The wave's maximum run-up amounted to 10.4 m.

The 1998 tsunami that occurred in the region of Papua New Guinea gave rise to particular interest among specialists. A relatively small earthquake of magnitude $M_w = 7.1$ resulted in an unexpectedly large wave of height amounting to 15 m. The tsunami attacked the coast with three waves about 18 min after the earthquake. The area influenced was limited to part of the coastline 30 km long, where several fishing villages were destroyed and about 3,000 people lost their lives. The formation of such a gigantic wave was mainly due to the underwater slide caused by the earthquake, rather than to the earthquake itself.

The catastrophic tsunami of December 26, 2004 that occurred in the Indian Ocean was caused by an exceptionally strong earthquake of magnitude $M_w = 9.3$, the epicentre of which was near the northern extremity of Island Sumatra. Comparable magnitudes were exhibited during the past 100 years only by several seismic events [Aleutian Islands, 1946; Kamchatka, 1952; Aleutian Islands, 1957; Chile, 1960; Alaska, 1964]. The manifestation of the tsunami was of a global character. Besides the castastrophic consequences in the vicinity of the source (the coast of Sumatra), where the run-up amounted to 35 m, waves were registered all over the World Ocean. Tsuami waves of significant amplitudes were registered in remote parts both of the Pacific coast (Manzanillo, Mexico—0.5 m, New Zealand—0.5 m, Chile—0.5 m, Severo-Kurilsk, Russia—0.3 m, British Columbia, Canada—0.2 m, San Diego, California—0.2 m) and of the Atlantic coast (Halifax—0.4 m, Atlantic City—0.2 m, the Bermuda islands—0.1 m, San Juan, Puerto Rico—0.05 m). The worst hit were countries of the basin of the Indian Ocean: Indonesia, Thailand, India, Sri Lanka, Kenya, Somalia, South Africa and the Maldive Islands. The total number of victims exceeded 250,000 people, the damage was enormous and it still has to be estimated. The number of casualties makes this catastrophe the largest of all known catastrophes in the history of tsunamis.

Central Kuril Islands Tsunamis. An extremely strong earthquake of magnitude $M_w = 8.3$ took place on November 15, 2006, in the Central-Kuril segment of the Kuril-Kamchatka seismofocal zone. The epicentre of the earthquake was located in the Pacific Ocean at about 85 km from the northern extremity of Simushur Island. Before this event, the Central-Kuril segment was considered a 'seismic gap' zone, an earthquake of such strength was registered here for the first time in the history of seismic observations. Nearly 2 months later, on January 13, 2007, another earthquake of practically the same strength, $M_w = 8.1$, occurred in the same region. Both seismic events were accompanied by tsunami waves, noted over the entire area of the Pacific Ocean: Shikotan Isl., Malokurilsk—1.55(0.72) m, Kunashir Isl., Yuzhno-Kurilsk—0.55(0.11) m, Alaska, Shemya—0.93(0.69) m, Crescent City, California—1.77(0.51) m, Hawaii, Kahului—1.61(0.24) m, Peru, Callao—0.73(0.3) m, Chile, Talcahuano—0.96(0.23) m (the wave heights indicated in brackets correspond to the event of January 13, 2007). However, owing to the absence of mareographic stations and inhabitants on the Central Kuril Islands, no information on the wave heights in the immediate vicinity of the sources was available. During the period from July 1 to August 14, 2007, two seafaring expeditions were organized with one of their main tasks consisting in the investigation of the coasts of the islands so as to determine the tsunami run-up heights [Levin et al. (2008)]. The participants of the expedition were the first people to visit the islands after the tsunamis and to estimate the scale of the natural disaster. The time for the expedition depended on the complicated weather conditions in the area. Landing on the coasts of the islands earlier (before April–May) was practically impossible to realize. The highest tsunami run-ups (up to 20 m) were revealed on Matua Island. The tsunami strongly altered the morphology of the coast in the Ainu Bay (south-west of Matua Island) by washing away a section of the sea terrace 20–30 m wide. The maximum run-up height in Dushnaya Bay (north-east part of Simushir Island) amounted to 19 m; here the tsunami left numerous scours. Besides erosion on the coasts investigated, accumulation was also observed everywhere. Tsunami deposits consisted of marine sand, pebbles, boulders and floating debris shifted toward the land. The vegetation on steep slopes was partly destroyed, and the soil washed away. If waves of such strength were to hit a densely populated coast, casualties could certainly not be avoided. The only reason the tsunamis of November 15, 2006, and of January 13, 2007, did not become an awful tragedy was the total absence of population on the Central Kuril Islands. These two events can rightfully be considered the strongest tsunamis that were not accompanied by human casualties.

Table 1.1 presents several examples of recent catastrophic tsunamis.

Table 1.1 Recent catastrophic tsunamis

No	Date	M^a	h^bmax (m)	Number of casualties	Location of event
1	12/12/1992	7.5	26	1,000[c]	Flores Is., Indonesia
2	12/07/1993	7.7	31	330[c]	Okushiri Is., Japan
3	02/06/1994	7.8	14	223	Java, Indonesia
4	04/10/1994	8.1	10	11[c]	Shikotan Is., Russia
5	09/10/1995	8.0	11	1	Manzanillo, Mexico
6	17/02/1996	8.1	7.7	110	Irian Jaya, Indonesia
7	17/07/1998	7.1	15	2,200	Papua New Guinea
8	23/06/2001	8.1	10	26	Peru
9	26/12/2004	9.3	36	250,000[c]	Indian Ocean, Sumatra
10	15/11/2006	8.3	20	0	Central Kuril Is., Russia

[a] Earthquake magnitude
[b] Maximum wave height
[c] May include earthquake victims

1.3 Tsunami Magnitude and Intensity

Estimation of the degree of tsunami danger for one or another coast (long-term tsunami forecast) is primarily based on the statistical analysis of events, that occurred in the past. Tsunamis evidently vary in strength within wide limits: from weak waves, that can be registered only with the aid of instruments, up to terrible catastrophic events devastating the coast along hundreds of kilometers. How can one estimate the strength of a tsunami? The point is that without the introduction of some quantitative characteristic of this strength it is not only impossible to perform any statistical analysis, but also to speak of estimating the degree of danger. The determination of such a quantitative characteristic is quite a non-trivial problem, the ultimate resolution of which has not yet been achieved. Similar difficulties are encountered by seismologists determining the strength of an earthquake. On the one hand, an earthquake is characterized by objective physical parameters showing the energy emitted by the source, or the released seismic moment. These parameters are measured quantitatively, and the scale of earthquake magnitudes is made to correspond to them. On the other hand, there exists a descriptive scale of earthquake intensities, which is related to the so-called macroseismic data, based on the results of in situ studies. Clearly, in practice, it is precisely the intensity scale that is important, but contrary to the magnitude scale it is not rigorous, from a physical standpoint.

Going back to tsunamis, we note that this phenomenon is also characterized, on the one hand, by objective and quantitatively measurable parameters (energy, amplitude, period, etc.), and on the other hand—by subjective descriptions, reflecting the scale and degree of the destructions caused by the wave or the character of its manifestations on the coast. Like in the case of earthquakes, for estimation of the tsunami danger precisely these subjective descriptions are more important than abstract physical parameters. The inhabitants of coastal regions are not interested in the energy of the approaching wave in joules, but they are interested in whether the wave is dangerous to their lives, what damage may be done and how it can be avoided. And, until further modelling is realized of the entire process starting from the actual formation of a wave up to its run-up onto the shore, such a situation will remain intact.

The first attempt at classification of tsunamis was made by Sieberg, who introduced a six-point scale of tsunami intensities by analogy with the scale of earthquake intensities [Sieberg (1927)]. This scale was not related to the measurement of physical parameters (wave heights, run-up lengths, etc.); it was based on the description of macroscopic effects, revealing the degree of destruction. Subsequently, the Sieberg scale was somewhat modified [Ambraseys (1962)].

The Sieberg–Ambraseys tsunami intensity scale

1. **Very light.** Waves can only be registered by special tide gauges (mareographs).
2. **Light.** Waves noticed by those living along the shore. On very flat shores waves are generally noticed.

3. **Rather strong.** Waves generally noticed. Flooding of gently sloping coasts. Light sailing vessels carried away on shore. Slight damage to light structures situated near the coasts. In estuaries reversal of the river flow some distance upstream.
4. **Strong.** Significant flooding of the shore. Buildings, embankments, dikes and cultivated ground near coast damaged. Small and average vessels carried either inland or out to sea. Coasts littered with debris.
5. **Very strong.** General significant flooding of the shore. Quay-walls and solid structures near the sea damaged. Light structures destroyed. Severe scouring of cultivated land. Littering of the coast with floating items, fish and sea animals thrown up on the shore. With the exception of big ships all other type of vessels carried inland or out to sea. Bores formed in estuaries of rivers. Harbour works damaged. People drowned. Wave accompanied by strong roar.
6. **Disastrous.** Partial or complete destruction of manmade structures for some distance from the shore. Strong flooding of coasts. Big ships severely damaged. Trees uprooted or broken. Many casualties.

Numerous attempts were made in Japan to introduce a quantitative characteristic of the tsunami strength. Imamura introduced, and Iida further improved, the concept of tsunami magnitude [Imamura (1942), (1949); Iida (1956), (1970)]. A proposal was made to estimate the magnitude by the formula

$$m = \log_2 H_{\max},$$

where $H_{\max}$ is the maximum wave height in metres, observed on the shore or measured by a mareograph. In practice, the Imamura–Iida scale is a six-point scale (from -1 up to 4).

In attempts at improving the Imamura–Iida scale S. L. Soloviev introduced the following tsunami intensity:

$$I = \frac{1}{2} + \log_2 H,$$

where H is the average tsunami height on the coast closest to the source. At present such a definition of the tsunami intensity is widespread, and the corresponding scale is conventionally termed the 'Soloviev–Imamura tsunami intensity scale'.

Note that the Imamura–Iida definition of magnitude is, generally speaking, unambiguous. It only requires knowledge of the maximum wave amplitude. The Solviev–Imamura definition of intensity is not mathematically rigorous and, consequently, provides for much 'freedom' in calculating the average height of tsunami waves. At any rate, both scales are not very sensitive to small errors in the determination of wave heights, since it is the logarithms of these quantities that count. It is also important to note that in the case of numerous historical events and, more so, of prehistoric events (paleotsunamis) the only available information comprises estimates of wave heights at a single point or at several points along the coast. Thus, both scales are quite convenient and will still be applied in practice for a long time. Anyhow, as a base characteristic to be measured in calculating the magnitude or intensity one may consider the flooded area, instead of the wave height. This characteristic may turn out to be a successful and promising alternative to the wave heights on the coast. A clear advantage of the flooded area consists not only in that it can be conveniently measured by remote means (from satellites, airplanes, etc.), but also in that this characteristic automatically reflects the scale of the catastrophe that took place.

Abe and Hatori proposed to modify the magnitude scale so as to take into account the weakening of waves, as the distance from the source increases [Abe (1979), (1981), (1985), (1989); Hatori (1986)],

$$M_t = a \log h + b \log \Delta + D,$$

where h is the maximum wave amplitude on the coast measured from the foot up to the crest in meters, Δ is the distance from the earthquake epicentre to the point of measurement in kilometers, a, b and D are constants. Such a definition resembles the definition of magnitude in seismology.

An essentially different approach to the definition of tsunami magnitude was put forward in [Murty and Loomis (1980)]. Here, the calculation of magnitude is based on estimation of the tsunami's potential energy E (in ergs),

$$\mathrm{ML} = 2\left(\log E - 19\right).$$

The definition of magnitude based on the wave energy is, naturally, the most adequate definition, from a physical point of view. However, it is not always possible to calculate the wave energy. At any rate, at the present-day stage calculations can be based on the potential energy of the initial elevation of the water surface, considering it to be identical to the residual displacements of the sea-floor. These displacements are calculated from the earthquake parameters by the Okada formulas [Okada (1985)].

It must be noted that the Imamura–Iida magnitude or the Soloviev–Imamura intensity gives an idea of the wave height on the coast and, consequently, permit to judge the scale of destructions. But, although the Murty–Loomis tsunami magnitude ML is a physically correct quantity, it cannot be unambiguously related to the manifestation of a tsunami on the coast.

Recently, a new detailed 12-point descriptive tsunami intensity scale was proposed in [Papadopoulos, Imamura (2001)]. Its elaboration was based on the more than 100-years-long experience, accumulated by seismologists in drawing up earthquake intensity scales. This scale is not related to any quantitative physical parameters (wave amplitudes, energy and so on), it is organized in accordance with the following three features:

(A) Its influence upon people
(B) Its impact on natural and artificial objects, including boats of different sizes
(C) The damage caused to buildings

Therefore, a tsunami of large amplitude that hits a weakly inhabited coast may be assigned a low intensity in accordance with the Papadopoulos–Imamura scale.

1.3 Tsunami Magnitude and Intensity

And, contrariwise, a tsunami of moderate amplitude that hits a densely populated coast may be characterized by quite a high intensity.

It is useful to present the Papadopoulos–Imamura intensity scale here completely. A consistent and systematic description of tsunami manifestations on the coast provides quite a full picture of the phenomenon.

The Papadopoulos–Imamura tsunami intensity scale

I. Not felt[1]

(a) Not felt even in most favourable circumstances
(b) No effect
(c) No damage

II. Scarcely felt

(a) Felt by some people in light boats. Not observed on the shore
(b) No effect
(c) No damage

III. Weak

(a) Felt by most people in light boats; observed by some people on the shore
(b) No effect
(c) No damage

IV. Largely observed

(a) Felt by all people in light boats and some on large vessels; observed by most people on shore
(b) Some light boats are slightly carried onto the shore
(c) No damage

V. Strong

(a) Felt by all people on large vessels; observed by all people on shore; some people are frightened and runup elevations.
(b) Many light vessels are carried inland over significant distances, some of them collide with each other or are overturned. The wave leaves layers of sand in places with favourable conditions. Limited flooding of cultivated land along the coast.
(c) Limited flooding of coastal structures, buildings and territories (gardens etc.) near residential houses.

VI. Slightly damaging

(a) Many people are frightened and run up elevations.
(b) Most light vessels are carried inland over significant distances, undergo strong collisions with each other or are overturned.

---

[1] Registered only by special instruments.

(c) Some wooden structures are destroyed and flooded. Most brick buildings have survived.

VII. Damaging

(a) Most people are frightened and try to run away onto elevations.
(b) Most light vessels are damaged. Some large vessels undergo significant vibrations. Objects of varying dimensions and stability (strength) are overturned and shifted from their positions. The wave leaves layers of sand and accumulates pebbles. Some floating structures are washed away to sea.
(c) Many wooden structures are damaged, some are totally wiped away or carried out to sea by the wave. Destructions of first degree and flooding of some brick buildings.

VIII. Heavily damaging

(a) All people run up elevations, some are carried out to sea by the wave.
(b) Most light vessels are damaged, many are carried away by the wave. Some large vessels are carried upshore and undergo collisions with each other. Large objects are washed away. Also Erosion and littering of the coast, widespread flooding and insignificant damage in antitsunami plantations of trees. Many floating structures are carried away by the wave, some are partially damaged.
(c) Most wooden structures are carried away by the wave or completely wiped of the earth's surface. Destructions of second degree in some brick buildings. Most concrete buildings are not damaged, some have undergone destruction of first degree and flooding.

IX. Destructive

(a) Many people are carried away by the wave.
(b) Most light vessels are destroyed and carried away by the wave. Many large vessels are carried inland over large distances, and some are destroyed. There is also Broad erosion and littering of the coast, local subsidence of the ground and partial destruction of antitsunami plantations of trees. Most floating structures are carried away, and many are partially damaged.
(c) Destructions of third degree in many brick buildings. Some concrete buildings have undergone destructions of second degree.

X. Very destructive

(a) General panic. Most people are carried away by the wave.
(b) Most large vessels are carried inland over large distances, many are destroyed or have undergone collisions with buildings. Small rocks (pebbles, stones) have been carried onshore from the sea-floor. Vehicles are overturned and displaced. Petroleum is spilt, there are fires and widespread subsidence of ground.
(c) Destructions of fourth degree in many brick houses, some concrete buildings have undergone destructions of third degree. Artificial dams (embankments) are destroyed and harbour wavebreakers damaged.

XI. Devastating

(b) Vital communications are destroyed. There are widespread fires. Reversed flows of water wash away to sea vehicles and other objects. Large rocks of different kinds are carried onshore from the sea-floor;

(c) Destructions of fifth degree in many brick buildings. Some concrete buildings suffer damage of fourth degree, and many of third degree.

XII. Completely devastating

(c) Practically all brick buildings are wiped out. Most concrete buildings have suffered destructions of degrees not lower than third.

1.4 Tsunami Warning Service: Principles and Methods

The extremely long and sad experience of Japan's population with many thousands of lives lost to tsunamis and earthquakes is expressed in the short inscription on the stone stellae often found near the coastline. The hieroglyphs on the stellae say the following:

Don't forget about earthquakes. If you feel an earthquake, don't forget about tsunamis. If you see a tsunami, run up a high slope.

The following legend is told by the inhabitants of the city of Wakayama, situated not far from Kyoto, the former capital of Japan and a most beautiful city. The major of Wakayama once felt an earthquake. He understood he had no time to warn the people on the shore of the tsunami danger, so he run-up the slope to the rice fields, where the rice had been harvested, and set the granaries on fire. People, seeing the burning supplies of rice, hurried up to put the fire out and, thus, they happily evaded the lethal strike of the tsunami wave against the coast. The grateful inhabitants of the city erected a monument to the wise ruler.

By the 1960s many countries of the Pacific region had organized national tsunami warning systems. The tsunami service organizations include a whole network of seismic and hydrometeorological stations, special systems for operative alert transmission, administrative organs for adopting resolutions and regional organizations for implementing evacuation plans of the population.

In past years the work of a tsunami warning service (TWS) was based on routine and/or urgent dispatches from operators on duty at seismic stations with round-the-clock tsunami services. If a nearby strong earthquake (of magnitude $M>7$) is registered, the operator had, within 10 min, to determine the distance to its epicentre, the earthquake's magnitude and the approximate region of its location. The operator had, then, to transmit the signal 'TSUNAMI warning' to the administrative organ, to the tsunami headquarters and to the meteostation. The oceanology on duty at the meteostation applied additional information to decide whether to announce the warning or not. The all-clear signal was announced by the tsunami headquarters upon agreement with specialists.

In modern TWS this technology is automized. However, the main physical principles of operative tsunami forecasting remain the same. The possibility itself of warning is based on the propagation velocity of seismic waves being many times larger than the velocity of a tsunami wave. A warning is announced, when registration occurs of an underwater earthquake of magnitude exceeding a threshold value.

In Russia, the Far East tsunami service is implemented by seismic stations of the Geophysical service of the Russian Academy of Sciences and by meteostations of the Administrations of Hydrometeorological Services (AHMS) subordinate to the Committee for Hydrometeorology of the Russian Federation. The tsunami service relies on seismostations of Petropavlovsk-Kamchatskyi and Ust-Kamchatsk (Kamchatka), Yuzhno-Sakhalinsk (isl. Sakhalin), Severo-Kurilsk (isl. Paramushir), Kurilsk (isl. Iturup) and Yuzhno-Kurilsk (isl. Kunahsir). The meteostations of the Kamchatka AHMS and the Sakhalin AHMS, as well as the tsunami centres of these Administrations are on round-the-clock duty for implementation of operative tsunami warning service and preparation of routine and/or urgent dispatches and reports.

A well-developed Tsunami Warning System has been organized in the USA within the National Oceanic and Atmospheric Administration (NOAA). It includes several hundreds of seismic and mareographic stations. All these stations, as well as several large oceanic buoys and sea-floor sensors reporting the ocean level transmit the information obtained in a real-time mode to the common servers of two centres: ATWC in Palmer, Alaska and PTWC in Honolulu, Hawaii. The information is freely available to all Internet users. The tsunami service in Japan, created significantly earlier than the others, is subordinate to the Japanese Meteorological Agency (JMA) and is noted for its very high level of organization.

An important success, achieved in operative tsunami prognosis, consists in the possibility of rapid (real-time) calculation, with a precision and reliability sufficient for practical purposes, of the arrival time of a wave at a given (protected) point of the coast. Such a calculation can be performed by applying simple ray theory. To this end it is only necessary to know the location of the tsunami source and the distribution of depths in the basin considered. We recall that the tsunami propagation velocity depends on the ocean depth, $c = \sqrt{gH}$. Data on the bathymetry of the World Ocean are free for a grid with steps of 1×1 angular minutes, and for many regions even with a significantly improved spatial resolution.

The situation concerning calculation of a tsunami run-up height at a given point of the coast is much worse. The calculation precision and speed required for practical purposes in resolving this problem have not been achieved yet. On the one hand, this is due to the enormous volume of calculations to be performed in estimating the evolution of a wave starting from its rise at the source up to its run-up to the shore. On the other hand, in the real-time mode it is impossible to calculate what has happened at the tsunami source with necessary precision. The time required for the reliable determination of sea-floor deformations, due to an earthquake, essentially exceeds minutes or even hours available for operative forecasting. In those cases, when underwater landslides participate in the tsunami generation, operative resolution of the problem turns out to be practically impossible.

There exist several promising ways for resolving the problem. One of them is realized in the Japanese system of operative tsunami prognosis [Tatehata(1998); Handbook for Tsunami Forecast (2001)]. The method is based on tsunami sources exhibiting the property of recurrence. Therefore the problem, requiring long-time calculations, has been resolved beforehand. The results of calculations are presented in a special database. When a real underwater earthquake takes place, then in accordance with its magnitude and epicentre location necessary data are extracted from the database and used for calculating the possible run-up heights applying the interpolation method.

The second way consists in making use of deep-water sensors of tsunamis established far from the coast (for instance, DART, JAMSTEC). The actual idea and its first realization are related to the name of S. L. Soloviev. There exist various possibilities for using such systems. One of them is based on the fact that timely registration of a wave permits to measure its characteristic (amplitude) with precision and to correct inaccuracies in the calculation of the wave generation at the source.

The catastrophic tsunami that occurred in December 2004 in the Indian Ocean was registered by a radio-altimeter established on the satellite JASON-1. Thus, the good prospects became evident of methods involving satellite monitoring of tsunamis.

Regretfully, the modern tsunami service is mainly based on regional principles. Analysis of the actions taken by national services on December 26, 2004, revealed their 'zones of responsibility' to be limited only to the sectors of the coast under their control. Taking into account the restricted capabilities of certain developing countries to provide operative tsunami warning at a modern level and, also, the scale of such catastrophes, it seems expedient to create a global monitoring system of the ocean's surface to function under international control. The authors of this book fully support the International Tsunami Information Center (ITIC, NOAA USA) in its efforts aimed at the rapid as possible creation of a worldwide tsunami warning system.

1.5 Databases and Tsunami Statistics

There exist several different informational resources containing the main information on tsunamis. One of the most effective and in greatest demand is the historical database for tsunamis in the Pacific Ocean, which was created in the Tsunami laboratory of the Institute of Computational Mathematics and Mathematical Geophysics, SB RAS (Novosibirsk) with support of UNESCO and the Russian Foundation for Basic Research [Gusiakov (2001)]. The Internet version of the database is available at http://tsun/sscc.ru/htdbpac.

The database contains information on approximately 1,500 tsunamigenic events that occurred in the Pacific region (within the geographical boundaries 60° S–60° N, 80° E–100° W) during the entire historical period of observations (starting from the year 47 BC up to now). This database includes a large volume of coastal observations of tsunamis (about 9,000 records) as well as various auxiliary information on regional bathymetry, seismicity, tectonics, vulcanism and on settlements in coastal regions, and, also, about the regional network of mareographic observations.

An important advantage of this informational resource is a specialized graphical shell, the construction of which is based on the technological principles of state informational networks and which provides convenient means for users to select, visualize and process data, and, also, to analyse the quality and completeness of historical catalogues. The cartographic shell includes means for working with raster images of the earth's surface and of the sea-floor, which permits the user to create digital maps of the area of interest and to subsequently superimpose observational data on them.

For application of the shell in tsunami warning services it has a special option—the 'New event' mode, permitting to realize selection of historical data within a circular vicinity of the event, undergoing operative processing. The built-in subsystem for estimation of the tsunami hazard permits to obtain estimates of the long-term tsunami hazard for coastal areas of the Pacific aquatorium, for which there exist sufficient observations of tsunami heights. The fields in which the created database is applied comprise scientific research and developments in the field of marine geophysics, seismology and oceanology, related to studies of natural catastrophes, the seismicity of the World Ocean and seismoproof building projects in coastal regions.

An example of graphical presentation of material is given in Fig. 1.4 by the distribution of sources of tsunamigenic earthquakes in the Pacific region that occurred during the period from 47 BC up to 2004.

An idea of the tsunami recurrence frequency can be obtained from Fig. 1.5, in which it is shown how the number of tsunamis (per decade) varied between 1800 and 2005. All the events are divided into two categories: the grey lines indicate significant tsunamis of intensities (the Soloviev-Imamura scale) $I \geqslant 1$, the dark lines

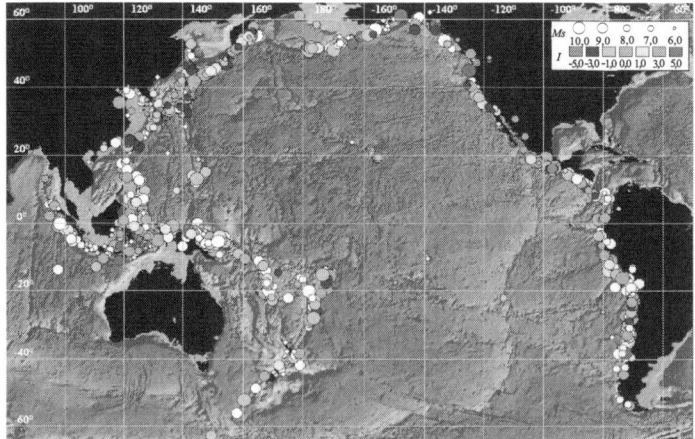

Fig. 1.4 Distribution of tsunami sources in the Pacific region within the period from 47 BC up to 2004. The sizes of the circles correspond to earthquake magnitudes and their colours to the tsunami intensities (see also Plate 1 in the Colour Plate Section on page 309)

1.5 Databases and Tsunami Statistics

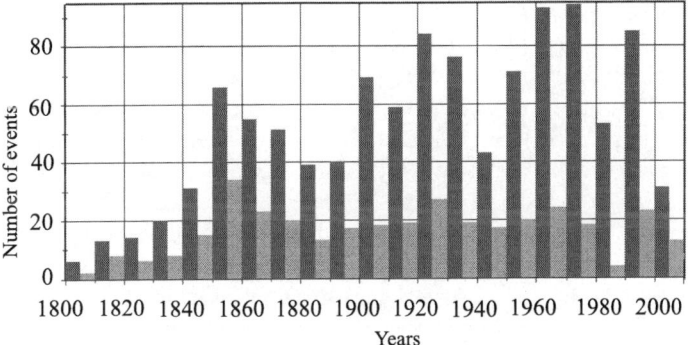

Fig. 1.5 Recurrence of tsunamis (number of events per decade) in the Pacific region between 1800 and 2005. The dark colour shows all the known tsunamis, the grey colour indicates tsunamis of intensity $I \geqslant 1$ according to the Soloviev-Imamura scale

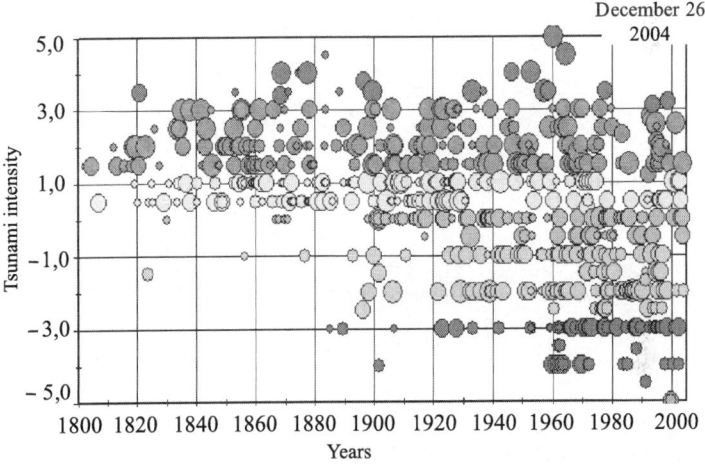

Fig. 1.6 Tsunamis of the Pacific region in the 'intensity–time' plane (see also Plate 2 in the Colour Plate Section on page 310)

show all known tsunamis. It can be seen that the recurrence of significant tsunamis remains approximately at the same level (about two events per year). The total number of tsunamis, here, shows a tendency of increasing, which is related to the progress in registering weak waves. Similar comments can also be addressed to Fig. 1.6, in which the tsunami intensity is plotted as a function of time. It must be stressed, here, that it would be wrong to conclude, on the basis of the presented data, that the tsunami recurrence frequency has increased during past centuries. The recurrence frequency of tsunamis can vary noticeably only over geological times.

The database can be applied for resolving a broad class of problems. Thus, for example, analysis of measurements of run-up heights of tsunami waves making use of approaches pertaining to dimensionality theory [Zav'yalov et al. (2005)] has permitted to reveal hidden information on the symmetry of a tsunami source and on peculiarities of the emission and damping of wave energy. It has been established that in the case of a large number of registered events, as the distance from the source increases, the weakening of waves follow a power law $r^{-\alpha}$, where the exponent α varies within the limits from 0.5 to 0.66. This corresponds to values of the source's symmetry parameter between 1 and 2, i.e. the source is an elongated ellipse. Similar estimates can be found in the book by Pelinovsky [Pelinovsky (1982)]. Note that precisely such a shape is attributed to the source of a tsunami in many publications dealing with the resolution of inverse hydrodynamic problems. In the case of continental earthquakes, the region of maximum shakings reconstructed in problems of macroseismics also exhibits a strongly elongated oval shape.

Statistical treatment of the material in the tsunami database has recently revealed an interesting periodicity in the appearance of tsunami sources [Levin, Sasorova (2002)]. The time sequence of events generated alternately in the Earth's northern and southern hemispheres is characterized by a 6-year period. Strong perturbations on Earth, caused by astronomical processes and the mutual arrangement of bodies in the Earth–Moon–Sun system arise with approximately the same periodicity. Further development of databases and of computational methods will permit future revelation of new laws in the processes of tsunami and underwater earthquake generation.

1.6 Seaquakes: General Ideas

Every year approximately 10,000 earthquakes are registered on Earth, of which about 4,000 events are perceptible earthquakes, for which the velocity of motion of particles in the wave or the mass velocity amounts to over 0.1 m/s. Of this amount the largest part are underwater earthquakes. Strong earthquakes result in the appearance near the coasts of gigantic devastating tsunami waves, while in the region of the earthquake epicentre unusual hydrodynamic phenomena are observed that are known to seafarers by the term seaquakes. In certain cases, the terms tide race and wave crowd are also used.

The transverse dimension of the perturbated region of the sea surface during a seaquake usually exceeds 50–100 km, while the duration of a strong seaquake may amount to 10 min. During a seaquake sets of very steep standing waves form on the surface of the aquatorium, individual vertical columns of water and solitary water formations arise and strong acoustic effects are noted. Spray sultans may be observed, as well as cavitational layers of water separating from each other and flying apart. A ship that happens to be in the zone of influence of a seaquake turns out to be surrounded by giant standing waves filling up the entire visible space. Terrible thunderous rumbling and howling are enhanced by sharp blows to the bottom, the most strong shaking of the vessel and the destruction of deck structures that had in the past endured more than a few storms.

We shall present the average quantitative characteristics of the phenomenon:
- Height of waves and of surges of water—over 10 m
- Velocity of surface motion up to 10 m/s
- The acceleration of water particles may amount to 10 m/s^2

1.6 Seaquakes: General Ideas

- The amplitude of pressure in the underwater wave of compression—up to 2 MP (20 atm)

Such effects may result in the destruction of structures on the ship, demoralization of the crew, the rise of critical and emergency situations on the vessel and the death of sea animals and fish. We shall point out the main parameters of a seaquake that are important for developing a physical model of the phenomenon:

1. The ratio of the water depth H and the characteristic horizontal dimension of the source L lies within the interval between 0.001 and 0.1.
2. The range of frequencies of the sea-floor oscillations is 0.001–10 Hz.
3. The acceleration amplitude of motions of the sea-floor amounts to $10 \, m/s^2$.

Modern catalogues, articles and scientific publications contain over 250 descriptions of seaquakes in various regions of the World Ocean. Among the recent events one must note the strong damage of eight fishing boats in the region of the South Kuril islands after the Shikotan earthquake of 1994 and the destruction of the huge tanker 'Exxon' in the Gulf of Alaska resulting from a seaquake in 1988. Below follows a scenario of the development of a 'generalized' seaquake, based on materials of individual descriptions due to witnesses [Levin (1996)].

Synthesized description of a seaquake

On a totally breezeless day the smooth mirror-like sea surface became all of a sudden covered with bumps. These aquatic bumps, that looked like waves, did not move away in any direction, but they did not remain motionless, either. They grew rapidly up to a height of about 8 m and then shrank down forming deep craters in the place of the recent bumps.

The oscillations were rapid, we were dazzled by these unusual boiling waves filling up all the visible sea space. The aquatic surface seethed and jumped up and down, as if it were in a red-hot kettle with salt brine. The boat was thrown up and down, and it rocked ominously on these jumping waves. They were as steep as the most ferocious storm waves, but no longer than 20 m. The keel rocking was so strong, that the propeller was several times seen to be completely dry in air, and the wheel of the ship compass fell off its pivot.

All the passengers and crew poured out on deck. The bright sunshine and total calm only enhanced the tension of this terrifying spectacle of a sea gone crazy.

Less than a minute passed, but no willpower was left to resist this monstrous galloping, which once in a while weakened, then strengthened again. Hands clinging to the ship's sides felt how unreliable this plaintively creaking vessel was in front of the mysterious and incomprehensible sea catastrophe.

The aquatic bumps started to become smaller, while the frequency of their blinking increased. At the same time, from somewhere out of the depth a low thunderous rumble arose, that supressed all willpower and reason. People started thrashing around the ship full of panicky fear. Many passengers, and even sailors, could not bear such torture and, having evidently lost their minds, started to jump overboard. Against the background of these blinking waves there started to appear very high jets of water, that collapsing created a strange rustling sound.

All of a sudden the ship was shaken by a most strong blow. Several persons were thrown overboard. The blows to the bottom of the ship came one after another. The ship seemed to batter the rocky sea floor, although the depth of the water exceeded 100 m. The impression was that enormous barrels full of water were jumping up and down in the hold and that the sheathing was on the brink of breaking. The shrouds trembled, the handrails of the ladder broke down, the windows of the deck cabin crumbled, the deck superstructures started to move and fall to pieces.

The vessel prepared for its unavoidable death.

Suddenly, the din stopped. The sea continued to vibrate, gradually calming down. The vessel, that had suffered in 2 mimutes more than from the most monstrous storm, was rocking quietly on the breezeless sunlit surface of the sea. If the seaquake had continued for half a minute more, then it would have surely led to the appearance of still another 'Flying Dutchman', abandoned by its crew, or to the mysterious disappearance in the ocean of one more vessel together with its crew.

In most parts of the World Ocean there exists a very clearly expressed and stable temperature stratification. Cold depth waters (a layer $\sim 10^3$ m thick) are separated from the atmosphere by a relatively thin ($\sim 10^1$–10^2 m) warm layer, which contains a mixed layer and a thermocline. The evidence, reflected in the synthesized description of seaquakes due to witnesses, permits to assume underwater earthquakes to be capable of causing so intense displacements of water that they are accompanied by sharp, but short, enhancement of the vertical exchange in the ocean (seismogenic upwelling). The concrete physical mechanisms of enhancement of the vertical exchange may be due to non-linear flows or to turbulence, the development of which in a layer of water results from strong seismic motions of the ocean bottom.

Seismogenic upwelling may affect many processes in the ocean. Sea water changing colour or becoming turbid as a result of ocean-floor sediments being carried up toward the surface and suspended represents the most noticeable manifestation of seismogenic upwelling. Moreover, the formation is possible of extensive anomalies in the sea surface temperature (SST) and, consequently, unusual weather phenomena may develop. The arrival of biogenes in the surface layer, usually depleted of such substances should be accompanied by explosive development of phytoplankton. Evolution of the zone with violated, and therefore unstable, stratification should create a powerful system of internal waves.

Some results of the research in seismogenic upwelling, under way since 1993 with participation and under the leadership of the authors of this book will be presented in Chap. 7.

1.7 Hydroacoustic Signals in the Case of Underwater Earthquakes

The capability of underwater earthquakes to excite hydroacoustic signals has been known long ago [Ewing et al. (1950); Soloviev et al. (1968); Kadykov (1986)]. Hydroacoustic waves propagating in the ocean from the epicentral zone of an earthquake are called the T-phase. Investigation of this phenomenon traditionally pertains to the scientific activity of seismologists. The term T-phase originated from

1.7 Hydroacoustic Signals in the Case of Underwater Earthquakes

seismological classification, since this wave is registered as the third phase (tertiae) after the appearance of phases P (primae) and S (secondae). The range of frequencies, usual for the T-phase, is 1–100 Hz. The lower boundary of this range, most likely, depends on the conditions for propagation of the signal along the SOFAR (low velocity waveguide).

Seismic deformations of the bottom are also capable of causing low-frequency (~ 0.1 Hz) elastic oscillations of the water layer. These oscillations exist near the tsunami source. Their nature is related to the multiple reflection of the elastic wave from the water–air and the water–bottom surfaces. These effects are dealt with in detail in Chap. 3.

Registration of the T-phase is possible not only with the aid of seismographs, but also with hydrophones. The latter method, for instance, is actively used in the American system SOSUS (SOund SUrveillance System) [Fox, Hammond (1994)], operating from the middle of the 1950s and initially intended for searching for submarines. The system represents a set of hydrophones, connected with the coastal services by a cable line. Registration of a T-phase signal by the SOSUS system permits to successfully determine the coordinates of epicentres of underwater earthquakes, which serves as a successful alternative to traditional seismological methods (http://www.pmel.noaa.gov/vents). Similar hydroacoustic systems were also created some time ago in the USSR [From the history (1998)].

In Russia, several recent years saw the revival of research aimed at making use of hydroacoustic signals from underwater earthquakes for tsunami warning [Sasorova et al. (2002)]. If a tsunami is excited by a nearby earthquake, the modern tsunami warning system has very little chance of providing a timely alert signal, since the time provided by nature for reacting (the time interval between the arrival of the seismic signal and the first tsunami wave) amounts to less than 5 min. At present, the only promising way of withstanding local tsunamis consists in making use in good time of available information on the preparatory stages of a developing underwater earthquake.

Analysis of the records of oceanic hydroacoustic noises, obtained by the Russian multi-purpose antenna AGAM within the framework of the international programme ATOC (Acoustic Thermometry of the Ocean's Climate) between 1998 and 1999 revealed promising results. The set of hydrophones established on the Pacific shelf of Kamchatka registered hydroacoustic signals of seismic origin in the 3–70 Hz frequency range, which appeared much earlier than the first blow from the earthquake (from hours down to several minutes) [Lappo et al. (2003)]. The signals were generated by microearthquakes in the preparation area of a strong earthquake and were evidence of the development of the event's critical stage.

It must be noted that signals of a similar type, caused by microdestructions of rock (acoustic emission and so on) and propagating in ground and rock dampen very rapidly and are practically imperceptible by land stations already at a distance of several kilometers from the source. The amplitude of an acoustic signal drops exponentially with the distance, and the exponent is proportional to the signal's frequency. The damping factor in water for a signal of frequency 100 Hz amounts to 0.0006 dB/km, in magmatic rock it is approximately 0.01 dB/km, in sedimentary rock and sand of the order of 0.1 and 0.5 dB/km, respectively. A signal of frequency

30 Hz dies out completely in sand at a distance of 2 km from the source, and in consolidated sedimentary rock at a distance of 10 km. In water such a signal is reliably registered at distances of up to 1,000 km.

Hence follows the important conclusion that hydroacoustic monitoring of the preparation process of an oceanic earthquake may lead to success in resolving the difficult problem of revealing in good time a nearby earthquake at its preparation stage and of issuing a timely and effective warning of the possible rise of a local tsunami.

During strong underwater earthquakes the ocean bottom in the epicentral area is deformed, and the deformation not only has a horizontal component, but a vertical one as well. If the motion is directed vertically upward, then a wave of compression forms in the water and propagates towards the surface; if the motion is downward, then a decompression wave forms. When reflected from a free surface of water, an elastic wave changes its polarity, therefore independently of the sign of the deformation, there may always be realized a wave of decompression, which tends to 'tear apart' the liquid. The amplitude of pressure variations related to elastic waves can be calculated by the formula

$$p_d = \rho c U,$$

where ρ is the water density, c is the velocity of sound in the water and U is the velocity of motion of the ocean bottom. If the bottom moves with a velocity of 1 m/s, the pressure amplitude will amount to $p_d = 1.5$ MPa.

Besides variations of pressure due to elastic waves, in the water layer there exists a hydrostatic pressure, the increase of which with the depth is approximately linear,

$$p_{st} = p_{atm} + \rho g z.$$

The total pressure $p_d + p_{st}$ at large depths is always positive, but in the layer near the surface (for $U = 1$ m/s down to $z \approx 140$ m) a situation may arise, when the dynamic pressure exceeds the hydrostatic pressure in absolute value, so the total pressure turns out to be negative.

The limit strength of water under tension is known to be about 0.25 MPa. Therefore, in the subsurface layer, where the total pressure $p_d + p_{st}$ exceeds in absolute value the limit strength of water, violation of the water continuity is possible, and it is called cavitation. The influence of the described mechanism taking place above the epicentre of an underwater earthquake results in the formation of a zone of cavitating (partly foaming) water. This zone has a reflection coefficient (albedo) differing from the reflection coefficient of all the remaining surface of the aquatorium. In this case the perturbated zone of the water surface can be registered by remote methods (from satellites, airplanes, etc.). Note that cavitation effects in the subsurface layer are observed in the case of underwater explosions.

1.8 Killer Waves in the Ocean

The killer-wave phenomenon is in no way related to the seismic activity of the sea-floor; however, the authors considered it expedient to present this rare and catastrophic oceanic event in the book. It is a striking example of the origination

1.8 Killer Waves in the Ocean

of rapidly proceeding high-energy wave phenomena, representing a significant danger for seafaring and a large interest for investigators of the physics of the ocean.

The problem of studying waves of anomalously large heights and of unusual shapes on the surface of the ocean has recently started to occupy the central part of intense theoretical and application research. The phenomenon of anomalously high waves (sometimes called 'killer waves', 'rogue waves' or 'freak waves') consists in the sudden appearance of enormous waves that are two or more times higher than the most significant agitation. The suddenness and large energy of such waves have repeatedly resulted in accidents of ships and to casualties (Fig. 1.7). The description of a series of such cases can be found, for example, in the book by Kurkin and Pelinovsky [Kurkin, Pelinovsky (2005)].

Fig. 1.7 Photographs of destructions of ships after encounters with killer waves. Modified after [Kurkin, Pelinovsky (2005)]

The following possible mechanisms for killer-wave formation have been proposed: linear, or with account of non-linear corrections, focusing of wave groups resulting from the influence of dispersion; variable conditions of wave propagation (variable bathymetry, variation of direction and the capture by currents); the action of variable atmospheric conditions; non-linear modulatory instability and interaction of solitonlike wave groups.

Killer waves have been repeatedly observed at the eastern coasts of Africa, where the combination of hydrophysical peculiarities of the narrow shelf zone and a strong wind influence provide favourable conditions for the rise of such waves. Recently, such waves have been registered quite often at the coasts of the Atlantic, of the North Sea and in the deep part of large water basins. The 'Newyear wave', registered on an oil rig in the North Sea on January 1, 1995, has become a popular object of research.

We here present a description of the killer wave that originated in the Black Sea on November 22, 2001. A 'Directional Waverider Buoy', made by the Dutch company DATAWELL, was established in 1996 in the vicinity of the town of Gelenjik. The Buoy was intended for direct measurement of parameters of the wind waves (including the direction of propagation), for initial data processing and transmission of the accumulated material to the coastal receiving device. The coordinates of the point where the buoy was established were: $44°30'40$ N, $37°58'70$ E, and the depth of the site was 85 m.

During the entire period of instrumental measurements of the wind waves, caused by wind in the north-eastern part of the Black Sea (1996–2003), three records were revealed with anomalously large waves. The record of displacements of the surface with time, registered by the buoy on November 22, 2001, including a wave of anomalous height, 10.3 m and a significant height of waves amounting to 2.6 m over the set, is presented in Fig. 1.8.

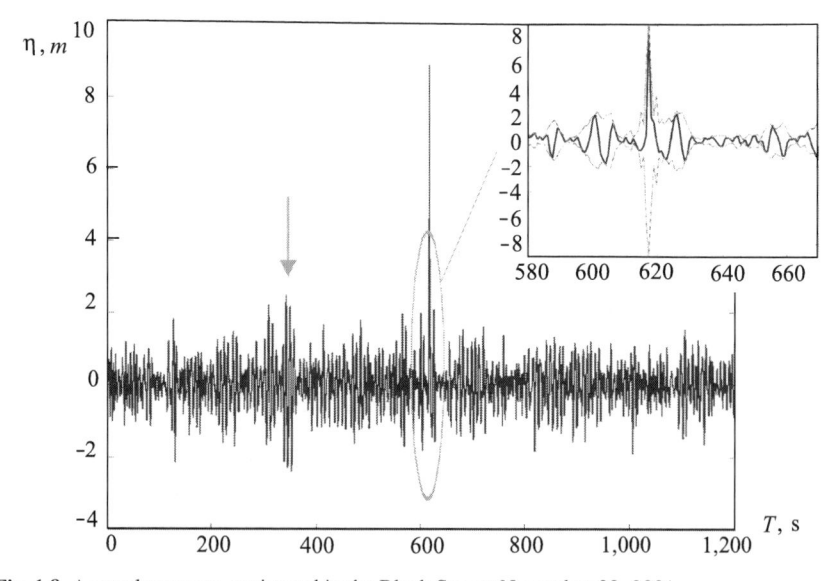

Fig. 1.8 Anomalous wave, registered in the Black Sea on November 22, 2001

The record shows an anomalous wave registered at a moment of time 617 s after registration started (the wave is shown in a larger scale in the insert); the wave's main characteristics are the following:

- Height $h_{max} = 10.32$ m (which exceeds the significant height $h_s = 2.6$ m by a factor of ~ 4)
- $h_{max}/h^- = 4.86$, $h_{max}/h^+ = 2.84$, where h^- and h^+ are the respective wave heights before and after the anomalous wave
- $C/h_{max} = 0.86$, where C is the height of the crest

Analysis of the maximum and minimum of displacements of the wave surface (of crests and troughs) as functions of the coordinate (determined during the entire registration time) revealed that the characteristic scale of localization of the peak wave is approximately 20 m. The wave covers this distance in 4.3 s. Thus, formation of the wave is sudden. Outside this region the wave field has no noticeable peaks.

The analysis performed of the record in [Divinsky et al. (2004)] and the search for long-lived non-linear solitary wave groups (applying the method of the inverse scattering problem in the approximation of the non-linear Schrödinger equation) revealed that the anomalous wave arose against a background of intense waves with amplitudes superior to the characteristic amplitudes of the surrounding waves. A group of intense waves with similar parameters is also present in the record at another moment of time, but it exhibits no anomalous behaviour.

References

Abe K. (1979): Tsunami propagation on a seismological fault model of the 1952 Kamchatka earthquake. Bull. Nippon Dental Univ. (8) 3–11

Abe K. (1981): Physical size of tsunamigenic earthquakes of the northwestern Pacific. Phys. Earth Planet Int. **27** 194–205

Abe K. (1985): Quantification of major earthquake tsunamis of the Japan Sea. Phys. Earth Planet Int. **38** 214–223

Abe K. (1989): Quantification of tsunamigenic earthquakes by the Mt scale. Tectonophysics **166** 27–34

Ambraseys N. N. (1962): Data for investigation of the seismic sea-waves in the eastern Mediterranean. Bull. Seismol. Soc. Am. **52** 895–913

Divinsky B. V., Levin B. V., Lopatukhin L. I., Pelinovsky E. N., Slyunyaev A. V. (2004): A freak wave in the Black Sea: Observations and simulation. Doklady Earth Sciences **395**(3) 438–443

Dotsenko S. F., Kuzin I. P., Levin B. V., Solov'eva O. N. (2000): Tsunami in the Caspian sea: Seismic sources and features of propagation. Oceanology **40**(4) 474–482

Ewing W. M., Tolstoy I., Press F. (1950): Proposed use of the T phase in tsunami warning systems, Bull. Seism. Soc. Am. **40** 53–58

Fox C. G., Hammond S. R. (1994): The VENTS Program T-Phase Project and NOAA's role in ocean environmental research. MTS Journal **27**(4) 70–74

From the history of our home hydroacoustics (in Russian) (1998): In: Collection, edited by Ya. S. Karlik. A. N. Krylov. TsNII Publishing house, St. Petersburg

Gusiakov V. K. (2001): Basic Pacific Tsunami Catalog and Database, 47 BC–2000 AD: results of the First Stage of the Project. In: ITS 2001 Proceedings, Session 1, No 1–2 263–272

Handbook for Tsunami Forecast in the Japan Sea. (2001): Earthquake and Tsunami Observation Division, Seismological and Volcanological Department, Japan Meteorological Agency, 22

Hatori T. (1986): Classification of tsunami magnitude scale. Bull. Earthquake Res. Inst. Univ. Tokyo **61** 503–515 (in Japanese with English abstract)

Iida K. (1956): Earthquakes accompanied by tsunamis occurring under the sea off the islands of Japan. J. Earth Sciences Nagoya Univ. **4** 1–43

Iida K. (1970): The generation of tsunamis and the focal mechanism of earthquakes. In: Tsunamis in the Pacific Ocean, edited by W. M. Adams, pp. 3–18. East–West Center Press, Honolulu

Imamura A. (1942): History of Japanese tsunamis. Kayo-No-Kagaku (Oceanography), **2** 74–80 (in Japanese)

Imamura A. (1949): List of tsunamis in Japan. J. Seismol. Soc. Japan, **2** 23–28 (in Japanese)

Kadykov I. F. (1986): The acoustics of submarine earthquakes (in Russian). Nauka, Moscow

Kurkin A. A., Pelinovsky E. N. (2005): Freak waves: facts, theory and modelling (in Russian). Publishing house of Nizhegorod. State Techical University, N. Novgorod

Lander J. F., Whiteside L. S., Lockridge, P. A. (2002): A brief history of tsunami in the Caribbean Sea. Sci. Tsunami Hazards **20**(2)

Lappo S. S., Levin B. W., Sasorova E. V., et al. (2003): Hydroacoustic location of an oceanic earthquake origin area. Doklady Earth Sci. **389**(2)

Levin B. W. (1996): Tsunamis and seaquakes in the ocean (in Russian). Priroda (5) 48–61

Levin B. W., Sasorova E. V. (2002): On the 6-year tsunami periodicity in the Pacific. Izvestiya, Physics of the Solid Earth **38**(12) 1030–1038

Levin B. W., Nosov M. A. (2008): On the possibility of tsunami formation as a result of water discharge into seismic bottom fractures. Izvestiya, Atmo. Ocean. Phys. **44**(1) 117–120

Murty T. S., Loomis H. G. (1980): A new objective tsunami magnitude scale. Mar. Geod. **4** 267–282

Nikonov, A. A. (1997): Tsunami occurrence on the coasts of the Black Sea and the Sea of Azov. Izv. Phys. Solid Earth **33** 72–87

Okada Y. (1985): Surface deformation due to shear and tensile faults in a half-space. Bull. Seismol. Soc. Am. **75**(4) 1135–1154

Papadopoulos G. A., Imamura F. (2001): A Proposal for a new tsunami intensity scale. In: Proceedings ITS, pp. 569–577

Pelinovsky E. N. (1982): Nonlinear dynamics of tsunami waves (in Russian), Institute of Applied Physics, USSR AS. Gorky

Sasorova E. V., Didenkulov I. N., Karlik Ya. S., Levin B. W., et al. (2002): Underwater earthquakes near shorelines: acoustic methods for identifying the preparation process of an earthquake and the prospects of their application for tsunami warning systems (in Russian). In: Collection of articles 'Local tsunamis: warning and risk mitigation', Yanus-K, Moscow, pp. 167–180

Sieberg A. (1927): Geologische, physikalische und angewandte Erdbebenkunde. Verlag von Gustav Fischer, Jena

Soloviev S. L. (1968): The tsunami problem and its significance for the Kamchatka and the Kuril islands (in Russian). In: The tsunami problem, pp. 7–50. Nauka, Moscow

Soloviev S. L., Ferchev M. D. (1961): Compilation of data on tsunamis in the USSR (in Russian). Bull. Seismol. Counc. (9) 43–55

Soloviev S. L., Voronin P. S., Voronina S. I. (1968): Seismic hydroacoustic data on the T wave (review of the literature) (in Russian). In: The tsunami problem, pp. 142–173. Nauka, Moscow

Soloviev S. L., Go C. N. (1975): Catalogue of tsunamis on the eastern coast of the Pacific Ocean (1513–1968) (in Russian). Nauka, Moscow

Soloviev S. L., Go C. N. (1974): Catalogue of tsunamis on the western coast of the Pacific Ocean (173–1968) (in Russian). Nauka, Moscow

Soloviev S. L., Go C. N., Kim Kh. S. (1986): Catalogue of tsunamis in the Pacific Ocean, 1969–1982 (in Russian). Izd. MGK, USSR AS, Moscow

Soloviev S. L., Go C. N., Kim Kh. S., et al. (1997): Tsunamis in the Mediterranean Sea, 2000 BC–1991 AD (in Russian), Nauchnyi mir, Moscow

References

Tatehata H. (1998): The new tsunami warning system of the Japan Meteorological Society. Sci. Tsunami Hazards **16**(1) 39–49

UNESCO-IOC. Tsunami Glossary. IOC Information document No. 1221. Paris, UNESCO, 2006

Zav'yalov P. O., Levin B. V., Likhacheva O. N. (2006): Statistical relations of the run-up height of tsunami waves to the distance and energy of the source. Oceanology **46**(1) 10–16

Chapter 2
Physical Processes at the Source of a Tsunami of Seismotectonic Origin

Abstract Modern ideas are presented concerning the source of an earthquake and the seismotectonic source of a tsunami. The main physical processes taking place at a tsunami source are described. Estimation is performed of the role of secondary effects: displacements of the bottom, occurring in its own plane, Coriolis force, and density stratification of the water. The Okada formulae are presented, and the technique is exposed for calculating residual bottom deformations caused by an underwater earthquake. Within the framework of linear potential theory of an incompressible liquid in a basin of fixed depth, the general analytical solution is constructed for the two-dimensional (2D) and three-dimensional (3D) problems of tsunami generation by bottom deformations of small amplitudes. The solution of the 3D problem is constructed in both Cartesian and cylindrical coordinates. For a series of model bottom deformation laws (piston, membrane and running displacements, bottom oscillations and alternating-sign displacement) physical regularities are revealed that relate the amplitude, energy and direction of tsunami wave emission to peculiarities of the bottom deformation at the source. In some cases, the theoretical regularities, obtained within potential theory, are compared with dependences following from the linear theory of long waves and, also, with the results of laboratory experiments.

Keywords Tsunami source · tsunami generation · earthquake · seismic moment · magnitude · tsunami earthquake · fault · residual deformation · initial elevation · bottom displacement · time scale · period of tsunami · duration of earthquake · Coriolis force · vortex · kinetic energy · potential energy · internal waves · Burger's vector · strike-slip fault · dip-slip fault · tensile fault · Lame constants · strike angle · dip angle · rake(slip) angle · slip distribution · linear potential theory · velocity potential · Laplace transformation · Fourier transformation · analytical solution · dispersion · directional diagrams

The following three stages are traditionally distinguished in the life of a tsunami: generation of the wave, its propagation in open ocean and its interaction with the coast (its uprush or run-up). Such a division is related to the existence of essential differences in the physical processes controlling one or another stage. Naturally, the description of all the stages is based on general principles of the mechanics of continuous media; however, application of complete three-dimensional (3D) equations for describing concrete tsunamis is not only irrational, but also just impossible at the present-day stage of development of computational technologies. The only possible way consists in the development of an inter-related complex of models, each of which will adequately describe a certain stage in the evolution of the wave. Strong underwater earthquakes are the most widespread cause for the rise of tsunami waves. Part of the energy of a seismic source is captured by the water column and is transferred, primarily, to various wave motions. The number of works devoted to investigation of the tsunami formation mechanism by a seismotectonic source is incredibly large. Without claiming to present a full list, we shall only mention several publications: [Takahasi (1934, 1963); Miyoshi (1954); Kajiura (1963, 1970); Van Dorn (1964); Kanamori (1972); Hammack (1973); Abe (1978); Levin, Soloviev (1985); Kowalik, Murty (1987); Dotsenko, Soloviev (1988); Nosov (1999); Ohmachi et al. (2001); Yagi (2004); Okal, Synolakis (2004); Satake et al. (1995)]. The physical processes taking place at a seismotectonic tsunami source and at an earthquake focus represent a unique whole. Therefore, studies of tsunami and of earthquake sources mutually complement and enrich each other. In spite of the significant progress achieved in this direction in the past decades, both tsunami and earthquake sources still remain 'terra incognita'.

2.1 Seismotectonic Source of a Tsunami: The Main Parameters and Secondary Effects

2.1.1 The Main Parameters

According to modern ideas, an earthquake is the abrupt release of strain accumulated in the Earth's crust, resulting from the relatively slow motion of lithosphere plates [Kanamori, Brodsky (2004)]. The source of an earthquake can be represented as a displacement that occurs owing to a fault along one or several planes. In the case of large shallow events the rupture speed amounts to 75–95% of the velocity of S-waves. An earthquake is characterized by the seismic moment

$$M_0 = \mu DS [\text{N} \cdot \text{m}],$$

where μ is the rigidity coefficient of the medium, D is the displacement amplitude between the opposite edges of the fault, and S is the area of the fault surface. The earthquake's magnitude is related to the seismic moment by the following relationship:

$$M_w = \frac{\log_{10} M_0}{1.5} - 6.07.$$

2.1 Seismotectonic Source of a Tsunami

Some seismic events (e.g. Sanriku, 1896, the Aleutian earthquake, 1946) caused tsunamis of intensities higher, than could be expected from the available seismic data. Kanamori [Kanamori (1972)] termed such earthquakes 'tsunami earthquakes' and presumed them to occur, when the process at the earthquake source underwent unusually slow development. This case is characterized by a low emission efficiency of the high-frequency component of seismic waves, which is not so important for the process of tsunami generation.

Figure 2.1 presents the relationship between the tsunami intensity (Soloviev–Imamura scale) and the earthquake magnitude for the Pacific region constructed by means of the Pacific tsunami database (see Sect. 1.5). The large spread between the data signifies that the relationship between tsunamis and earthquakes is complex

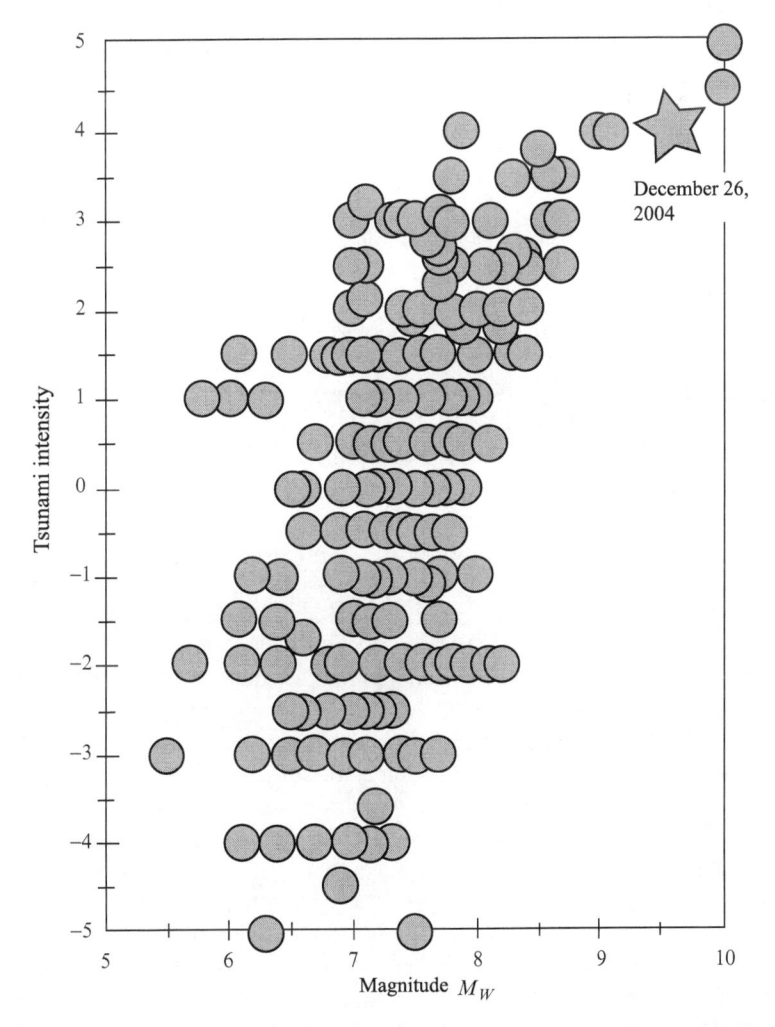

Fig. 2.1 Dependence of the tsunami intensity, according to the Soloviev–Imamura scale, upon the earthquake magnitude M_w for the Pacific region

and ambiguous. Besides the earthquake magnitude, a tsunami intensity may depend on many parameters: the hypocentre depth, shape and orientation of the fault area, duration of processes at the earthquake source, ocean depth, etc. It is seen that success in the investigation of tsunami generation is related not only to resolution of the hydrodynamic part of the problem, but also to progress in resolving such a difficult problem as description of the earthquake source. It must be noted that the large spread is also due to the tsunami intensity not being a rigorously defined physical quantity like, energy. At any rate, a certain positive correlation within the dependence under consideration can be identified: earthquakes of higher magnitudes are generally accompanied by tsunamis of higher intensities. The dependence presented is a good illustration of the magnitude criterion applied in the tsunami warning system. It is seen that the formation of practically all significant tsunamis ($I > 2$) was due to earthquakes of magnitudes $M_w > 7$.

The process of tsunami generation has been studied relatively weakly, which largely due to the fact that no measurements have hitherto been performed at tsunami sources. Indeed, all the information on processes proceeding at a tsunami source has been obtained by remote measurements done with mareographs (coastal and deep-water devices), hydroacoustic systems or seismogrpahs. Evidence provided by witnesses of underwater earthquakes is quite scarce, and it naturally concerns phenomena that took place at the ocean surface. In principle, it became possible to investigate a tsunami formation at its source in 1996, when a set of measuring devices comprising several sensors of bottom pressure (JAMSTEC, Japan Agency for Marine-Earth Science and Technology) was established on the continental slope close to the Japanese islands. The Tokachi-Oki earthquake of 2003 was the first strong seismic event with its epicentre located in the immediate vicinity of the JAMSTEC sensors. In Sect. 3.1.6 data will be analysed on variations of the bottom pressure, registered by the JAMSTEC sensors at the source of the 2003 Tokachi-Oki earthquake.

In simulating a tsunami of seismic origin a convenient method is usually applied that permits not to deal with the description of the generation process in a straightforward manner. The 'roundabout manoeuvre' consists in the following. An earthquake is considered to suddenly cause residual deformations of the ocean bottom (actually the duration of the process at the source may amount to 100 s and more). The residual deformations of the bottom are deduced from the parameters of the earthquake source. The calculation technique will be presented in detail in Sect. 2.1.3. Then, the assumption is made that the displacement of the bottom is simultaneously accompanied by formation at the surface of the ocean of a perturbation, the shape of which is fully similar to the residual deformations of the bottom. The perturbation of the water surface (the initial elevation), thus obtained, is then applied as the initial condition in resolving the problem of tsunami propagation.

It is interesting that the possibility to transfer sea-floor perturbations up to the surface is based on the actual structure of the equations for shallow water requiring the sole condition that the sea-floor deformation process be rapid. If, contrariwise, one applies, for instance, potential theory, then, even if the process is instantaneous, the perturbation of the liquid's surface and the residual deformation will differ from each other.

2.1 Seismotectonic Source of a Tsunami

In general, it is evidently not correct, from a physical point of view, to transfer sea-floor deformations up to the surface. In the case of deformation of the sea-floor, lasting for a long time, i.e. when a long wave has time to propagate over a noticeable distance, as compared with the horizontal dimension of the source, elevation of the surface will at no particular moment of time coincide with the residual displacements of the sea-floor. But this effect could still be taken into account within the framework of the long-wave theory. If, on the other hand, the duration of the deformation is small, then the motion of the water layer must be described within the framework of the theory of a compressible liquid. Here, the theory of long waves turns out to be totally inapplicable. In the case of high-speed displacement of the sea-floor an additional contribution to the tsunami wave can also be given by non-linear effects.

Note the paradoxical effect, manifested when tsunami generation is considered a process proceeding in an incompressible liquid. For definiteness we shall assume an earthquake resulting in area S ($\sqrt{S} \gg H$) of the sea-floor being displaced vertically with a constant velocity by a quantity η_0 during a time interval τ. According to the theory of an incompressible liquid, practically all the water layer immediately above the moving part of the sea-floor acquires a vertical velocity $\eta_0 \tau^{-1}$, and, consequently, the kinetic energy

$$W_k = \frac{\rho S H \eta_0^2}{2\tau^2}. \tag{2.1}$$

The displacement results in a perturbation forming on the water surface (we shall consider it identical to the deformation of the sea-floor), which contains the potential energy

$$W_p = \frac{\rho S g \eta_0^2}{2}. \tag{2.2}$$

The paradox consists in that the kinetic energy involved in the process has a fixed value, but immediately after its completion the kinetic energy disappears without leaving a trace. The paradox is readily resolved, naturally, if the condition $W_p \gg W_k$ is applied. But in reality the kinetic energy may not only be comparable to the potential energy, but even significantly exceed it. Indeed, from formulae (2.1) and (2.2) we have

$$\frac{W_k}{W_p} = \frac{\tau_0^2}{\tau^2},$$

where $\tau_0 = (H/g)^{1/2}$ is the propagation time of a long gravitational wave over a distance equal to the depth of the ocean ($\tau_0 \approx 20\,\mathrm{s}$ for $H = 4{,}000$ m). In many cases $\tau < \tau_0$, and, consequently, $W_k > W_p$. An accurate resolution of the said paradox is possible within the framework of the theory of compressible liquids.

For an adequate mathematical description of the processes occurring when waves are generated it is necessary to have a clear idea of the characteristic values of the main parameters defining the problem. The range of tsunami wave periods has already been indicated above. The depth of the ocean in area of a tsunami source may vary from several kilometres to zero (when the area of the sea-floor deformation extends onto the land). The horizontal size of the tsunami source usually amounts to tens and even hundreds of kilometres. The empirical dependence that relates the mean radius R_{TS} [km] of the tsunami source and the earthquake magnitude M is known as

$$\lg R_{TS} = (0.50 \pm 0.07)M - (2.1 \pm 0.6). \qquad (2.3)$$

Note that real tsunami sources, naturally, do not exhibit a circular, but instead a more complex, as a rule, elongated shape. At any rate, the boundary of a tsunami source is a concept that is essentially conventional. The source of a tsunami of seismic origin can be defined as the area, within which an earthquake has resulted in noticeable residual deformations of the sea-floor or within which significant seismic oscillations have occurred. From records of waves made by the method of inverse isochrones it is possible to reconstruct the tsunami source region. It is interesting that a source reconstructed in this manner usually exhibits a reasonable correspondence to the area of aftershock manifestations. It must also be stressed that, as a rule, residual deformations are bipolar, i.e. elevation of the sea-floor takes place in one part of the source and it is subsided in another part. Figure 2.2, adapted from [Satake, Imamura (1995)], presents the example of the reconstruction of the Tokachi-Oki 1968 tsunami source.

Figure 2.3 shows the areas of the fault surface at the earthquake source (solid line) and of the tsunami source (dotted line) as functions of the earthquake seismic moment (magnitude). The area of the tsunami source was calculated as the area of a circle with a radius determined by formula (2.3). The area of the tsunami source can be seen to be several times larger than the area of the fault at the earthquake source, which is quite reasonable from a physical point of view. It is interesting to note that the said dependencies are practically parallel.

Another essential parameter characterizing tsunami generation by an earthquake is the displacement amplitude ξ_0 [m] of the oceanic surface at the source. This quantity approximately follows the vertical residual deformations of the ocean bottom. The corresponding regression estimate exhibits the following form:

$$\lg \xi_0 = (0.8 \pm 0.1)M - (5.6 \pm 1.0). \qquad (2.4)$$

Formulae (2.3) and (2.4) were derived in [Dotsenko, Soloviev (1990)] for magnitudes within the range of $6.7 < M < 8.5$ by analysis of the wave field at the source, reconstructed from measurements at the coast. The estimates for intervals correspond to an 80% probability. Note that formula (2.4) seems to yield overestimated values of residual displacements in the case of large magnitudes. The catastrophic tsunamigenic earthquake that occcurred on December 26, 2004, and the magnitude of which was $M_w = 9.3$ exhibited maximal vertical residual displacements of 8.6 m for the elevation area and of 3.8 m for the depression area [Grilli et al. (2007)].

The duration of processes at the tsunami source also represents an important parameter of the problem. Here, one must distinguish among several characteristic quantities. Earlier, we already introduced the timescale $\tau_0 = (H/g)^{1/2}$ peculiar to problems involving surface gravitational waves. Besides, there also exists the propagation time of a long gravitational wave over a distance, equal to the horizontal

2.1 Seismotectonic Source of a Tsunami   37

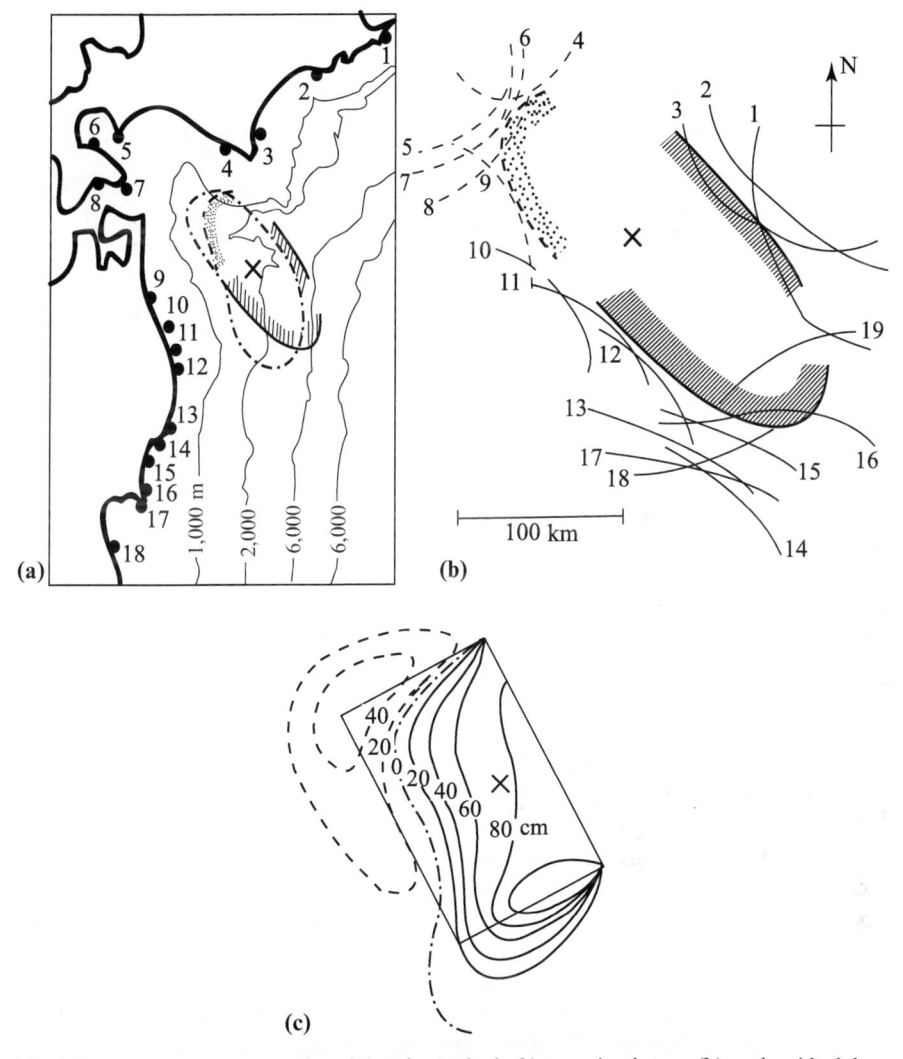

Fig. 2.2 Tsunami source restored applying the method of inverse isochrones (**b**), and residual deformations of the sea-floor (**c**) for the Tokachi-Oki 1968 earthquake. The figures are the numbers of mareographs, the locations of which are shown in the map (**a**). The solid and dotted curves correspond to the positive and negative leading wave, respectively. Adapted from [Satake, Imamura (1995)]

extension of the source, $T_{TS} = R_{TS}(gH)^{-1/2}$. Note that the order of the tsunami wave period depends precisely on the quantity T_{TS}. In a similar manner one can also introduce the propagation time of a hydroacoustic wave along the source, $T_S = R_{TS}/c$, where c is the speed of sound in water. The maximum period of normal elastic oscillations of a water layer, $T_0 = 4H/c$, is also related to hydroacoustic waves. And, ultimately, there exists a time that characterizes the duration of a process occurring

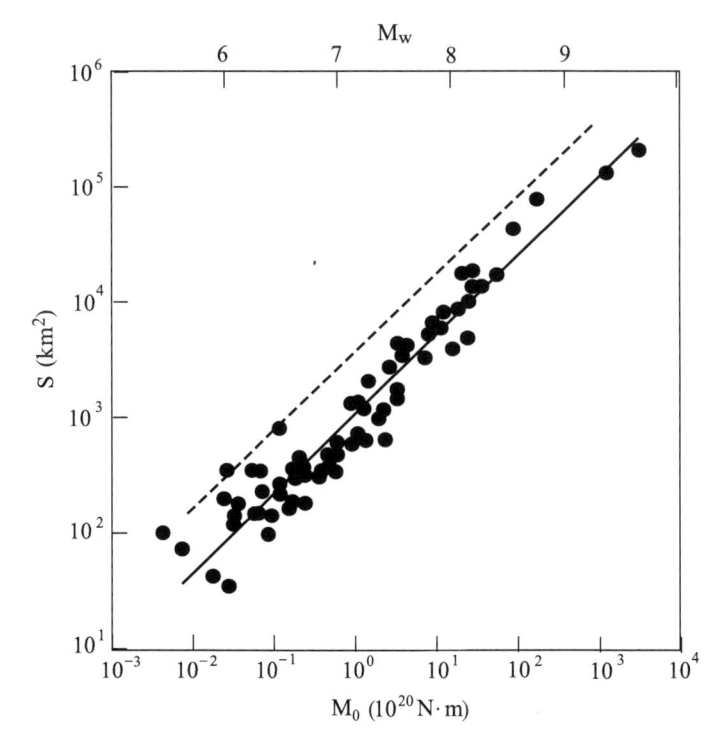

Fig. 2.3 The area of the fault at the source of the earthquake (solid line) versus the seismic moment (magnitude). Adapted from [Kanamori, Brodsky (2004)]. The dotted line represents an estimation of the area of the tsunami source in accordance with formula (2.3)

at an earthquake source, T_{EQ}. Note that deformation of the sea-floor (especially in the case of strong earthquakes) does not proceed simultaneously over the entire area of the tsunami source, but propagates horizontally following the fault that forms at the earthquake source. Therefore, the duration of the sea-floor deformation at a certain point may turn out to be significantly shorter than the quantity T_{EQ}. In the Harward seismic catalogue (http://www.seismology.harvard.edu/) a temporal characteristic termed 'half duration' is presented, which corresponds to half the duration of the process at an earthquake source. We shall denote this quantity by T_{hd} [s]. Analysis of all the earthquakes of magnitude $M_w > 7$, presented in the Harward catalogue for the period between January 1976 and March 2005 (370 events) permitted us to obtain the following regression relationship:

$$\lg T_{hd} = (0.42 \pm 0.02) M_w - (1.99 \pm 0.14). \tag{2.5}$$

Such a range of amplitudes was chosen, because significant tsunamis are excited by earthquakes with $M_w > 7$.

Figure 2.4 demonstrates the relation between the above temporal scales and the earthquake magnitude. In constructing the dependences we have applied formulae (2.3) and (2.5) and, besides, for definiteness, we have assumed the ocean depth to vary between 10^2 and 10^4 m.

2.1 Seismotectonic Source of a Tsunami

Fig. 2.4 Timescales of a tsunami source as functions of the earthquake magnitude. T_{TS} is the tsunami period, T_{hd} is the duration of the process at the earthquake source (the "half duration"), T_S is the propagation time of the hydroacoustic wave along the tsunami source, T_0 is the maximal period of normal elastic oscillations of the water layer, τ_0 is the time scale for gravitational waves. The ranges correspond to the interval of oceanic depths, 10^2–10^4 m

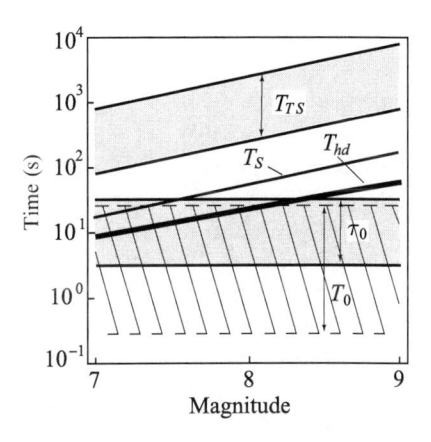

From Fig. 2.4 it can be seen that, as a rule, the duration of processes at the earthquake source, T_{hd}, is significantly inferior to the period of the tsunami wave, T_{TS}, that lies within the range 10^2–10^4 s. Therefore, the generation of waves is generally a relatively rapid process. The quantity τ_0 (within the considered range of magnitudes) is always smaller than the period of the tsunami wave, T_{TS}, however, in a number of cases this difference may turn out to be not so significant. In this connection, a tsunami can be considered a long wave, but with certain restrictions: in the case of small-size sources phase dispersion is certain to be manifested. Let us, now, turn to the quantity T_S, which always lies between the quantities T_{TS} and T_{hd}. This reflects the fact that the speed of hydroacoustic waves is always superior to the speed of long waves, but inferior to the speed, with which the fault opens up at the earthquake source. We further turn to elastic oscillations of the water layer. It is readily noted that the quantities T_0 and T_{hd} have very close values, so that effective excitation of elastic oscillations of the water layer is possible at the tsunami source. From the figure it is also seen that the maximal period T_0 of elastic eigen oscillations of the water layer is always smaller than the tsunami period T_{TS}, i.e. elastic oscillations and tsunami waves exist in ranges that do not intersect. This, however, does not mean that elastic oscillations cannot at all contribute to the energy of tsunami waves. Such a contribution can be realized by means of non-linear effects.

2.1.2 Secondary Effects

In setting boundary conditions on hard surfaces in hydrodynamic problems one conventionally distinguishes between the normal and tangential components of the flow velocity of the liquid. In the problem of tsunami generation such a hard surface is represented by the ocean bottom, which in the case of an earthquake can undergo motion both in its own plane, and in a perpendicular direction. We will term such displacements as tangential and normal. Actually, the surface of the ocean bottom has a complex structure, therefore the normal is conventionally constructed in a certain plane—the result of averaging either over the entire area of the tsunami source, or over a part of it. We shall consider the differences between this plane and the actual surface of the bottom to be irregularities.

We shall show that, for the excitation of motions in a water layer, normal displacements of the ocean bottom are essentially more effective than tangential ones. Let each point of the bottom surface at the tsunami source of area S undergo displacement over a distance η_0 during a time τ: once in the tangential direction and then in the normal direction. The normal to the bottom surface is at an angle α to the vertical direction. The slope of the surface of the oceanic bottom rarely exceeds 0.1, therefore the angle α can be considered small.

During tangential shifts the ocean bottom exerts a force on the water layer, equal to $\rho(u^*)^2 S$, where u^* is the friction velocity and ρ is the density of water. The energy transferred to the water layer by the ocean bottom undergoing motion can be estimated as the work performed by this force along the path η_0:

$$W_t = \rho(u^*)^2 S \eta_0. \qquad (2.6)$$

If one passes to the reference frame related to the moving ocean bottom, then one obtains the traditional problem of a logarithmic boundary layer, in which the quantity η_0/τ plays the part of the velocity of the average flow far from the boundary. The friction velocity is known to be essentially smaller than the velocity of the average flow, therefore, it is possible to write

$$W_t \ll \rho S \frac{\eta_0^3}{\tau^2}. \qquad (2.7)$$

We shall estimate the energy transferred to an incompressible layer of water by a normal displacement as the potential energy of the initial elevation above the water surface. We shall assume the horizontal dimensions of the source to essentially exceed the ocean depth $S^{1/2} \gg H$ and the displacement to be quite rapid, $\tau \ll S^{1/2}(gH)^{-1/2}$. In this case the entire volume of water dislodged by the displacement, $\eta_0 S$, will be distributed over an area $S\cos\alpha$ of the ocean surface. Thus, the amplitude of the initial elevation will amount to $\eta_0/\cos\alpha$. Taking into account the smallness of the angle α we obtain the following estimate for the potential energy of the intial elevation:

$$W_n = \rho g S \frac{\eta_0^2}{2}. \qquad (2.8)$$

Let us find the ratio between the energies transferred to the water layer by the normal and tangential displacements,

$$\frac{W_n}{W_t} \gg \frac{g\tau^2}{\eta_0}. \qquad (2.9)$$

If one assumes $\eta_0 = \xi_0$, $\tau = T_{hd}$ and applies formulae (2.4) and (2.5), then one can readily show that $g\tau^2/\eta_0 \approx 800 \gg 1$. Hence it follows that tangential motions of the ocean bottom can be neglected in the problem of tsunami generation.

2.1 Seismotectonic Source of a Tsunami

The real ocean is always stratified, and, moreover, owing to rotation of the Earth each moving particle of the water is under the influence of a Coriolis force. Therefore, tsunami generation is, generally speaking, accompanied by the formation of internal waves and vortical motions.

Let us estimate the effect due to rotation of the Earth, when vertical displacements of the ocean bottom are generated by a tsunami. We shall apply the linearized equations of shallow water written with account of the Coriolis force for a horizontally infinite ocean of depth H.

$$\frac{\partial u}{\partial t} = -g\frac{\partial \xi}{\partial x} + fv, \tag{2.10}$$

$$\frac{\partial v}{\partial t} = -g\frac{\partial \xi}{\partial y} - fu, \tag{2.11}$$

$$H\left(\frac{\partial u}{\partial x} + \frac{\partial v}{\partial y}\right) + \frac{\partial \xi}{\partial t} - \frac{\partial \eta}{\partial t} = 0, \tag{2.12}$$

where u, v are the components of the horizontal flow velocity, $f = 2\omega \sin\varphi$ is the Coriolis parameter, η represents small vertical deformations of the ocean bottom (deviations from the initial position), ξ is the displacement of the free surface from the equilibrium position. We differentiate equation (2.10) with respect to the coordinate y and equation (2.11) with respect to the coordinate x, and, then, we subtract one from the other. With account of the continuity equation (2.12) we ultimately obtain an evolution equation for the vertical curl component of the velocity

$$\frac{\partial}{\partial t}(\text{rot}_z \mathbf{v}) = \frac{f}{H}\left(\frac{\partial \xi}{\partial t} - \frac{\partial \eta}{\partial t}\right). \tag{2.13}$$

We shall assume no motion to exist in the water layer at the time moment $t = 0$ and the surfaces of the water and ocean bottom to be in an unperturbed state ($\mathbf{v} = 0$, $\eta = 0$, $\xi = 0$). We shall further assume deformation of the ocean bottom, arbitrary in space and time, but quite rapid ($\tau \ll R(gH)^{-1/2}$), to take place within a circular area of radius R, which will result in the formation of certain residual displacements. For simplicity we shall consider the residual displacements to differ from zero only inside the circular area of radius R, where they assume the fixed value η_0. The ocean bottom displacement results in formation of a wave perturbation of the surface, which after a sufficiently long period of time ($T \gg R(gH)^{-1/2}$) will leave the area of the source and the water surface will return to its initial unperturbed state.

The said assumptions make it possible to integrate equation (2.13) over time in the time interval from 0 up to T.

$$(\text{rot}_z \mathbf{v})|_{t=T} = -\frac{f}{H}\eta_0. \tag{2.14}$$

Expression (2.14) permits to conclude that influence of the Earth's rotation manifested at the tsunami source area, considering residual displacements of the ocean bottom to form at the site, must result in the formation of a certain vortical structure. Usually, bipolar deformation of the ocean bottom occurs at real tsunami sources, therefore, it may be assumed that several vortical structures are formed with different directions of rotation.

Let us estimate the energy of the vortical structure formed by the circular residual deformation. To this end we integrate expression (2.14) over the area of a circle of radius $r \leqslant R$, the centre of which coincides with the centre of the source. Applying the known Stokes formula, we pass in the left-hand part of the obtained expression to circulation of the velocity. With account of the radial symmetry of the problem we obtain for the velocity of vortical motion at a distance r from the centre,

$$V(r) = -\frac{f}{2H}\eta_0 r. \tag{2.15}$$

Note that, when $r > R$, the velocity $V = 0$. Knowledge of the velocity distribution readily permits to calculate the kinetic energy of the vortex,

$$W_k = \frac{\pi \rho f^2 \eta_0^2 R^4}{16H}. \tag{2.16}$$

Let us, now, compare the energy of the vortex with the energy of the tsunami wave, which we estimate as the potential energy of the initial elevation, similar in shape to the residual deformation of the ocean bottom (a circular area of radius R and height η_0),

$$W_p = \frac{\pi \rho g R^2 \eta_0^2}{2}. \tag{2.17}$$

Comparison of formulae (2.16) and (2.17) reveals the ratio of the energy of the vortex, formed at the tsunami source and due to rotation of the Earth, and the energy of the tsunami wave itself to be given by the following expression:

$$\frac{W_k}{W_p} = \frac{f^2 R^2}{8gH} \sim 10^{-2} - 10^{-4}. \tag{2.18}$$

The part of the energy due to vortical motion is seen to increase quadratically with the horizontal dimension of the source and to decrease as the ocean depth increases. But, in any case, the contribution of this energy does not exceed 1% of the energy of the tsunami wave. Note that such an estimate is correct for medium or high latitudes; for equatorial regions, where the Coriolis parameter is small, it will be significantly overestimated.

Let us, now, estimate the energy contribution of internal waves that are due to ocean bottom displacements. We shall consider the model of an ocean consisting of two layers: the upper layer of thickness h_1 with a free surface, and the lower layer of thickness h_2. The density of the upper layer is ρ_1 and of the lower layer is ρ_2 ($\rho_2 > \rho_1$). In this case it is convenient to base estimations on the one-dimensional (1D) (along the horizontal coordinate) model, constructed within the framework of the linear theory of long waves. We shall consider a segment of the ocean

2.1 Seismotectonic Source of a Tsunami

bottom of length L to undergo a vertical displacement η_0 during a time interval $\tau \ll L(g(h_1+h_2))^{-1/2}$. Such a displacement represents an impulse not only for surface waves, but also for internal waves, since the propagation velocity of the latter is significantly smaller. The displacement results in the formation of initial elevations both on the water surface and on the boundary surface separating the two layers; we shall consider these elevations to be similar in shape to the deformation of the ocean bottom. In principle, it should be possible already at this stage of reasoning to compare the energies of internal, W_{int} and surface, W_{sur} tsunami waves by comparison of the potential energies of the initial elevations. This ratio is evidently given by the formula

$$\frac{W_{int}}{W_{sur}} \approx \frac{\rho_2 - \rho_1}{\rho_2} \sim 10^{-3}. \tag{2.19}$$

But such a value is actually strongly overestimated. The point is that the evolution of initial elevations gives rise to two sets of waves, each of which consists of perturbations on the water surface and on the jump of density [Hammack (1980)]. One of the sets of waves propagates rapidly with the velocity of surface waves, the other one is essentially slower and propagates with the velocity of internal waves. As the initial elevation evolves, the water particles on the free water surface in the vicinity of the source are shifted downward. The maximum of this displacement, equal to η_0, corresponds to the free surface. At the ocean bottom, owing to there being no flow, the displacement equals zero. Assuming the displacement to depend linearly on the vertical coordinate, we obtain the displacement at the level of the density jump, $\Delta \eta = \eta_0 h_2/(h_1+h_2)$. The evolution of the elevation on the free surface is seen to result in the initial elevation at the density jump being reduced by the quantity $\Delta \eta$, while its height becomes equal to $\eta_{int} = \eta_0 h_1/(h_1+h_2)$. Naturally, the potential energy that is proportional to the square height of the initial elevation, also, decreases here. A more correct estimation yields the following relationship between the energies of the internal and surface tsunami waves:

$$\frac{W_{int}}{W_{sur}} \approx \frac{\rho_2 - \rho_1}{\rho_2} \left(\frac{h_1}{h_1+h_2}\right)^2 \sim 10^{-5}. \tag{2.20}$$

Estimations reveal that stratification of the ocean and rotation of the Earth cannot significantly influence the process of tsunami generation by an earthquake. But a small part of the earthquake's energy is transferred both to baroclinic motions and to vortical fields.

A complete physical formulation of the problem of tsunami generation by an earthquake should, generally speaking, consider a layer of viscous compressible stratified liquid on an elastic semispace in the gravitational field with account of the Earth's rotation. The above reasoning makes it possible to essentially simplify formulation of the problem. As a first approximation, we shall consider the process of tsunami generation by an earthquake to be a phenomenon occurring in a homogeneous (nonstratified) perfect incompressible liquid in the gravitational field in an inertial (without rotation) reference frame. Deformations of an absolutely rigid ocean bottom of finite duration and small amplitude ($A \ll H$) serve as the source of waves. Owing to tsunami waves being subject to dispersion, it is expedient to resolve the problem within the framework of potential theory.

In conclusion, we shall briefly dwell upon one more possible mechanism of tsunami formation in the case of underwater earthquakes. Experience of the investigation of catastrophic and strong seismic events shows that numerous seismic cracks of lengths exceeding tens of kilometers and widths amounting to 5–15 m arise at the epicentral zone. Dilatant changes of the state of rock in the same area develop, enhancement of the specific volume of the medium takes place, as well as revelation of microcracks and growth of its permeability. In the case of underwater earthquakes such processes should clearly take place in the rock of the ocean bottom. Rapid opening of the cracks at the ocean bottom should lead to an impetuous drainage of water.

Evidence provided by witnesses of the 1999 Izmit earthquake revealed that one of the shallow regions of the Sea of Marmara was dried up by the exclusive drainage of water through cracks in the sea-floor; large areas of the sea-floor were completely uncovered. In scientific literature, such phenomena are conventionally termed the Moses effect, in memory of the biblical Exodus through the Red Sea. Naturally, the dried areas of the sea-floor remain for a short time, until the water fills up the entire volume formed by the created set of cracks.

The impetuous drainage of water into cracks results in a local lowering of the ocean level. Such an initial perturbation is also capable of generating tsunami waves. The first results of mathematical simulation of the formation mechanism of a tsunami, caused by a fault opening up in the bottom, are presented in [Levin, Nosov (2008)].

2.1.3 Calculation of Deformations of the Ocean Bottom

For simulating tsunami waves of seismotectonic origin it is necessary to have realistic data concerning the residual deformations of the ocean bottom, resulting from an underwater earthquake. Residual deformations can be calculated on the basis of seismic data, making use of the analytical solution for the stationary problem of elasticity theory, presented in [Okada (1985)].

In this chapter, formulae are presented for surface displacements due to inclined shear and tensile faults in an isotropic homogeneous elastic half-space. The expressions have been carefully checked to be free from any singularities and misprints.

The Yoshimitsu Okada formulae are quite cumbersome and contain numerous variables. Therefore, in this section, in order to avoid errors, instead of our traditional notation, we shall accurately follow [Okada (1985)] and apply the original notation adopted therein.

We take the Cartesian reference system as it is shown in Fig. 2.5. The elastic medium occupies the region of $z \leqslant 0$. The $0x$ axis is taken to be parallel to the strike direction of a finite rectangular fault of length L and width W. Burger's vector $\mathbf{D} = (U_1, U_2, U_3)$ shows the movement of the hanging-wall side block relative to

2.1 Seismotectonic Source of a Tsunami

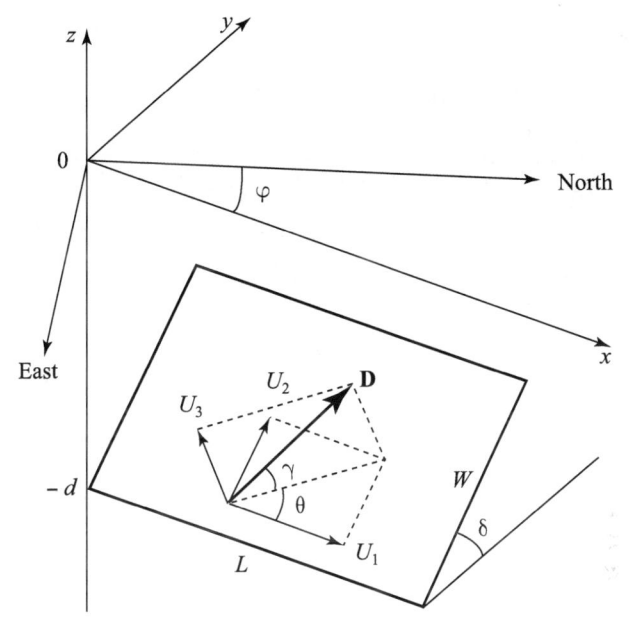

Fig. 2.5 Geometry of the source model (length L, width W, Burger's vector $\mathbf{D}$, dip angle δ, rake angle θ, angle between Burger's vector $\mathbf{D}$ and the fault plane γ)

the foot-wall side block. Elementary dislocations U_1, U_2, and U_3 are defined so as to correspond to strike–slip, dip–slip and tensile components of arbitrary dislocations. The tensile component U_3 is normal to the fault plane.

A dislocation is determined by four angles: the strike angle φ (clockwise from North), the dip angle δ, the rake (slip) angle θ and the angle γ between Burger's vector $\mathbf{D}$ and the fault plane. Elementary dislocations U_1, U_2 and U_3 are linked to Burger's vector in the following way: $U_1 = |\mathbf{D}|\cos\gamma\cos\theta$, $U_2 = |\mathbf{D}|\cos\gamma\sin\theta$, $U_3 = |\mathbf{D}|\sin\gamma$.

The final results are condensed into compact forms using Chinnery's notation $\|$ to represent the substitution

$$f(\xi,\eta)\| = f(x,p) - f(x,p-W) - f(x-L,p) + f(x-L,p-W). \quad (2.21)$$

For strike–slip

$$u_x = -\frac{U_1}{2\pi}\left[\frac{\xi q}{R(R+\eta)} + \arctan\left(\frac{\xi\eta}{qR}\right) + I_1 \sin\delta\right]\bigg\|,$$

$$u_y = -\frac{U_1}{2\pi}\left[\frac{\tilde{y}q}{R(R+\eta)} + \frac{q\cos\delta}{R+\eta} + I_2 \sin\delta\right]\bigg\|, \quad (2.22)$$

$$u_z = -\frac{U_1}{2\pi}\left[\frac{\tilde{d}q}{R(R+\eta)} + \frac{q\sin\delta}{R+\eta} + I_4 \sin\delta\right]\bigg\|.$$

For dip–slip

$$u_x = -\frac{U_2}{2\pi}\left[\frac{q}{R} - I_3 \sin\delta\cos\delta\right]\bigg|\bigg|,$$

$$u_y = -\frac{U_2}{2\pi}\left[\frac{\tilde{y}q}{R(R+\xi)} + \cos\delta\arctan\left(\frac{\xi\eta}{qR}\right) - I_1\sin\delta\cos\delta\right]\bigg|\bigg|, \quad (2.23)$$

$$u_z = -\frac{U_2}{2\pi}\left[\frac{\tilde{d}q}{R(R+\xi)} + \sin\delta\arctan\left(\frac{\xi\eta}{qR}\right) - I_5\sin\delta\cos\delta\right]\bigg|\bigg|.$$

For tensile fault

$$u_x = \frac{U_3}{2\pi}\left[\frac{q^2}{R(R+\eta)} - I_3\sin^2\delta\right]\bigg|\bigg|,$$

$$u_y = \frac{U_3}{2\pi}\left[\frac{-\tilde{d}q}{R(R+\xi)} - \sin\delta\left\{\frac{\xi q}{R(R+\eta)} - \arctan\left(\frac{\xi\eta}{qR}\right)\right\} - I_1\sin^2\delta\right]\bigg|\bigg|, \quad (2.24)$$

$$u_z = \frac{U_3}{2\pi}\left[\frac{\tilde{y}q}{R(R+\xi)} + \cos\delta\left\{\frac{\xi q}{R(R+\eta)} - \arctan\left(\frac{\xi\eta}{qR}\right)\right\} - I_5\sin^2\delta\right]\bigg|\bigg|,$$

where

$$I_1 = -\frac{\mu}{\lambda+\mu}\left[\frac{\xi}{(R+\tilde{d})\cos\delta}\right] - I_5\tan\delta,$$

$$I_2 = -\frac{\mu}{\lambda+\mu}\ln(R+\eta) - I_3,$$

$$I_3 = \frac{\mu}{\lambda+\mu}\left[\frac{\tilde{y}}{(R+\tilde{d})\cos\delta} - \ln(R+\eta)\right] + I_4\tan\delta, \quad (2.25)$$

$$I_4 = \frac{\mu}{\lambda+\mu}\frac{1}{\cos\delta}\left[\ln(R+\tilde{d}) - \sin\delta\ln(R+\eta)\right],$$

$$I_5 = \frac{\mu}{\lambda+\mu}\frac{2}{\cos\delta}\arctan\left(\frac{\eta(X+q\cos\delta)+X(R+X)\sin\delta}{\xi(R+X)\cos\delta}\right),$$

and if $\cos\delta = 0$,

$$I_1 = -\frac{\mu}{2(\lambda+\mu)}\frac{\xi q}{(R+\tilde{d})^2},$$

$$I_3 = \frac{\mu}{2(\lambda+\mu)}\left[\frac{\eta}{R+\tilde{d}} + \frac{\tilde{y}q}{(R+\tilde{d})^2} - \ln(R+\eta)\right], \quad (2.26)$$

$$I_4 = -\frac{\mu}{\lambda+\mu}\frac{q}{R+\tilde{d}},$$

$$I_5 = -\frac{\mu}{\lambda+\mu}\frac{\xi\sin\delta}{R+\tilde{d}},$$

2.1 Seismotectonic Source of a Tsunami

$$p = y\cos\delta + d\sin\delta,$$
$$q = y\sin\delta - d\cos\delta,$$
$$\tilde{y} = \eta\cos\delta + q\sin\delta,$$
$$\tilde{d} = \eta\sin\delta - q\cos\delta, \quad (2.27)$$
$$R^2 = \xi^2 + \eta^2 + q^2,$$
$$X^2 = \xi^2 + q^2.$$

The Lame constants λ and μ enter into expressions (2.25) and (2.26) in the form of a combination, which for practical calculations is conveniently expressed via the respective velocities of longitudinal and transverse seismic waves, c_p and c_s,

$$\frac{\mu}{\lambda+\mu} = \frac{c_s^2}{c_p^2 - c_s^2}.$$

Usually, this ratio varies within the range from 0.3 to 0.5. More precise information for a concrete region can be obtained, for instance, from the Reference Earth Model (http://mahi.ucsd.edu/Gabi/rem.html).

Under special conditions some terms in formulas (2.22)–(2.26) become singular. To avoid all singularities, the following rules should be obeyed:

(i) When $q = 0$, set $\arctan(\xi\eta/qR) = 0$ in equations (2.22)–(2.24)
(ii) When $\xi = 0$, set $I_5 = 0$ in equation (2.25)
(iii) When $R + \eta = 0$, set all the terms which contain $R + \eta$ in their denominators to zero in equations (2.22)–(2.26), and replace $\ln(R+\eta)$ by $-\ln(R-\eta)$ in equations (2.25) and (2.26).

To assist the development of a computer program based on expressions (2.21)–(2.27), several numerical results, permitting to check it, are listed in Table 2.1. A medium is assumed to be $\lambda = \mu$ in the all cases, and the results are presented in units of U_i.

Table 2.1 Checklist for numerical calculations. Adapted from [Okada (1985)]

	u_x	u_y	u_z
Case 1: $x=2$; $y=3$; $d=4$; $\delta=70°$; $L=3$; $W=2$			
Strike	$-8.689E-3$	$-4.298E-3$	$-2.747E-3$
Dip	$-4.682E-3$	$-3.527E-2$	$-3.564E-2$
Tensile	$-2.660E-4$	$+1.056E-2$	$+3.214E-3$
Case 2: $x=0$; $y=0$; $d=4$; $\delta=90°$; $L=3$; $W=2$			
Strike	0	$+5.253E-3$	0
Dip	0	0	0
Tensile	$+1.223E-2$	0	$-1.606E-2$
Case 3: $x=0$; $y=0$; $d=4$; $\delta=-90°$; $L=3$; $W=2$			
Strike	0	$-1.303E-3$	0
Dip	0	0	0
Tensile	$+3.507E-3$	0	$-7.740E-3$

When applying formulae (2.21)–(2.27) in geophysics, one should bear in mind that the effect of the Earth's curvature is negligible for shallow events at distances of less than 20°, but that vertical stratification or lateral inhomogeneity can sometimes considerably influence the deformation field.

For calculation of the bottom deformation the following parameters of the earthquake source are necessary: coordinates of the epicentre, the hypocentre depth, the seismic moment or the moment-magnitude and the strike, dip, rake(slip) angles. All these parameters are to be found in earthquake catalogues. The geometrical dimensions of the fault area (rectangular $L \times W$) and the dislocation (or mean $|\mathbf{D}|$) can be estimated by empirical formulae [Handbook for Tsunami Forecast (2001)]:

$$\log_{10} L[\text{km}] = 0.5 M_w - 1.9,$$
$$\log_{10} W[\text{km}] = 0.5 M_w - 2.2, \qquad (2.28)$$
$$\log_{10} D[\text{m}] = 0.5 M_w - 3.2.$$

Note that formulae (2.28) can be obtained from the definition of the seismic moment $M_0 = \mu D L W$, the relationship between the earthquake moment and moment-magnitude $M_w = \log_{10} M_0 / 1.5 - 6.07$ and then following empirical relationships: $L/W = 2, D/L = 5 \cdot 10^{-5}$ (scaling law [Kanamori, Anderson (1975)]). The rigidity of the crustal rock is assumed to be $\mu \approx 3 \cdot 10^{10} Pa$ [Kanamori, Brodsky (2004)].

In recent years methods have been developed that permit to determine, how the fault at an earthquake source developed in time, and to reveal its space structure [Ji et al. (2002); Yagi (2004)]. In this case, the fault surface is divided into a finite number (usually several hundreds) of rectangular elements, for each of which Burger's vector $\mathbf{D}$ is determined. The bottom deformation, caused by each of these rectangular elements, is calculated by formulae (2.21)–(2.27). Then, the contributions of all elements are summed up. Digitized data on the structure of fault surfaces for certain strong earthquakes (slip distribution) are presented on the site http://earthquake.usgs.gov/regional/world/historical.php.

In Fig. 2.6 (see colour section), the example is presented of bottom deformation calculations for the tsunamigenic Central Kuril Islands earthquake of November 15, 2006. According to USGS NEIC data the moment-magnitude of the earthquake amounted to $M_w = 7.9$ (8.3 CMT), its hypocentre location: 46.616°N, 153.224°E, 26.7 km (depth). The dip and the strike angles were determined to be 14.89° and 220.23°, respectively. The fault plane dimension was 400 km (along the strike) by 137.5 km, which was further divided into 220 subfaults (20 km by 12.5 km). The maximum slip was 8.9 m. Data on the slip distribution are available from http://earthquake.usgs.gov/eqcenter/ eqinthenews/2006/usvcam/finite_fault.php.

Figure 2.7a (see colour section) demonstrates the space distribution of the horizontal projection length of the bottom deformation vector. Figure 2.7b shows the vertical component of the deformation vector. In both cases, the step, with which the isolines are drawn, is 0.2 m. According to calculations, the maximum horizontal bottom deformation amounted to about 3.8 m. The calculated maximum and minimum vertical bottom deformations amounted to 2.7 m and −0.6 m, respectively.

2.2 Problem of Excitation of Gravitational Waves

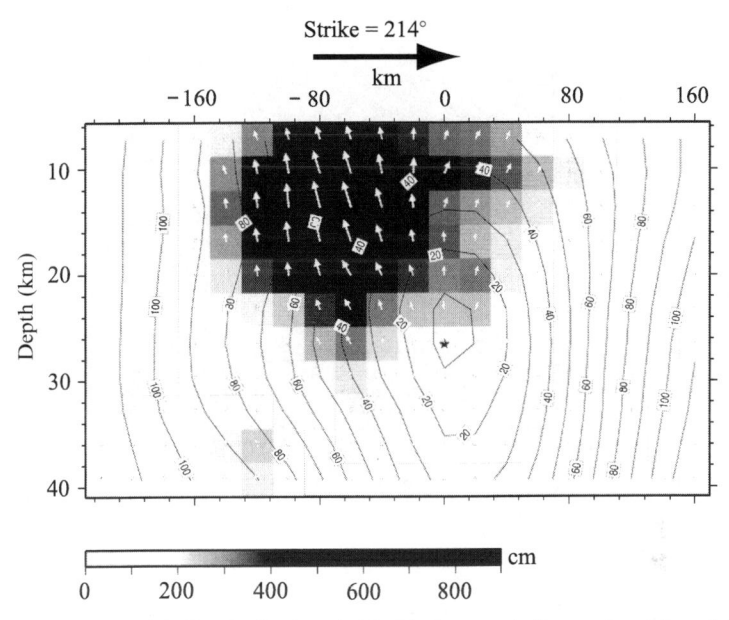

Fig. 2.6 Cross section of slip distribution. A big black arrow indicates the strike of the fault plane. The gray level shows the amplitude of dislocations and white arrows represent the motion of the hanging wall relative to the footwall. Contours show the rupture initiation time in seconds and the black star indicates the hypocentre location. The figure is adapted from http://earthquake.usgs.gov/eqcenter/eqinthenews/2006/usvcam/finite fault.php (see also Plate 3 in the Colour Plate Section on page 311)

2.2 General Solution of the Problem of Excitation of Gravitational Waves in a Layer of Incompressible Liquid by Deformations of the Basin Bottom

2.2.1 Cartesian Coordinates

The goal of this section is the construction of a mathematical model describing the motion of a layer of homogeneous incompressible liquid in the case of deformation of the basin bottom, proceeding in accordance with a certain given space–time law. The liquid is limited from above by its free surface and is in a gravitational field, characterized by the acceleration of gravity, g. We shall only deal with the case of a basin of constant depth H—such an approach will permit to obtain an analytical solution of the problem. We shall consider the amplitude of motions of the basin bottom, η_0, a small quantity as compared to the depth, $\eta_0 \ll H$. In practice, this condition is actually always satisfied (the average depth of the ocean is $H \sim 4{,}000$ m, while $\eta_0 < 10$ m even in the case of catastrophic earthquakes). The amplitude of gravitational surface waves, A, excited by one or another motion of the basin bottom with the amplitude η_0, will clearly be of the same order

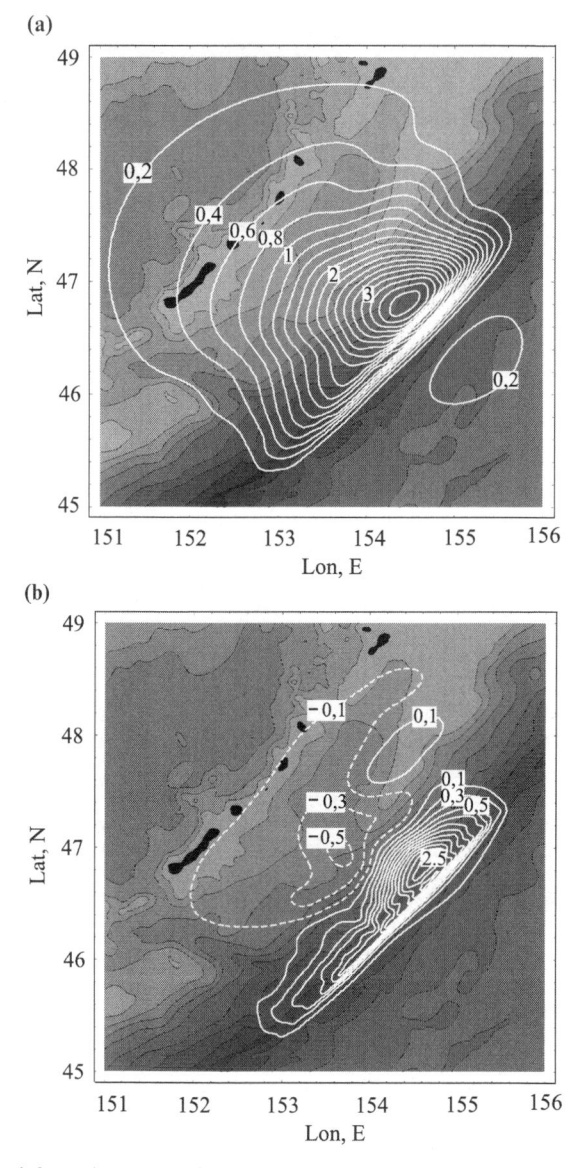

Fig. 2.7 Bottom deformations (a—is the length of the horizontal component, b—the vertical component) due to the Central Kuril Islands earthquake of November 15, 2006. The isolines are drawn in steps of 0.2 m. Calculations are performed applying Yoshimitsu Okada's formulae in accordance with USGS NEIC slip distribution data (see also Plate 4 in the Colour Plate Section on page 312)

of magnitude: $A \sim \eta_0$. The amplitude of the wave being small in comparison with its length $A \ll \lambda$ makes it possible to apply linear theory. The motion of the liquid will be considered potential. Consider a layer, infinite in the Oxy plane, of an ideal incompressible homogeneous liquid of constant depth H in the field of gravity. We

2.2 Problem of Excitation of Gravitational Waves

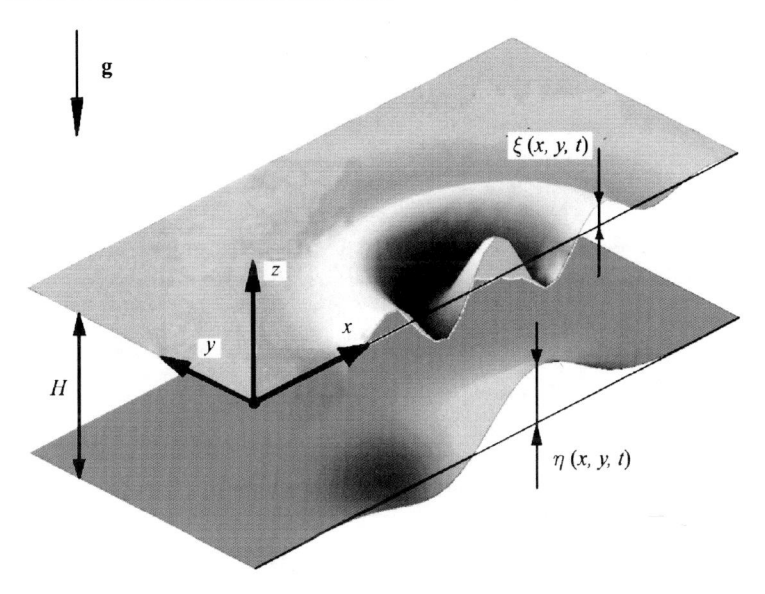

Fig. 2.8 Mathematical formulation of the 3D problem

shall put the origin of the Cartesian reference frame, $Oxyz$, in the unperturbed free surface and direct the Oz axis vertically upward (Fig. 2.8). The liquid is at rest until the time moment $t = 0$. To find the wave perturbation $\xi(x, y, t)$, formed on the surface of the liquid, and the velocity field, $\mathbf{v}(x, y, z, t)$, throughout the thickness of the layer in the case of motions of the basin floor, occurring in accordance with the law $\eta(x, y, t)$, we shall resolve the problem with respect to the velocity potential $F(x, y, z, t)$ [Landau, Lifshits (1987)]:

$$\frac{\partial^2 F}{\partial x^2} + \frac{\partial^2 F}{\partial y^2} + \frac{\partial^2 F}{\partial z^2} = 0, \tag{2.29}$$

$$g\frac{\partial F}{\partial z} = -\frac{\partial^2 F}{\partial t^2}, \quad z = 0, \tag{2.30}$$

$$\frac{\partial F}{\partial z} = \frac{\partial \eta}{\partial t}, \quad z = -H. \tag{2.31}$$

The physical meaning of the boundary condition (2.30) consists in the pressure on the free surface of the liquid being constant. The boundary condition (2.31) signifies equality of the vertical component of the flow velocity to the velocity of motion of the basin floor (a no-flow condition). Displacement of the free surface and the flow velocity vector are related to the potential of the flow velocity by the following known formulae:

$$\xi(x, y, t) = -\frac{1}{g}\frac{\partial F}{\partial t}\bigg|_{z=0}, \tag{2.32}$$

$$\mathbf{v}(x,y,z,t) \equiv \{u(x,y,z,t), v(x,y,z,t), w(x,y,z,t)\} = \nabla F(x,y,z,t). \quad (2.33)$$

The Laplace equation (2.29) is resolved by the standard method of separation of variables. We shall omit elementary calculations and write out the general solution of the problem in the form of Laplace and Fourier expansions over the time and space coordinates:

$$F(x,y,z,t) = \int_{s-i\infty}^{s+i\infty} dp \int_{-\infty}^{+\infty} dm \int_{-\infty}^{+\infty} dn \exp\{pt - imx - iny\}$$
$$\times (A(p,m,n)\cosh(kz) + B(p,m,n)\sinh(kz)), \quad (2.34)$$

where $k^2 = m^2 + n^2$.

Substitution of the general solution (2.34) into the boundary condition on the surface (2.30) yields the relationship between the coefficients:

$$B(p,m,n) = -A(p,m,n)\frac{p^2}{gk}. \quad (2.35)$$

Applying the formulae for the direct and inverse Laplace and Fourier transformations we obtain the integral representation for the laws, satisfied by motion of the basin floor:

$$\eta(x,y,t) = \frac{1}{8\pi^3 i} \int_{s-i\infty}^{s+i\infty} dp \int_{-\infty}^{+\infty} dm \int_{-\infty}^{+\infty} dn \exp\{pt - imx - iny\} H(p,m,n), \quad (2.36)$$

where

$$H(p,m,n) = \int_0^{\infty} dt \int_{-\infty}^{+\infty} dx \int_{-\infty}^{+\infty} dy \exp\{-pt + imx + iny\} \eta(x,y,t). \quad (2.37)$$

Substituting expression (2.34), written with the aid of formula (2.35), into the boundary condition on the basin floor (2.31), one can calculate the coefficient $A(p,m,n)$. As a result, one obtains the following expression for the potential of the flow velocity, which corresponds to motions of the basin floor, satisfying the law $\eta(x,y,t)$.

$$F(x,y,z,t) = -\frac{1}{8\pi^3 i} \int_{s-i\infty}^{s+i\infty} dp \int_{-\infty}^{+\infty} dm \int_{-\infty}^{+\infty} dn$$
$$\times \frac{p \exp\{pt - imx - iny\} \cosh(kz)\left(gk - p^2 \tanh(kz)\right)}{k \cosh(kH)\left(gk \tanh(kH) + p^2\right)} H(p,m,n). \quad (2.38)$$

Applying formulae (2.32) and (2.33), we obtain expressions describing the behaviour of the free surface,

2.2 Problem of Excitation of Gravitational Waves

$\xi(x,y,t)$

$$= \frac{1}{8\pi^3 i} \int_{s-i\infty}^{s+i\infty} dp \int_{-\infty}^{+\infty} dm \int_{-\infty}^{+\infty} dn \frac{p^2 \exp\{pt - imx - iny\}}{\cosh(kH)(gk \tanh(kH) + p^2)} H(p,m,n), \quad (2.39)$$

horizontal, $u(x,y,z,t)$, $v(x,y,z,t)$ and vertical, $w(x,y,z,t)$ components of the flow velocity,

$$u(x,y,z,t) = \frac{\partial F}{\partial x} = \frac{1}{8\pi^3} \int_{s-i\infty}^{s+i\infty} dp \int_{-\infty}^{+\infty} dm \int_{-\infty}^{+\infty} dn$$

$$\times \frac{mp \exp\{pt - imx - iny\} \cosh(kz) \left(gk - p^2 \tanh(kz)\right)}{k \cosh(kH)(gk \tanh(kH) + p^2)} H(p,m,n); \quad (2.40)$$

$$v(x,y,z,t) = \frac{\partial F}{\partial y} = \frac{1}{8\pi^3} \int_{s-i\infty}^{s+i\infty} dp \int_{-\infty}^{+\infty} dm \int_{-\infty}^{+\infty} dn$$

$$\times \frac{np \exp\{pt - imx - iny\} \cosh(kz) \left(gk - p^2 \tanh(kz)\right)}{k \cosh(kH)(gk \tanh(kH) + p^2)} H(p,m,n) \quad (2.41)$$

$$w(x,y,z,t) = \frac{\partial F}{\partial z} = -\frac{1}{8\pi^3 i} \int_{s-i\infty}^{s+i\infty} dp \int_{-\infty}^{+\infty} dm \int_{-\infty}^{+\infty} dn$$

$$\times \frac{p \exp\{pt - imx - iny\} \cosh(kz) \left(gk \tanh(kz) - p^2\right)}{\cosh(kH)(gk \tanh(kH) + p^2)} H(p,m,n) \quad (2.42)$$

In principle, expressions (2.39)–(2.42) provide an exhaustive solution of problem (2.29)–(2.31), but obtaining concrete results requires the calculation of sixfold integrals, which represents quite a realistic, but extremely labour-consuming (from the point of view of the volume of calculations) and irrational task. To be able to perform part of the calculations analytically it is necessary to set the concrete form of function $\eta(x,y,t)$.

2.2.2 Cylindrical Coordinates

In a number of cases, when the model displacement of the basin floor exhibits appropriate symmetry, it may turn out to be convenient to apply a cylindrical reference system, which we shall introduce in a standard manner with respect to the Cartesian system, described in Sect. 2.2.1. In this case the Laplace equation (2.29) assumes the following form:

$$\frac{1}{r}\frac{\partial}{\partial r}\left(r\frac{\partial F}{\partial r}\right) + \frac{1}{r^2}\frac{\partial^2 F}{\partial \varphi^2} + \frac{\partial^2 F}{\partial z^2} = 0, \qquad (2.43)$$

while the boundary conditions (2.30) and (2.31) remain intact.

For resolving equation (2.43) we apply the traditional method of variable separation, i.e. we shall assume that

$$F(r,\varphi,z) = R(r)\,\Phi(\varphi)\,Z(z). \qquad (2.44)$$

Substitution of expression (2.44) into equation (2.43) results in the following set of ordinary differential equations:

$$r^2\frac{\partial^2 R}{\partial r^2} + r\frac{\partial R}{\partial r} + (r^2 - n^2)R = 0, \qquad (2.45)$$

$$\frac{\partial^2 \Phi}{\partial \varphi^2} + n^2 \Phi = 0, \qquad (2.46)$$

$$\frac{\partial^2 Z}{\partial z^2} - k^2 Z = 0. \qquad (2.47)$$

The Bessel equation (2.45) is written with account of the substitution of variable $r* = rk$ (the asterisk '*' is dropped). The solutions of equations (2.45)–(2.47) are well known and can be written as follows:

$$R(rk) = C_1 J_n(kr) + C_2 Y_n(kr),$$

$$\Phi(\varphi) = C_3 \cos(n\varphi) + C_4 \sin(n\varphi),$$

$$Z(z) = C_5 \cosh(kz) + C_6 \sinh(kz),$$

where J_n and Y_n are Bessel functions of the first and second kinds and of the nth order, C_i are arbitrary constants.

Functions $\Phi(\varphi)$ must satisfy the periodicity condition:

$$\Phi(\varphi) = \Phi(\varphi + 2\pi),$$

from which it follows that parameter n is an integer, $n = 0, \pm 1, \pm 2, \ldots$ The condition, that function $R(rk)$ be limited at $r = 0$ requires the coefficient of the Bessel function of the second kind to be equal to zero: $C_2 = 0$.

Thus, it is expedient to seek for the general solution of the problem in the form of a Fourier expansion and of Laplace and Fourier–Bessel transformations [Nikiforov, Uvarov (1984)]:

$$F(r,\varphi,z,t) = \int_0^\infty dk \int_{s-i\infty}^{s+i\infty} dp$$

$$\times \exp\{pt\} J_0(kr) \frac{C_3^0}{2} \left(C_5^0(p,k) \cosh(kz) + C_6^0(p,k) \sinh(kz) \right)$$

2.2 Problem of Excitation of Gravitational Waves

$$+ \int_0^\infty dk \int_{s-i\infty}^{s+i\infty} dp \exp\{pt\} \sum_{n=1}^\infty J_n(kr) \left(C_3^n \cos(n\varphi) + C_4^n \sin(n\varphi)\right)$$

$$\times \left(C_5^n(p,k) \cosh(kz) + C_6^n(p,k) \sinh(kz)\right). \quad (2.48)$$

Substitution of equation (2.48) into the boundary condition on the surface (2.30) yields the relationship between the coefficients,

$$C_6^n(p,k) = -C_5^n(p,k) \frac{p^2}{gk}. \quad (2.49)$$

We shall now write the integral representation for the function describing the space–time law of motion of the basin floor, $\eta(r,\varphi,t) = \eta^r(r)\,\eta^\varphi(\varphi)\,\eta^t(t)$

$$\eta(r,\varphi,t) = \int_0^\infty dk \int_{s-i\infty}^{s+i\infty} dp \exp\{pt\} J_0(kr) k \frac{A_0}{2} H^0(p,k) + \int_0^\infty dk \int_{s-i\infty}^{s+i\infty} dp$$

$$\times \exp\{pt\} \sum_{n=1}^\infty J_n(kr) k \left(A_n \cos(n\varphi) + B_n \sin(n\varphi)\right) H^n(p,k), \quad (2.50)$$

where

$$A_n = \frac{1}{\pi} \int_{-\pi}^\pi \eta^\varphi(\varphi) \cos(n\varphi)\,d\varphi,$$

$$B_n = \frac{1}{\pi} \int_{-\pi}^\pi \eta^\varphi(\varphi) \sin(n\varphi)\,d\varphi,$$

$$H^n(p,k) = \frac{1}{2\pi i} \int_0^\infty dt \int_0^\infty dr \eta^r(r)\,\eta^t(t) \exp\{-pt\} r J_n(kr).$$

Substitution of formulae (2.48) and (2.50) into the boundary condition on the basin floor (2.31) reveals that equality of the left-hand and right-hand parts is possible, only when the following three conditions are fulfilled:

$$C_3^n = A^n, \quad C_4^n = B^n,$$

$$C_5^n(p,k) = -\frac{p H^n(p,k)}{k\left(\sinh(kH) + \dfrac{p^2}{gk}\cosh(kH)\right)}.$$

It is now possible to write out the resultant expression for the potential, which is the solution of equation (2.43) with the boundary conditions (2.30) and (2.31)

$$F(r,\varphi,z,t) = -\int_0^\infty dk \int_{s-i\infty}^{s+i\infty} dp \exp\{pt\} \frac{p\left(\cosh(kz) - \frac{p^2}{gk}\sinh(kz)\right)}{\left(\sinh(kH) + \frac{p^2}{gk}\cosh(kH)\right)}$$

$$\times \left(J_0(kr)\frac{A_0}{2}H^0(p,k) + \sum_{n=1}^\infty J_n(kr)\Big(A_n\cos(n\varphi)\right.$$

$$\left. + B_n\sin(n\varphi)\Big) H^n(p,k)\right). \quad (2.51)$$

Making use of expression (2.51), it is not difficult to obtain formulae for calculation of the displacement of the surface and of the velocity components $v_r = \frac{\partial F}{\partial r}$, $v_\varphi = \frac{1}{r}\frac{\partial F}{\partial \varphi}$, $v_z = \frac{\partial F}{\partial z}$, the explicit expressions for which will not be written out here, because they are too cumbersome.

Below we shall turn to the case, when the source of waves exhibits axial symmetry. The solution of the problem, here, will be of the following form:

$$F(r,z,t)$$

$$= -\int_0^\infty dk \int_{s-i\infty}^{s+i\infty} dp \exp\{pt\} J_0(kr) \frac{p\left(\cosh(kz) - \frac{p^2}{gk}\sinh(kz)\right)}{\left(\sinh(kH) + \frac{p^2}{gk}\cosh(kH)\right)} X(p,k), \quad (2.52)$$

where

$$X(p,k) = \frac{1}{2\pi i}\int_0^\infty dt \int_0^\infty dr \exp\{-pt\} J_0(kr)\, r\,\eta(r,t).$$

2.3 Plane Problems of Tsunami Excitation by Deformations of the Basin Bottom

In this section two-dimensional models (in the vertical plane) are dealt with. Solution of the plane problem permits to demonstrate clearly many important peculiarities of the physical processes taking place during tsunami generation. A significant part of the results, obtained within the framework of the two-dimensional model, remains valid in the three-dimensional case also. The 2D⇒3D transition for problems of the type considered actually permits to investigate only two new points: the direction of wave irradiation and changes in their characteristics, as the distance from the source increases.

2.3 Plane Problems of Tsunami Excitation by Deformations of the Basin Bottom

2.3.1 Construction of the General Solution

We shall consider (Fig. 2.9) a layer of ideal incompressible homogeneous liquid, infinite along the $0x$ axis, of constant depth H, and in the field of gravity. We shall put the origin of the Cartesian reference system, $0xz$, on the unperturbed free surface, the $0z$ will be directed vertically upward. To find the perturbation of the free surface, $\xi(x,t)$, and the field of flow velocities $\mathbf{v}(x,z,t)$, arising in the layer of liquid, when the basin floor undergoes motion in accordance with the law $\eta(x,t)$, we shall resolve the problem with respect to the potential of the flow velocity, $F(x,z,t)$:

$$\frac{\partial^2 F}{\partial x^2} + \frac{\partial^2 F}{\partial z^2} = 0, \quad (2.53)$$

$$g\frac{\partial F}{\partial z} = -\frac{\partial^2 F}{\partial t^2}, \qquad z=0, \quad (2.54)$$

$$\frac{\partial F}{\partial z} = \frac{\partial \eta}{\partial t}, \qquad z=-H. \quad (2.55)$$

Without dwelling on the details of resolving the problem (2.53)–(2.55) that were exposed above for the three-dimensional case, we shall present the resultant formulae.

$$F(x,z,t) = -\frac{1}{4\pi^2 i} \int_{s-i\infty}^{s+i\infty} dp \int_{-\infty}^{+\infty} dk$$
$$\times \frac{p \exp\{pt-ikx\} \cosh(kz)\left(gk-p^2\tanh(kz)\right)}{k\cosh(kH)\left(gk\tanh(kH)+p^2\right)} H(p,k), \quad (2.56)$$

$$\xi(x,t) = \frac{1}{4\pi^2 i} \int_{s-i\infty}^{s+i\infty} dp \int_{-\infty}^{+\infty} dk \frac{p^2 \exp\{pt-ikx\}}{\cosh(kH)\left(gk\tanh(kH)+p^2\right)} H(p,k), \quad (2.57)$$

$$u(x,z,t) = \frac{\partial F}{\partial x} = \frac{1}{4\pi^2} \int_{s-i\infty}^{s+i\infty} dp \int_{-\infty}^{+\infty} dk$$
$$\times \frac{p\exp\{pt-ikx\}\cosh(kz)\left(gk-p^2\tanh(kz)\right)}{\cosh(kH)\left(gk\tanh(kH)+p^2\right)} H(p,k), \quad (2.58)$$

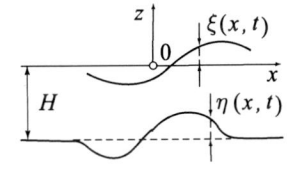

Fig. 2.9 Mathematical formulation of the 2D problem

$$w(x,z,t) = \frac{\partial F}{\partial z} = -\frac{1}{4\pi^2 i} \int_{s-i\infty}^{s+i\infty} dp \int_{-\infty}^{+\infty} dk$$

$$\times \frac{p \exp\{pt - ikx\} \cosh(kz) \left(gk \tanh(kz) - p^2\right)}{\cosh(kH) \left(gk \tanh(kH) + p^2\right)} H(p,k), \quad (2.59)$$

where

$$H(p,k) = \int_0^\infty dt \int_{-\infty}^{+\infty} dx \exp\{-pt + ikx\} \eta(x,t).$$

In the case of arbitrary motion of the basin floor the solution of the problem involves a cumbersome procedure—the calculation of a fourfold integral. Therefore, for physical interpretation of the obtained integral representations it is expedient to select several concrete versions of function $\eta(x,t)$. This will permit to calculate a large part of the integrals analytically.

Consider the following three types of deformation of the basin floor:

1. A linear (in time) displacement

$$\eta_L(x,t) = \eta_0 \left(\theta(x+a) - \theta(x-a)\right) \theta(t) t \tau^{-1}, \quad (2.60)$$

2. Running displacement

$$\eta_R(x,t) = \eta_0 \left(\theta(x) - \theta(x-b)\right) \left(1 - \theta(x-vt)\right), \quad (2.61)$$

3. Harmonic oscillations of the basin floor

$$\eta_{osc}(x,t) = \eta_0 \left(\theta(x+a) - \theta(x-a)\right) \sin(\omega t) \quad (2.62)$$

where η_0 is the amplitude of the basin floor displacement, θ is the Heaviside function, $2a$ and b are the horizontal dimensions of the source. In all cases we consider the rectangular distribution of deformations of the basin floor. The scheme of motions of the basin floor in the case of a running displacement is shown in Fig. 2.10.

For the tsunami problem the linear displacement itself has no physical significance—it is useful only as a mathematical model. But from the function

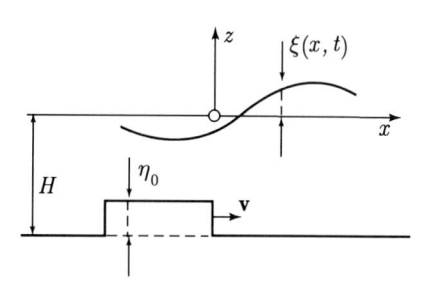

Fig. 2.10 Model of running displacement of basin floor

2.3 Plane Problems of Tsunami Excitation by Deformations of the Basin Bottom

$\eta_L(x,t)$ it is possible to 'construct' the two principal model laws of deformation of the basin floor at the tsunami source: a motion of the basin floor involving residual displacement

$$\eta_1(x,t) = \eta_L(x,t) - \eta_L(x,t-\tau) \tag{2.63}$$

and a motion of the basin floor without residual displacement

$$\eta_2(x,t) = 2\eta_L(x,t) - 4\eta_L(x,t-0.5\tau) + 2\eta_L(x,t-\tau). \tag{2.64}$$

Complying with the terminology proposed in [Dotsenko, Soloviev (1990)], we call the two indicated types of motion 'piston' and 'membrane' displacements.

The problem considered is linear, therefore, the solutions for the piston and membrane displacements can be expressed through the solution for the linear displacement, making use of the superposition principle:

$$F_1(x,z,t) = F_L(x,z,t)\,\theta(t) - F_L(x,z,t-\tau)\,\theta(t-\tau), \tag{2.65}$$

$$F_2(x,z,t) = 2F_L(x,z,t)\,\theta(t) - 4F_L(x,z,t-0.5\tau)\,\theta(t-0.5\tau)$$
$$+ 2F_L(x,z,t-\tau)\,\theta(t-\tau), \tag{2.66}$$

where $F_L(x,z,t)$ is the solution of the problem (2.53)–(2.55) in the case of $\eta(x,t) = \eta_L(x,t)$. The perturbation of the free surface and of the velocity component corresponding to the piston or membrane displacements is obviously calculated by formulae, similar to (2.65) and (2.66). Only a formal substitution of ξ_L, u_L or w_L for F_L is required.

Calculation of the intermediate integrals, performed applying residue theory, results in the following expressions for the linear displacement:

$$F_L(x,z,t) = -\frac{1}{2\pi\tau}\int_{-\infty}^{+\infty} dk$$
$$\times \frac{\exp\{-ikx\}\cosh(kz)\,(1-(1+\tanh(kH)\tanh(kz))\cos(t\,p_0))}{k\sinh(kH)}X(k). \tag{2.67}$$

$$\xi_L(x,t) = \frac{1}{2\pi\tau}\int_{-\infty}^{+\infty} dk\,\frac{\exp(-ikx)\sin(t\,p_0)}{p_0\cosh(kH)}X(k), \tag{2.68}$$

$$u_L(x,z,t) = \frac{i}{2\pi\tau}\int_{-\infty}^{+\infty} dk$$
$$\times \frac{\exp\{-ikx\}\,(\cosh(kz)-[\cosh(kz)+\tanh(kH)\sinh(kz)]\cos(t\,p_0))}{\sinh(kH)}X(k), \tag{2.69}$$

$$w_L(x,z,t) = -\frac{1}{2\pi\tau}\int_{-\infty}^{+\infty} dk$$

$$\times \frac{\exp\{-ikx\}(\sinh(kz) - [\sinh(kz) + \tanh(kH)\cosh(kz)]\cos(t\,p_0))}{\sinh(kH)} X(k), \quad (2.70)$$

where $p_0 = (gk\tanh(kH))^{1/2}$, $X(k) = \eta_0 2\sin(ka)/k$.

In the case of a running displacement of the basin floor the solution is given by the following formulae:

$$F_R(x,z,t) = -\frac{\eta_0}{4\pi}\int_{-\infty}^{+\infty} dk \frac{\exp\{-ikx\}}{k\cosh(kH)}$$

$$\times \frac{(gk\cosh(kz) + p_0^2\sinh(kz))}{p_0}\left(\exp\{-ip_0 t\}\frac{\exp\left\{ib\left(k+\frac{p_0}{v}\right)\right\} - 1}{k+\frac{p_0}{v}}\right.$$

$$\left. - \exp\{ip_0 t\}\frac{\exp\left\{ib\left(k-\frac{p_0}{v}\right)\right\} - 1}{k-\frac{p_0}{v}}\right), \quad (2.71)$$

$$\xi_R(x,t) = \frac{\eta_0}{4\pi i}\int_{-\infty}^{+\infty} dk \frac{\exp\{-ikx\}}{\cosh(kH)}\left(\exp\{-ip_0 t\}\frac{\exp\left\{ib\left(k+\frac{p_0}{v}\right)\right\} - 1}{k+\frac{p_0}{v}}\right.$$

$$\left. + \exp\{ip_0 t\}\frac{\exp\left\{ib\left(k-\frac{p_0}{v}\right)\right\} - 1}{k-\frac{p_0}{v}}\right), \quad (2.72)$$

$$u_R(x,z,t) = \frac{\eta_0 i}{4\pi}\int_{-\infty}^{+\infty} dk \frac{\exp\{-ikx\}}{\cosh(kH)}$$

$$\times \frac{(gk\cosh(kz) + p_0^2\sinh(kz))}{p_0}\left(\exp\{-ip_0 t\}\frac{\exp\left\{ib\left(k+\frac{p_0}{v}\right)\right\} - 1}{k+\frac{p_0}{v}}\right.$$

$$\left. - \exp\{ip_0 t\}\frac{\exp\left\{ib\left(k-\frac{p_0}{v}\right)\right\} - 1}{k-\frac{p_0}{v}}\right), \quad (2.73)$$

2.3 Plane Problems of Tsunami Excitation by Deformations of the Basin Bottom

$$w_R(x,z,t) = -\frac{\eta_0}{4\pi} \int_{-\infty}^{+\infty} dk \frac{\exp\{-ikx\}}{\cosh(kH)}$$

$$\times \frac{(gk\sinh(kz) + p_0^2 \cosh(kz))}{p_0} \left(\exp\{-ip_0 t\} \frac{\exp\left\{ib\left(k+\frac{p_0}{v}\right)\right\} - 1}{k + \frac{p_0}{v}} \right.$$

$$\left. - \exp\{ip_0 t\} \frac{\exp\left\{ib\left(k-\frac{p_0}{v}\right)\right\} - 1}{k - \frac{p_0}{v}} \right). \quad (2.74)$$

We stress that expressions (2.71)–(2.74) are valid only if the condition $t \geqslant b/v$ is fulfilled. At any rate, this fact gives rise to no essential complications in calculations for time periods inferior to b/v, since, from a physical point of view, the solution of the problem involving a running displacement at $t = t_0 < b/v$ is equivalent to the solution of a similar problem for $b = vt_0$.

2.3.2 *Piston and Membrane Displacements*

As it was already shown above, tsunami waves are generated by motions of the ocean bottom occurring along the normal to its surface (normal displacements). Motions of the ocean bottom in its own plane (tangential displacements) are not effective, from the standpoint of tsunami generation. The term 'vertical displacement' is often encountered in the literature. In the case of small slope angles of the ocean bottom the difference between vertical and normal displacements is, naturally, insignificant.

The goal of this section consists in the revelation of relationships between the main parameters of a tsunami wave and the characteristics of the source generating it—the deformation area of the ocean bottom. The wave parameters of interest to us comprise its amplitude, length and the energy of the wave perturbation. The source is characterized by the amplitude and duration of the ocean bottom deformation, as well as its horizontal extension.

The piston and membrane mechanisms of wave generation, both of impulse and finite duration, have been investigated analytically [Kajiura (1970); Murty (1977); Dotsenko et al. (1993), (1995)] and numerically [Marchuk et al. (1983)]. There also exists a small number of publications devoted to laboratory simulation of the generation process [Takahasi (1934, 1963); Hammack (1973); Nosov, Shelkovnikov (1997)]. A review of experimental works can be found in [Levin (1978)].

We shall first deal with elementary results that can be obtained within the framework of linear theory of long waves. The one-dimensional wave equation, describing displacements of the free surface, ξ, in the case of deformations η of the ocean bottom, exhibit the following form:

$$\frac{\partial^2 \xi}{\partial t^2} - gH\frac{\partial^2 \xi}{\partial x^2} = \frac{\partial^2 \eta}{\partial t^2}. \tag{2.75}$$

Let deformations of the ocean bottom be given by the formula

$$\eta(x,t) = \left(\theta(x+a) - \theta(x-a)\right)\eta(t), \tag{2.76}$$

where $\eta(t)$ represents an arbitrary law of motion of the ocean bottom. Note that the piston and membrane displacements, (2.63) and (2.64), respectively, are special cases of formula (2.76). Deformations of the ocean bottom of the form (2.76) result in the formation of two identical waves, travelling in opposite directions. In the one-dimensional case a long linear wave does not undergo transformation during propagation, so it suffices to know its characteristics at any single point, for instance, close to the right boundary of the generation area ($x = a + \varepsilon$, $\varepsilon > 0$, $\varepsilon \ll a$). The solution of equation (2.75) is readily found analytically. Thus, for example, at $x = a + \varepsilon$ the wave perturbation is described by the following simple formula:

$$\xi(t) = \frac{1}{2}\left(\eta(t) - \eta(t - 2a(gH)^{-1/2})\right). \tag{2.77}$$

In Fig. 2.11 examples are presented of the shapes of wave perturbations formed by piston and membrane displacements (solid lines). Calculations are performed in accordance with formula (2.77). A piston-like displacement always forms a sole

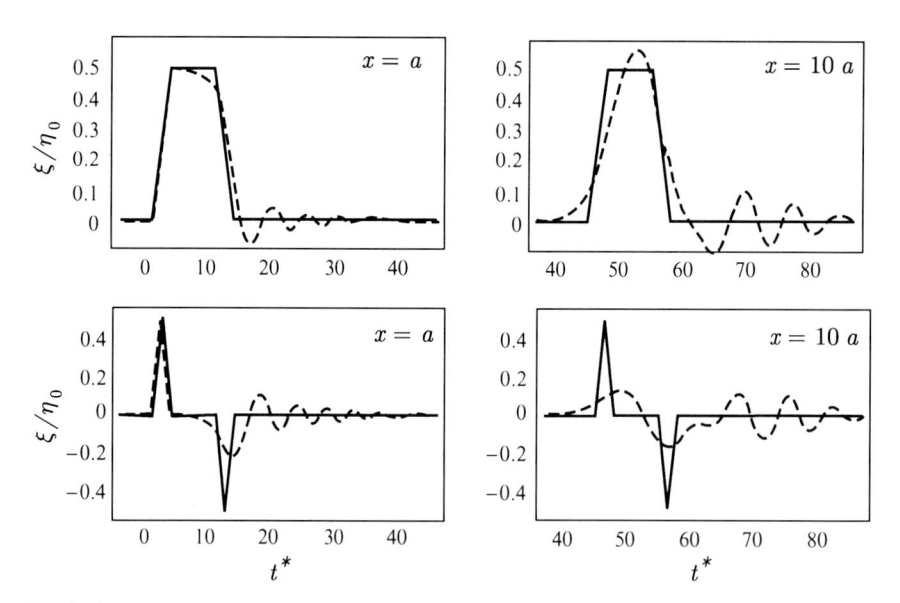

Fig. 2.11 Waves formed by piston (upper row) and membrane (lower row) displacements of duration $\tau^* = 3$ close to the right boundary of the generation area ($x = a$) and at a significant distance from it ($x = 10a$). The horizontal extension of the source $2a = 10$. The solid line represents the linear theory of long waves, the dotted one the linear potential theory

2.3 Plane Problems of Tsunami Excitation by Deformations of the Basin Bottom

wave of trapezoidal shape, the polarity of which coincides with the polarity of the seabed displacement. In the case of a membrane-like displacement a bipolar wave arises that comprises a crest and a trough. We shall present the formulae relating the main parameters of waves and the characteristics of a displacement:

- Wave amplitude in the case of piston-like displacement

$$A_{max}^1 = \eta_0 \begin{cases} 1/2, & \tau^* \leqslant 2, \\ 1/\tau^*, & \tau^* > 2, \end{cases} \tag{2.78}$$

- Crest and trough amplitude in the case of membrane-like displacement

$$A_{max}^2 = A_{min}^2 = \eta_0 \begin{cases} 1/2, & \tau^* \leqslant 4, \\ 2/\tau^*, & \tau^* > 4, \end{cases} \tag{2.79}$$

- Wave energy (We consider total energy of waves propagating in both positive and negative directions of the Ox axis) in the case of piston-like displacement

$$W_1 = ag\rho \eta_0^2 \begin{cases} 1 - \tau^*/6, & \tau^* \leqslant 2, \\ 2/\tau^* - 4/3(1/\tau^*)^2, & \tau^* > 2, \end{cases} \tag{2.80}$$

- Wave energy in the case of membrane-like displacement

$$W_2 = ag\rho \eta_0^2 \begin{cases} \tau^*/3 & \tau^* \leq 2, \\ \tau^*/3 - 4/3\left[(\tau^*/2)^{1/3} - (\tau^*/2)^{-2/3}\right]^3, & 2 < \tau^* \leq 4, \\ (8/\tau^*)(1 - 2/\tau^*), & \tau^* > 4. \end{cases} \tag{2.81}$$

- Period of wave perturbation for piston-like and membrane-like displacements

$$T_1 = T_2 = \frac{a(2+\tau^*)}{(gH)^{1/2}}, \tag{2.82}$$

- Wavelength of perturbation for piston-like and membrane-like displacements

$$\lambda_1 = \lambda_2 = a(2+\tau^*). \tag{2.83}$$

The formulae presented contain the dimensionless displacement duration $\tau^* = \frac{\tau}{a}(gH)^{1/2}$. Below, we shall make use of dimensionless time, determined by a similar formula, $t^* = \frac{t}{a}(gH)^{1/2}$. The energy of the wave (per unit 'channel' width) was calculated by the Kajiura formula [Kajiura (1970)]:

$$W = \rho g(gH)^{1/2} \int_0^T \xi^2 \, dt, \qquad (2.84)$$

where T is the duration of the wave perturbation. From formulae (2.80) and (2.81) it is seen that the wave energy is conveniently normalized to the quantity $W_0 = ag\rho\eta_0^2$, representing the potential energy of a free-surface rectangular elevation of length $2a$ and height η_0. Precisely such an elevation should arise on the water surface in the case of a impulse piston-like displacement of the seabed (if the process is described within the framework of the linear theory of long waves).

Owing to the problem considered being linear, the tsunami wave amplitude is proportional to the seabed (ocean bottom) deformation amplitude. In the case of short motions the amplitude is independent of the duration of the displacement or the horizontal size of the source and amounts to half the amplitude of the seabed deformation. When the displacements are longer in time ($\tau^* \gg 1$), the amplitude drops monotonously according to the law $(\tau^*)^{-1}$. The dependencies (2.78) and (2.79) are shown in Figs. 2.12 and 2.13, respectively, by broken lines.

From formulae (2.80) and (2.81) it follows that the energy of a tsunami wave is proportional to the square amplitude of the seabed deformation η_0 and to the horizontal dimension of the source a. The respective dependences are shown in Fig. 2.14 (curve 3). In the case of a piston-like displacement the wave energy decreases monotonously as the duration of the seabed deformation increases. In the case of

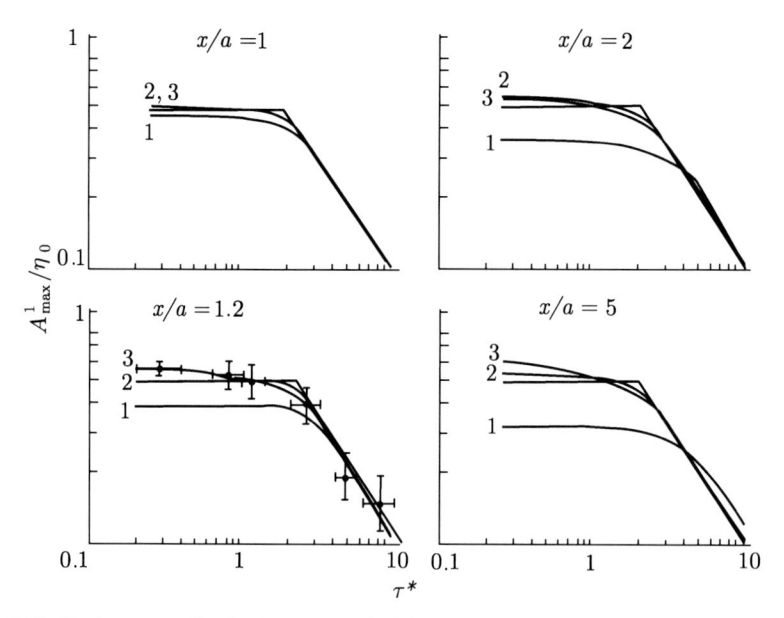

Fig. 2.12 Maximum amplitude of wave, excited by piston-like displacement of seabed, versus displacement duration for various distances from the generation area. Curves 1–3 correspond to values of parameter $a/H = 1, 3, 9$. The broken line represents linear theory of long waves

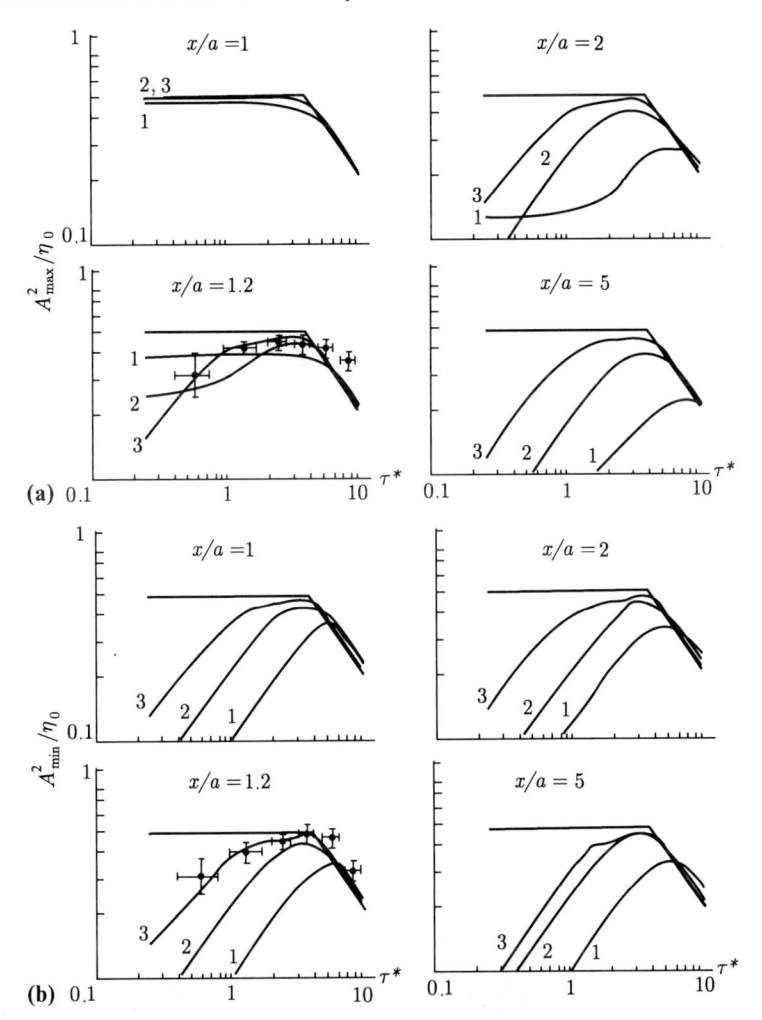

2.3 Plane Problems of Tsunami Excitation by Deformations of the Basin Bottom

Fig. 2.13 Maximum amplitude of first crest (**a**) and first trough (**b**) of wave, excited by membrane-like displacement of seabed, versus displacement duration for various distances from the generation area. Curves 1–3 correspond to values of parameter $a/H = 1, 3, 9$. The broken line represents linear theory of long waves ($a/H \gg 1$)

a membrane-like displacement the corresponding dependence is not monotonous: as the duration of the displacement increases, the energy starts to increase and, then, drops. The maximum corresponds to $\tau^* = 4$.

As to the wave periods and lengths, these quantities increase monotonously with the displacement duration, in accordance with formulae (2.82) and (2.83). Actually, $\tau^* \ll 1$, therefore both the period and length of a tsunami wave mostly depend on the horizontal dimension of the area of seabed deformation.

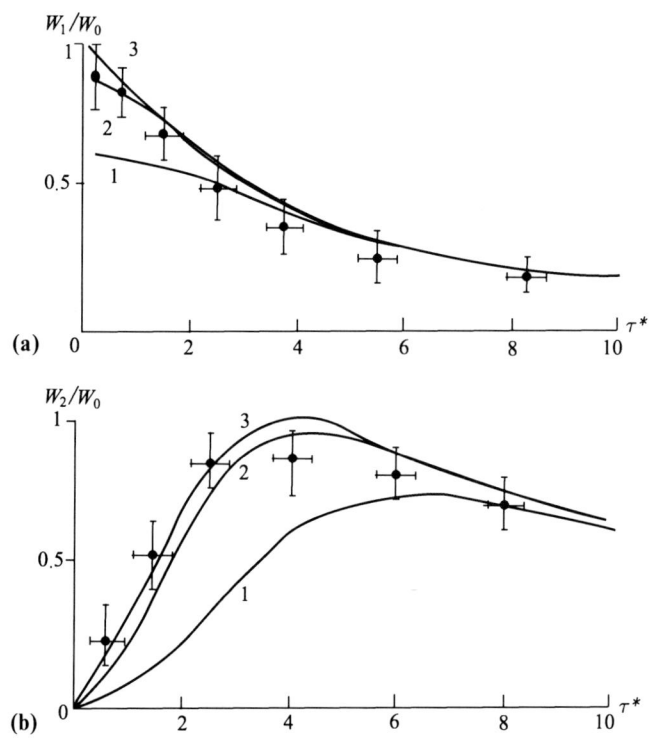

Fig. 2.14 Energy of wave, excited by piston-like (**a**) and membrane-like (**b**) displacements of seabed, versus displacement duration. Curves 1 and 2 correspond to values of parameter $a/H = 1, 3$. Curve 3 corresponds to linear theory of long waves ($a/H \gg 1$)

The results expounded above follow from the linear theory of long waves, which actually describes the process of wave generation and propagation not quite adequately. Figure 2.11 demonstrates waves formed by identical sources, but calculated within the frameworks of two different linear theories: long-wave and potential. In calculations we applied formulae (2.77), (2.65) and (2.66). Note that in the calculations presented in Fig. 2.11 use was made of quite an extended source, the length of which amounted to ten ocean depths.

In the case of a membrane-like displacement the wave shape is essentially different, especially at significant distances from the source. But in the case of a piston-like displacement one can see a noticeable difference in the wave shape and amplitude. In accordance with potential theory, the main perturbation is followed by an oscillating 'tail', due to phase dispersion. A small enhancement of the wave amplitude at large distances from the source (piston-like displacement) is also explained by dispersion. Actually, as the distance from the source increases, the wave amplitude first increases and only subsequently starts to decrease. The physical interpretation of this phenomenon consists in the following: the sharp wave front includes shortwave components travelling slower than the main wave, and,

2.3 Plane Problems of Tsunami Excitation by Deformations of the Basin Bottom

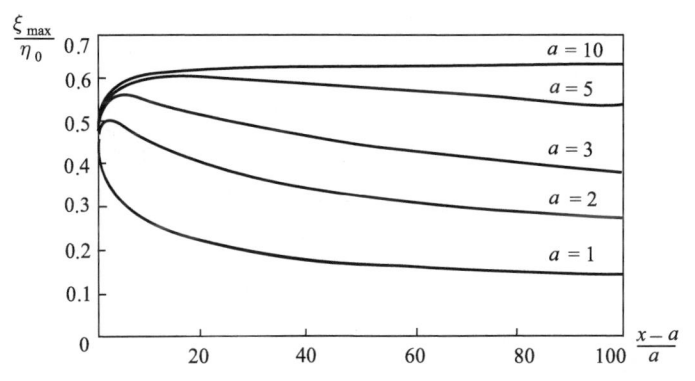

Fig. 2.15 Maximum amplitude of wave caused by piston-like displacement, $\tau = 1$, versus distance from the boundary of the source for different source sizes

therefore, as the wave propagates, the front 'overtakes' it, thus causing enhancement of the amplitude. Variation of the wave amplitude, as it travels away from the source, is shown in Fig. 2.15. Dispersive amplification can be seen to be capable of enhancing the wave amplitude by 25%, but it is not always present and exists only in such cases, when the size of the source is noticeably greater than the basin depth. The effect of dispersive tsunami amplification was first dealt with in [Mirchina, Pelinovsky (1987)] for volcanogenic tsunamis.

Figures 2.12 and 2.13 show the dependences of wave amplitudes upon the displacement duration (curves 1–3), calculated within the framework of potential theory. In the case of a piston-like displacement the behaviour of these dependencies does not differ very strongly from the broken line, corresponding to the long-wave theory. Significant differences are observed only in the case of small-size sources and at large distances from it. In the case of waves due to a membrane-like displacement, also, no noticeable difference exists at the boundary of the generation area between calculations performed by the long-wave and potential theories. But, already at a small distance from the source the dependence of the amplitude essentially changes in character and becomes monotonous. Such a character of the dependence is conserved for any horizontal dimensions of the source. It is important to note that rapid membrane-like displacements do not cause tsunami waves of significant amplitudes.

We shall now turn to the relationship between the wave energy and the source parameters, presented in Fig. 2.14. It can be seen that in the case of a large-size source or of significant displacement durations, the energy values calculated by potential and long-wave theories comply quite well with each other. The most essential difference is again observed in the case of short membrane-like displacements. Taking into account that in the case of real tsunami sources $\tau^* < 1$ clarifies the leading role of seabed motions with residual displacements in the excitation of strong tsunamis. A similar conclusion is made, for example, in [Dotsenko, Soloviev (1990)] from analysis of a source with axial symmetry.

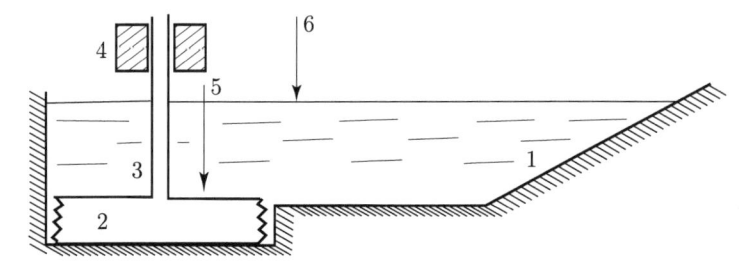

Fig. 2.16 Layout of laboratory set-up for simulating tsunami generation by deformations of the basin bottom. 1—wave damper (slope), 2—pneumatic wave generator, 3—main line of pressure supply, 4—guiding cylinder, 5—sensor of basin bottom motion, 6—IR wavegraph

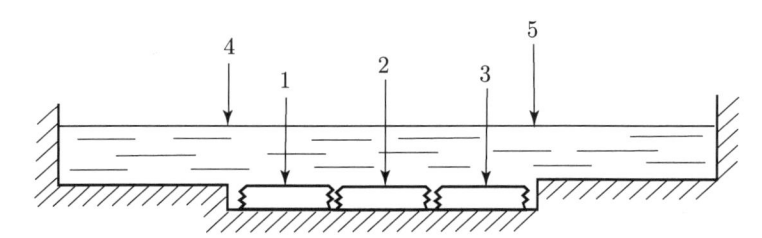

Fig. 2.17 Layout of laboratory set-up for simulating tsunami generation by a running displacement. 1, 2, 3—pneumatic wave generators, 4, 5—IR wavegraphs

In conclusion of this section we shall turn to experimental tests of the theoretical relationships found between wave parameters and source characteristics. We shall briefly describe the layout of laboratory experiments. The set-up was an open rectangular hydrocanal with transparent walls of organic glass of dimensions $0.15 \times 0.15 \times 3.3$ m (Fig. 2.16). As the source of waves, imitating vertical displacements of the basin bottom, use was made of a pneumatic generator representing a rectangular volume with rigid upper and lower sides, and elastic lateral sides. Model displacements of the basin bottom were registered by a sensor, representing a fixed inductance coil and a ferrite core, connected to the moving upper side. Several generators of the same type of dimensions 0.3×0.15 and 0.7×0.15 m were used. The inclined plane at the end of the canal served as a wave damper. The depth of the water varied between 0.04 and 0.1 m.

The described system permitted to simulate not only single piston-like and membrane-like displacements, but also oscillations of the basin bottom. For simulation of a running displacement use was made of three identical generators, driven sequentially (Fig. 2.17). The registration of waves on the free water surface was performed with the aid of optical sensors—IR wavegraphs [Nosov, Shelkovnikov (1991)]. Unlike traditional contact methods of wave measurements on a water surface, an IR wavegraph introduces no distortion in the surface at the point of measurement, therefore, it can measure waves of small amplitude (0.1 mm and less). Measurement of waves of such small amplitudes is essential in physical simulation of tsunamis in the open ocean. Observation of geometrical similarity [Basov et al. (1984)] requires conservation of the relationship between the wave amplitude

2.3 Plane Problems of Tsunami Excitation by Deformations of the Basin Bottom 69

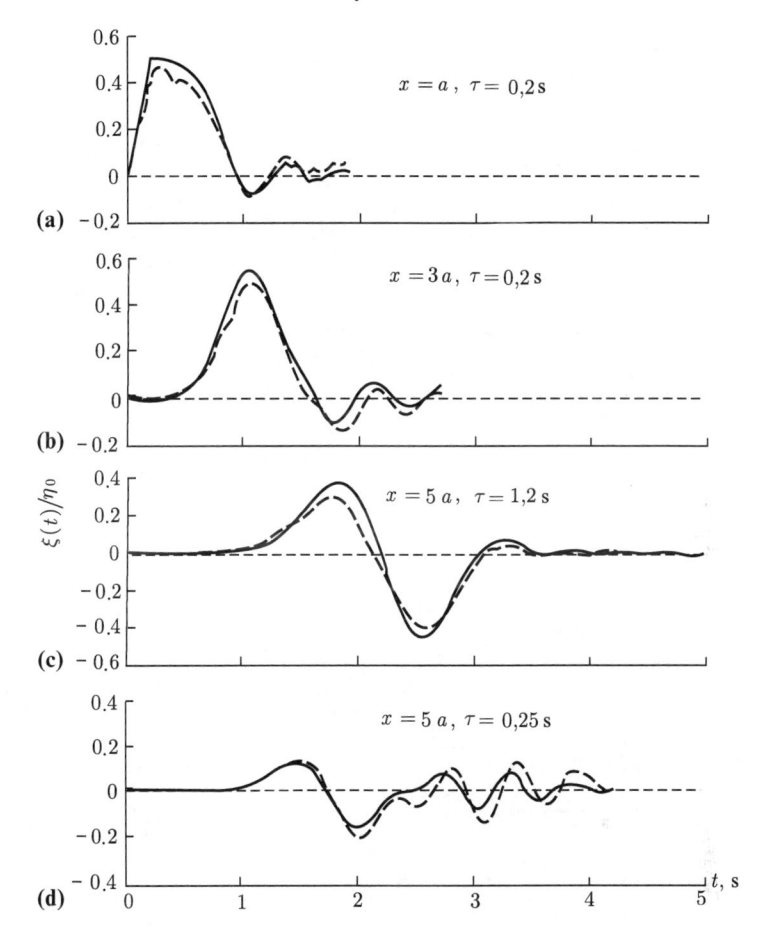

Fig. 2.18 Examples of time evolvents of waves generated by piston-like (**a, b**) and membrane-like (**c, d**) displacements of the basin bottom. Solid line—linear potential theory, dotted line—experiment; $a = 0.3$ m, $H = 0.1$ m

and the basin depth, $A/H \sim 10^{-3}$, while the depth and wave length are related as $H/\lambda \sim 10^{-2}$–10^{-1}. It is extremely difficult to establish such relationships in laboratory conditions. Owing to application of the IR wavegraph we have succeeded to perform the first investigation in the case of realistic relationships between the basin depth, wave length and amplitude.

Figure 2.18 presents examples of waves, registered in the experiment (dotted line) and calculated in accordance with linear potential theory (solid line). The theory is seen to describe the wave perturbations quite adequately. Explanation of the small discrepancy between experiment and theory consists in that the actual time dependences of the basin motion differed insignificantly from the theoretical dependences, the first time derivatives of which exhibit discontinuities. The experimental points shown in Figs. 2.12–2.14 were obtained as a result of analysing several hundreds of experiments. The experimental data are seen to confirm the main peculiarities of the obtained theoretical dependences.

Once again it must be stressed that not to take into account dispersion in describing the process of tsunami generation may result in significant errors in determining amplitude and energy characteristics of waves, especially in the case of displacements of the basin bottom not accompanied by residual deformations.

For comparison, Fig. 2.19 demonstrates the results of experiments and theoretical calculations, performed in [Hammack (1973)]. Note that Hammack only investigated waves excited by piston-like displacements. Moreover, he applied a somewhat different, smoother time law of the basin bottom deformation. The main characteristics of the dependences, obtained by Hammack and by us for the piston-like displacement, are identical.

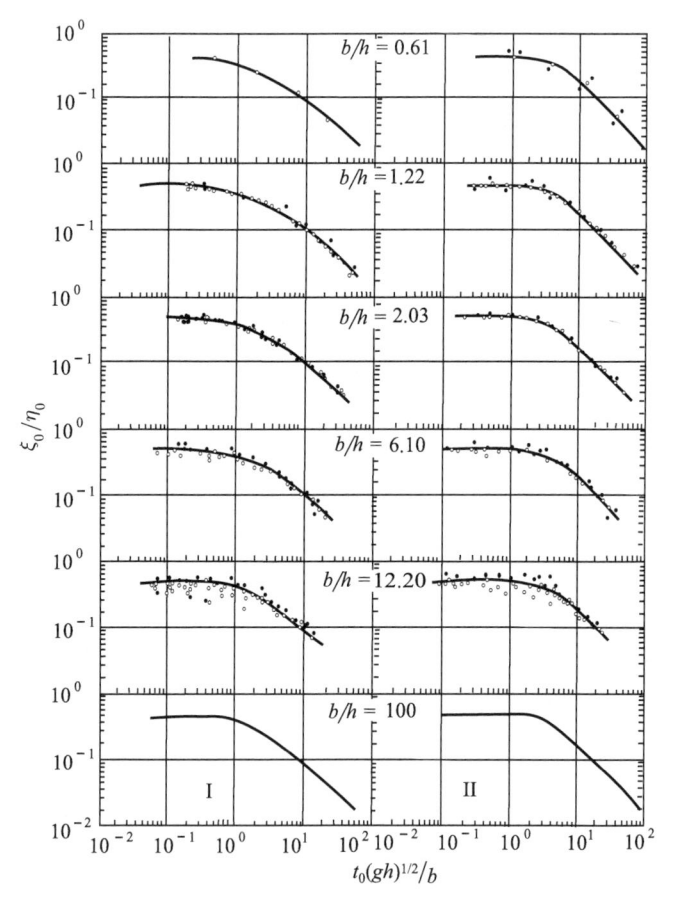

Fig. 2.19 Amplitude ξ_0 of wave, excited by piston-like basin bottom displacement of amplitude η_0 as function of the displacement duration τ. Adapted from [Hammack (1973)]

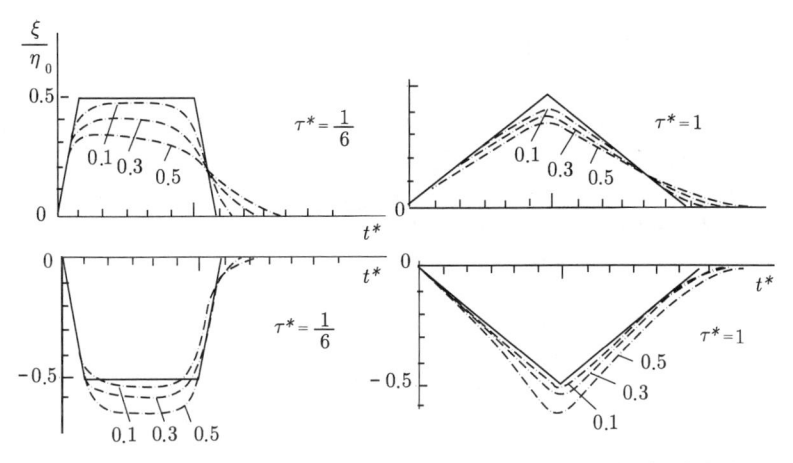

2.3 Plane Problems of Tsunami Excitation by Deformations of the Basin Bottom

Fig. 2.20 Waves formed by a piston-like displacement of large amplitude. Calculations are performed within the framework of long-wave theory: solid line—linear theory, dotted line—nonlinear theory. The numbers, indicating the curves, show the ratio of the basin bottom deformation amplitude and the basin depth

In his experiments Hammack applied the traditional method of wave registration on water, making use of a parallel-wire wavegraph. For this reason he had to excite waves of higher amplitudes. From Fig. 2.19 it is seen that the points, corresponding to positive (empty circles) and negative (full circles) displacements of the basin bottom, are stratified, i.e. lie respectively below and above the theoretical dependence obtained within the framework of linear theory. Note that the 'stratification' effect of experimental points is observed only in the case of large dimensions of the generation area (for instance, $b/h = 12, 20$ or $6, 10$). This is readily explained by the large relative dimensions of the generation area (the quantity b/h) being achieved by the choice of a small depth of water in the canal, h, owing to which the relative amplitude of the displacement became comparable to the depth. Large relative amplitudes were accompanied by manifestation of non-linear effects [Kostitsyna et al. (1992)]. Attention must also be drawn to the fact that in the case of prolonged displacements the 'stratification' effect became noticeably smaller, which was evidently related to the drop in the relative wave amplitude. Figure 2.20 clearly demonstrates the manifestation of non-linearity in the case of tsunami generation by piston-like displacements of large amplitude.

2.3.3 Running and Piston-like Displacements

The idea of the deformation of a basin bottom being a process taking place simultaneously throughout the entire active region is, naturally, far from reality, although it does serve as an illustrative model of tsunami generation. Actually, deformation of a basin bottom is a consequence of the fault at the earthquake source propagating along a certain plane. In the case of strong earthquakes the fault plane may extend over hundreds of kilometers, exhibiting a small angle to the horizontal plane. Therefore, a displacement of the basin bottom, as a rule, has a component that can be represented as a perturbation propagating in the horizontal direction. In the literature, such perturbations of the ocean bottom are conventionally termed 'running displacements' [Novikova, Ostrovsky (1979); Vasilieva (1981); Marchuk et al. (1983)]. Let us name several other natural prototypes of the running displacement. This role may be assumed by a non-simultaneous (sequential) displacement of blocks of the bottom [Lobkovsky, Baranov (1982)], a crack propagating over the basin bottom [Bobrovich (1988)], surface seismic waves [Belokon' et al. (1986)], the motion of an underwater landslide [Garder et al. (1993); Kulikov et al. (1998)], [Watts et al. (2001)]. Similar effects may be observed, also, in the case of wave generation by a moving area of low or elevated pressure [Pelinovsky et al. (2001)].

The interest in running displacements arose, because when the propagation velocity of a displacement coincides (even approximately) with the velocity of long waves, $(gH)^{1/2}$, a resonance pumping is realized of energy into the tsunami wave. Like in the preceding section, we shall first turn to the linear theory of long waves. We shall take advantage of the one-dimensional wave equation (2.75), describing perturbation of the surface, $\xi(x,t)$, that arises with deformation of the basin bottom, $\eta(x,t)$. Assume a deformation of the basin bottom, the shape of which is set by a certain function f, to propagate in the positive direction of the $0x$ axis with a constant velocity v: $\eta(x,t) = f(x-vt)$. Consider the motion to be established, therefore the solution of equation (2.75) will also have the form of a perturbation $\xi(x,t) = A_0 f(x-vt)$ running over the surface, where A_0 is a constant. Substituting the form of the solution, $\xi(x,t)$, and function $\eta(x,t)$ into the wave equation, we find the dependence of the constant A_0 upon the velocity of long waves and the propagation velocity of the perturbation,

$$\xi(x,t) = \frac{v^2}{v^2 - gH} f(x-vt). \tag{2.85}$$

From formula (2.85) it is seen that over a deformation of the basin bottom, travelling horizontally, there always exists a similar in shape perturbation of the water surface. Given the condition $v < (gH)^{1/2}$, the perturbations of the surface and of the basin bottom exhibit different polarities, while, when $v > (gH)^{1/2}$, their polarities coincide. The velocities of the perturbation and of long waves being close to each other result in a sharp enhancement of the amplitude of the surface perturbation.

If the problem of an established running displacement is considered within the framework of linear potential theory, then the main conclusion concerning resonance pumping of energy into the wave, when $v \approx (gH)^{1/2}$, does not change. At velocities $v > (gH)^{1/2}$ there will exist over the displacement a perturbation of similar polarity. But for velocities $v < (gH)^{1/2}$, besides the perturbation of opposite polarity located over the displacement, there also exists behind it a periodic in space and stationary in time perturbation with a wavelength determined by the velocity v. Standing waves, similar in nature, form when underwater obstacles are bypassed by the flow [Sretensky (1977)].

2.3 Plane Problems of Tsunami Excitation by Deformations of the Basin Bottom

Problems of established motion are doubtless expedient for understanding the peculiarities of physical processes taking place during wave generation by running displacements. But in reality a tsunami forms during a certain finite time interval. Therefore, we shall further consider models assuming deformations of the basin bottom to be limited in time.

It must be noted that practically all prototypes of the running displacement (with the exception of underwater landslides) exhibit velocities superior to the velocity of sound in water, therefore, the model of an incompressible liquid, considered here, is often not adequate for describing the process. Nevertheless, the solution of this problem is certainly not without significance, for the following reasons. Earlier, the running displacement as a tsunami generator was studied exclusively within the framework of the theory of long waves [Novikova, Ostrovsky (1979)], which occupies a lower position than potential theory in the hierarchy of models. The theory of incompressible liquids is a special case (and limit for $c \to \infty$) of the more general theory of compressible liquids. Consequently, the solution of the problem for an incompressible liquid will be a convenient benchmark in the construction of a more complex theory, and, moreover, the possibility arises of direct comparison of solutions of one and the same problem, obtained within the frameworks of different theories.

Making use of solutions (2.65) and (2.72), obtained within the framework of linear potential theory, we shall perform comparative analysis of dispersive tsunami waves excited by piston-like and running displacements of the basin bottom and subject to dispersion [Nosov (1996)]. We shall also compare such piston-like and running displacements that form identical residual deformations during the same time period, which, evidently, is expressed by the condition $b = v\tau$, where b is the horizontal size of the source, τ is the duration of the process at the source and v is the propagation velocity of a running displacement. In other words, we are attempting to compare the efficiency of wave excitation, when the area filled exhibits a rectangular shape and is adjacent to the basin bottom, by two methods: from below upward and from left to right. Figure 2.21 presents the profiles of waves calculated at time moment $t = 50(H/g)^{1/2}$ for the value of parameter $b = 10H$, which is characteristic of real tsunami sources. An ordinary piston-like displacement forms identical waves in the positive and negative directions of the Ox axis, while waves, excited by running displacements of the basin bottom, manifest an explicit asymmetry: a more intense train of waves runs in the direction of propagation of the displacement. The clearest asymmetry is revealed at propagation velocities of displacements, v, close to the velocity of long waves, $(gH)^{1/2}$. In the case of sufficiently large velocities v the profiles of waves, corresponding to piston-like and running displacements, actually become identical.

As a measure of the intensity of wave generation by the two mechanisms investigated, we shall take advantage of energy (per unit 'canal' width), calculated by the formula

$$W = \rho g \int \xi^2 \, dx. \tag{2.86}$$

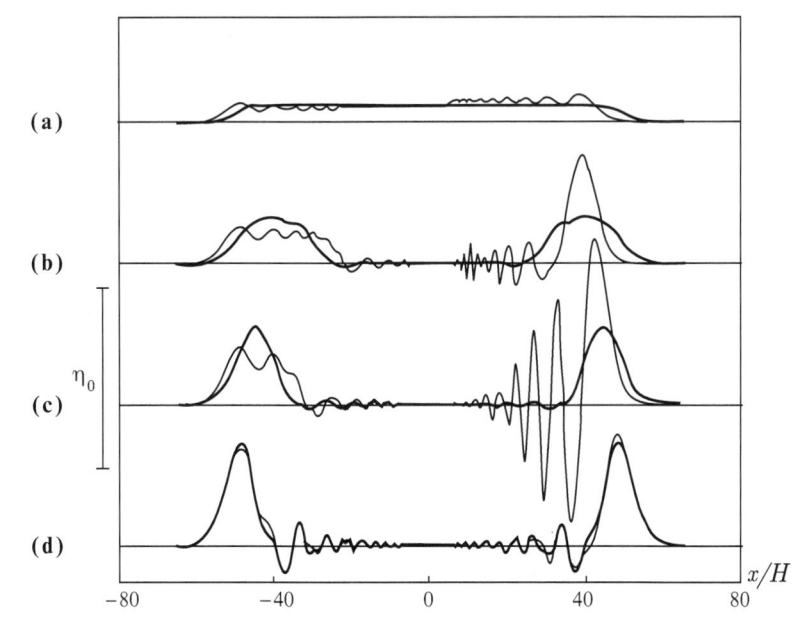

Fig. 2.21 Profiles of waves formed by running (thin line) and piston-like (thick line) displacements at time moment $t = 50$ for $L = 10$; (**a–d**) correspond to $v = 0.2, 0.5, 1$ and 10; $\tau = 50, 20, 10$ and 1

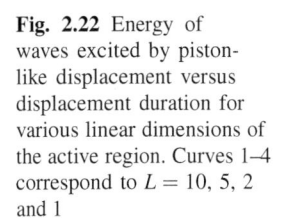

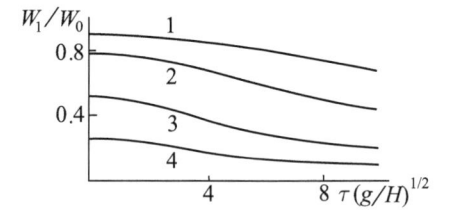

Fig. 2.22 Energy of waves excited by piston-like displacement versus displacement duration for various linear dimensions of the active region. Curves 1–4 correspond to $L = 10, 5, 2$ and 1

The quantity W equals twice the potential energy of the wave. The calculation of energy was performed for the time moment $t = 50(H/g)^{1/2}$, when energy redistribution between the potential and kinetic energies had been totally completed and the value of W no longer depended on time.

The results of calculations are presented in Fig. 2.22 as dependences of the energy of a wave excited by a piston-like displacement of the basin bottom, W_1, upon the displacement duration τ. The energy values are normalized to the quantity $W_0 = \rho g b \eta_0^2 / 2$, representing the specific potential energy of a rectangular elevation of height η_0 and length b of the free surface of a liquid. As the duration of a piston-like displacement increases the wave energy undergoes a monotonous decrease. Moreover, the energy depends essentially on the size of the generation area. Curves 1–4 in Fig. 2.22 correspond to values of parameter $b/H = 10, 5, 2$ and 1. Note that these results, naturally, do not contradict the data presented in Fig. 2.14.

2.3 Plane Problems of Tsunami Excitation by Deformations of the Basin Bottom

Figure 2.23 shows the dependence of the energy W_2 of a wave, excited by a running displacement, upon the velocity of the displacement propagation, v. When the parameter $b/H > 2$, the dependence exhibits a maximum, determined by the value of b/H, in the region of $v \sim (gH)^{1/2}$. The figure presents the relationship between the fractions of energy attributed to waves running along (W^+) and against (W^-) the direction of propagation of the running displacement versus the displacement velocity. This dependence also has a maximum in the vicinity of $v \sim 1$. When the parameter b/H decreases, the maximum is shifted noticeably towards smaller velocities. When the propagation velocity of the displacement increases, the curves in Fig. 2.23b asymptotically tend towards unity, independently of the value of b/H, which points to a loss of orientation by the energy emission at large values of v.

Figure 2.24 presents the dependence of the maximum possible energy of waves, excited by piston-like and running displacements of the basin bottom, upon the parameter b/H. In the case of $b/H < 2$ the type of displacement is seen to be irrelevant. When $b/H > 2$, a running displacement turns out to be capable of exciting waves more effectively than a piston-like displacement, and, while curve 1 tends asymptotically towards an evident long-wave limit equal to unity, the maximum energy of a wave, excited by a running displacement actually increases linearly with the parameter b/H. It must be noted that in this case the linear increase of dimensionless energy signifies a quadratic dependence of its dimensional value upon the size of the generation area, b.

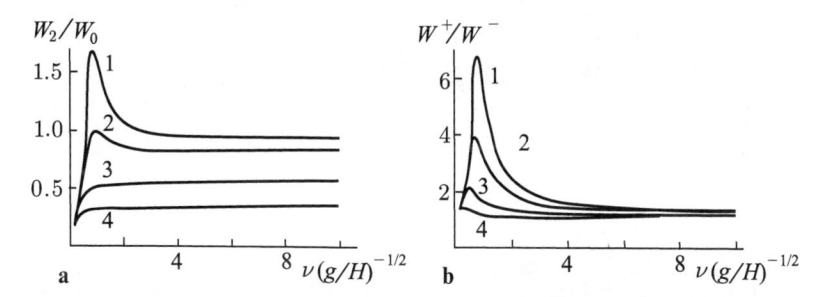

Fig. 2.23 Energy of waves (**a**), excited by a running displacement versus the propagation velocity of the displacement. **b**—The relationship between the fractions of energy attributed to waves running along (W^+) and against (W^-) the direction of propagation of the running displacement versus the displacement velocity for $L = 10, 5, 2$ and 1 (1–4, respectively)

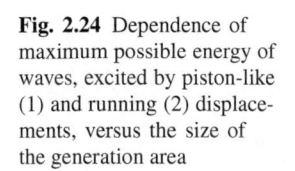

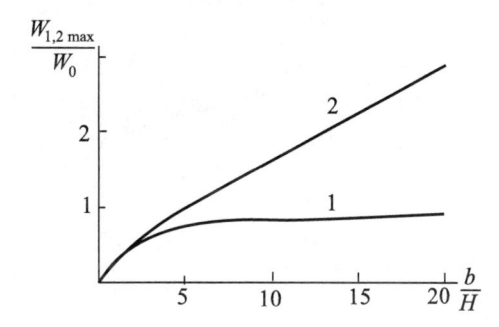

Fig. 2.24 Dependence of maximum possible energy of waves, excited by piston-like (1) and running (2) displacements, versus the size of the generation area

Laboratory simulation of a running displacement [Nosov, Shelkovnikov (1995)] was performed with the set-up depicted in Fig. 2.17. As the wave source use was made of three identical ocean bottom wave generators (of length $l = 0.3$ m) located at the centre of the hydrocanal and driven sequentially. The motion of each of the generators simulated a vertical displacement of the basin bottom, involving residual displacement, and was controlled by its individual sensor. The amplitude of motions of the generators did not exceed 2 mm. The duration of motion of each generator, τ, was chosen so as to have pulsed displacements: $\tau \ll l(gH)^{-1/2}$ (usually, ~ 0.2 s). The depth of the water in experiments amounted to 3, 5, 7 and 10 cm.

Perturbations of the free water surface were registered with the aid of two IR wavegraphs, located at the boundaries of the generation area. Records of the generator motions and of signals arriving from the wavegraphs were used in determining the maximum amplitude of the wave perturbations running along and against the direction of propagation of the displacement; the vertical displacement of each of the generators is η_i and the propagation velocity of the displacement, $v = (l/t_{12} + l/t_{23})/2$, where t_{12} and t_{23} are the time intervals between the connections of the first and second and of the second and third generators.

The results of experiments and of calculations, performed in accordance with formula (2.72), are presented in Fig. 2.25 as dependences of the maximum amplitude of the wave perturbation, A_{max}, upon the propagation velocity of the basin bottom displacement. The dependence is presented in dimensionless coordinates: the wave perturbation amplitude is normalized to the amplitude of the bottom displacement, $A_0 = (\eta_1 + \eta_2 + \eta_3)/3$, averaged for each given experiment, while the velocity v is normalized to the propagation velocity of long waves, $v_0 = (gH)^{1/2}$. The data on the maximum amplitude of the wave, running against the direction of propagation of

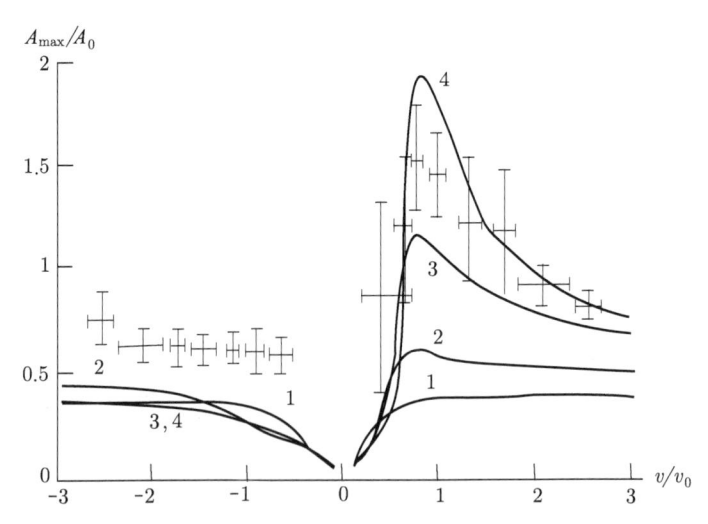

Fig. 2.25 Experimental and calculated dependences of the maximum wave amplitude at the boundary of the generation area at points $x = 0$ ($v/v_0 < 0$) and $x = b$ ($v/v_0 > 0$) versus the propagation velocity of the displacement. Curves 1–4 correspond to values of parameter $b = 1, 2, 5$ and 10

the displacement, correspond to negative values of the dimensionless velocity. The large spread of experimental data, due to the amplitudes of bottom displacements, η_i, not being strictly equal to each other, did not permit to separate the experimental dependences for different water depths, so the experimental points in Fig. 2.25 reflect the data averaged over all the indicated water depths H.

Motions of the basin bottom in laboratory and theoretical models somewhat differed from each other. Therefore, one cannot expect perfect coincidence of theory and experiment, which is particularly noticeable, when $v/v_0 < 0$. This is also related to the fact that the difference between displacement amplitudes of bottom wave generators could amount to 30%, while the maximum amplitude of the wave, running against the direction of propagation of the displacement, is determined by the amplitude of the largest of η_i. Consequently, in connection with the wave amplitude being normalized to the quantity $A_0 = (\eta_1 + \eta_2 + \eta_3)/3$, the dimensionless amplitude will certainly be overestimated as compared with the case of identical bottom displacement amplitudes.

Theory and experiment show that a running displacement can indeed serve as an effective mechanism for the excitation of tsunami waves. In the case of propagation velocities of bottom displacements close to the velocity of long waves, $(gH)^{1/2}$, sharp enhancement occurs of the amplitude and energy of waves running in the direction of the displacement propagation. It is known that a tsunami amplitude in the open ocean cannot exceed the amplitude of a piston-like displacement of the ocean bottom. Contrariwise, in the case of a running displacement the wave amplitude can significantly exceed the bottom displacement amplitude. In the case of identical residual deformations of the bottom, a running displacement may turn out to be many times more effective that a piston-like displacement. The energy transferred by a running displacement to gravitational waves, when $v = (gH)^{1/2}$, increases in proportion to the square distance covered by the displacement.

2.3.4 The Oscillating Bottom

In the case of established harmonic oscillations of the bottom we cannot directly take advantage of the general solution, obtained applying the Laplace transformation, since the oscillations take place at times $t < 0$. But in the case considered this is not necessary. For established oscillations it is possible to obtain a fully analytical solution, which does not require numerical calculation of integrals [Nosov (1992)]. Owing to the response of a linear system existing only at the frequency of inducing oscillations, we know the frequency of excited waves. Therefore, the solution of the problem is expediently sought in the following form:

$$F_{osc}(x,z,t) = \exp\{i\omega t\} \int_{-\infty}^{+\infty} dk \exp\{-ikx\} \left(A(\omega,k)\cosh(kz) + B(\omega,k)\sinh(kz)\right). \quad (2.87)$$

Taking advantage of the boundary conditions on the surface and on the bottom, we obtain:

$$F_{\text{osc}}(x,z,t) = \frac{\eta_0 \omega H}{2\pi} \exp\{i\omega t\} \int_{-\infty}^{+\infty} dk$$

$$\times \frac{(\exp\{-ik(x-a)\} - \exp\{-ik(x+a)\})\,(k\cosh(kz) + \omega^2 \sinh(kz))}{k^2(\omega^2 \cosh(k) - k\sinh(k))}. \quad (2.88)$$

Expression (2.88) contains dimensionless variables under the integral sign that were introduced in accordance with the formulae (the sign '*' has been dropped):

$$(x^*, z^*, a^*) = (x, z, a)H^{-1}; \qquad t^* = t g^{1/2} H^{-1/2};$$

$$\omega^* = \omega\, g^{-1/2} H^{1/2}; \qquad k^* = kH,$$

but the multiplier before the integral and the velocity potential itself are dimensional quantities.

To calculate the integral (2.88) it suffices to know the value of an integral of the following form:

$$\int_{-\infty}^{+\infty} dk \frac{\exp\{-ik\alpha\}\,(k\cosh(kz) + \omega^2 \sinh(kz))}{k^2(\omega^2 \cosh(k) - k\sinh(k))}, \quad (2.89)$$

where the parameter $\alpha = x \pm a$ may assume positive, negative and zero values.

Let us continue the integrand function in (2.89) analytically from the real axis onto the entire complex plane ($\{\text{Re}(k), \text{Im}(k)\}$). The integrand has two singular points on the real axis, $k = \pm k_0$, and an infinite number of singular points on the imaginary axis, $k = \pm i k_j$. The singular points are poles of the first order, and their positions are determined from the solutions of the two following transcendental equations:

$$\cosh(k)\,\omega^2 - k\sinh(k) = 0, \quad (2.90)$$

$$\cos(k)\,\omega^2 + k\sin(k) = 0. \quad (2.91)$$

The integrand function in (2.89) has no other singular points, which is readily demonstrated with the aid of the theorem on counting the number of zeros of an analytical function [Sveshnikov, Tikhonov (1999)].

Since the integrand function has poles on the real axis, the integral (2.89) must be understood in the sense of its principal value (v.p.), according to Cauchy. For its calculation the theorem of residues was applied. The ultimate expression, determining the velocity potential of a liquid flow in the case of established oscillations of a part of the bottom, has the following form:

2.3 Plane Problems of Tsunami Excitation by Deformations of the Basin Bottom

- For $|x| \leqslant a$

$$\frac{F_{\text{osc}}(x,z,t)}{H\eta_0 i\omega} = \exp\{i\omega t\}\left(\frac{1}{\omega^2}+z-2\sum_{j=1}^{\infty}Q\exp\{-k_j a\}\cosh(k_j x)\right)$$
$$+P\left(\exp\{i(\omega t+k_0(x-a))\}+\exp\{i(\omega t-k_0(x+a))\}\right), \quad (2.92)$$

- For $x \geqslant a$

$$\frac{F_{\text{osc}}(x,z,t)}{H\eta_0 i\omega} = \exp\{i\omega t\}\left(2\sum_{j=1}^{\infty}Q\exp\{-k_j x\}\sinh(k_j a)\right)$$
$$+P\left(-\exp\{i(\omega t-k_0(x-a))\}+\exp\{i(\omega t-k_0(x+a))\}\right) \quad (2.93)$$

- For $x \leqslant -a$

$$\frac{F_{\text{osc}}(x,z,t)}{H\eta_0 i\omega} = \exp\{i\omega t\}\left(2\sum_{j=1}^{\infty}Q\exp\{k_j x\}\sinh(k_j a)\right)$$
$$+P\left(\exp\{i(\omega t+k_0(x-a))\}-\exp\{i(\omega t+k_0(x+a))\}\right), \quad (2.94)$$

where

$$P = \frac{k_0\cosh(k_0 z)+\omega^2\sinh(k_0 z)}{k_0^2((\omega^2-1)\sinh(k_0)-k_0\cosh(k_0))},$$

$$Q = \frac{k_j\cos(k_j z)+\omega^2\sin(k_j z)}{k_j^2((\omega^2-1)\sin(k_j)-k_j\cos(k_j))}.$$

With knowledge of the velocity potential of the flow it is not difficult to obtain expressions for the displacement of a free surface and for the velocity components:

- For $|x| \leqslant a$

$$\xi_{\text{osc}}(x,t) = \eta_0\exp\{i\omega t\}\left(1-2\omega^2\sum_{j=1}^{\infty}Q\exp\{-k_j a\}\cosh(k_j x)\right)$$
$$+\eta_0\omega^2 P\left(\exp\{i(\omega t+k_0(x-a))\}+\exp\{i(\omega t-k_0(x+a))\}\right), \quad (2.95)$$

- For $x \geqslant a$

$$\xi_{\text{osc}}(x,t) = \eta_0 e^{i\omega t}2\omega^2\sum_{j=1}^{\infty}Q\exp\{-k_j x\}\sinh(k_j a)$$
$$+\eta_0\omega^2 P\left(-\exp\{i(\omega t-k_0(x-a))\}+\exp\{i(\omega t-k_0(x+a))\}\right), \quad (2.96)$$

- For $x \leqslant -a$

$$\xi_{osc}(x,t) = \eta_0 \exp\{i\omega t\} 2\omega^2 \sum_{j=1}^{\infty} Q \exp\{k_j x\} \sinh(k_j a)$$
$$+ \eta_0 \omega^2 P \left(\exp\{i(\omega t + k_0(x-a))\} - \exp\{i(\omega t + k_0(x+a))\}\right). \quad (2.97)$$

- For $|x| \leqslant a$

$$u(x,z,t) = \eta_0 i\omega \, \exp\{i\omega t\} \left(-2 \sum_{j=1}^{\infty} Q k_j \exp\{-k_j a\} \sinh(k_j x)\right), \quad (2.98)$$

$$w(x,z,t) = \eta_0 i\omega \, \exp\{i\omega t\} \left(1 - 2\sum_{j=1}^{\infty} \frac{\partial Q}{\partial z} \exp\{-k_j a\} \cosh(k_j x)\right), \quad (2.99)$$

- For $|x| \geqslant a$

$$u(x,z,t)$$
$$= \eta_0 i\omega \, \exp\{i\omega t\} \left(-2\,\text{sign}(x) \sum_{j=1}^{\infty} Q k_j \exp\{-k_j |x|\} \sinh(k_j a)\right), \quad (2.100)$$

$$w(x,z,t) = \eta_0 i\omega \, \exp\{i\omega t\} \left(2 \sum_{j=1}^{\infty} \frac{\partial Q}{\partial z} \exp\{-k_j |x|\} \sinh(k_j a)\right), \quad (2.101)$$

Note that, owing to the discontinuity exhibited by the function, describing the space distribution of oscillations of the basin bottom, expressions (2.98)–(2.101) do not yield adequate values of the flow velocity at the points with coordinates $\{x = \pm a, z = -1\}$. To obtain the exact velocity values in the immediate vicinity of the points indicated it is necessary to take into account quite a large number of terms of the expansion in j.

From the structure of the obtained formulae it is seen that perturbation of a liquid consists of induced oscillations, occurring in the immediate vicinity of the source (and exponentially dying away with the distance from it), and a series of progressive waves starting at points $x = \pm a$.

From the point of view of tsunami generation precisely the amplitude of progressive waves is important. From formulae (2.96) and (2.97) it is seen that this amplitude is largely determined by the quantity $|\omega^2 P|$, the dependence of which upon the cyclic frequency is shown in Fig. 2.26. A most important peculiarity of the response of the liquid to oscillations of a part of the ocean bottom consists in the existence of a certain boundary frequency, which, when surpassed, the efficiency of wave emission drops drastically. Thus, at high frequencies all the motions of the liquid is concentrated exclusively in the vicinity of the source and represents induced oscillations.

2.3 Plane Problems of Tsunami Excitation by Deformations of the Basin Bottom

Fig. 2.26 Dependence of the quantity $|\omega^2 P|$, determining the amplitude of a gravitational wave upon the cyclic frequency of oscillations of the basin bottom

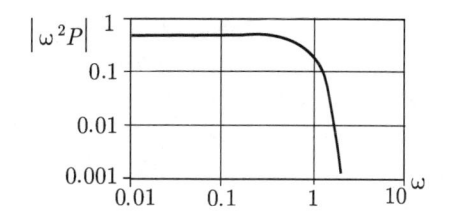

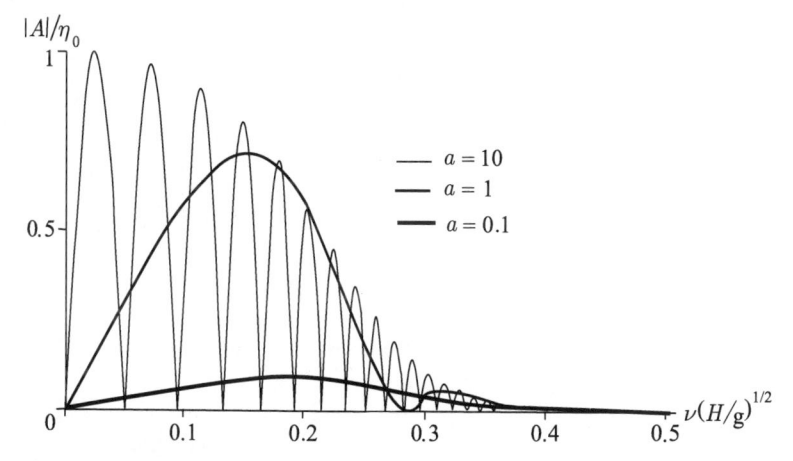

Fig. 2.27 Amplitude of progressive wave excited by oscillating area of ocean bottom versus the frequency of bottom oscillations for different sizes of the source, a

But the amplitude of emitted waves, A, does not depend only on the frequency of basin bottom oscillations, but on the horizontal extension of the oscillating area of the bottom also. In accordance with formula (2.96) we can write

$$A(\omega) = \eta_0 \omega^2 \frac{2 \sin(k_0 a) \cosh(k_0)}{k_0 [k_0 + \sinh(k_0) \cosh(k_0)]}, \quad (2.102)$$

where $\omega^2 = k_0 \tanh(k_0)$. The dependence of the absolute value of the amplitude upon the dimensionless oscillation frequency of the ocean bottom $(\nu(H/g)^{1/2})$ is presented in Fig. 2.27. Calculations were performed for three different values of parameter a. Surface manifestation of the oscillations of a part of the ocean bottom with linear dimensions, smaller than the depth of the layer of liquid, will be relatively weak. The existence of a set of frequencies, at which the amplitude of the emitted wave turns to zero, is related to the interference of waves forming at points $x = \pm a$. The automatic locking of the source is a consequence of the rectangular space distribution of the amplitude of bottom oscillations. Actually, manifestation of the automatic locking effect is extremely improbable.

The dependence (2.102) permits to reveal parameters determining the limits of the tsunami frequency spectrum, $\nu_{\min}$ and $\nu_{\max}$. We shall find the limit frequencies from the solution of equation $A(\nu) = \eta_0/10$. From Fig. 2.27 it is not difficult to conclude that $\nu_{\max} \sim 0.3$, while the quantity $\nu_{\max}$ does not depend on the size of the generation area, a.

We shall now determine v_{min}. At small oscillation frequencies of the ocean bottom (which also corresponds to small values of k_0), expression (2.102) is essentially simplified and assumes the following form: $A(\omega) = \eta_0 a \omega$. Thus, the quantity v_{min} can be estimated as $v_{min} \sim (20\pi a)^{-1}$. Passing to dimensional quantities, we obtain the following formulae for the limits of the tsunami frequency spectrum:

$$v_{max} \sim 0.3 \left(\frac{g}{H}\right)^{1/2}, \qquad (2.103)$$

$$v_{min} \sim \frac{(gH)^{1/2}}{20\pi a}. \qquad (2.104)$$

The lower frequency limit is seen to be related both to the ocean depth and to the horizontal dimension of the source, while the upper limit only to the depth. For a depth $H \sim 10^3$ m and size of the source equal to $a \sim 10^4$ m we obtain $v_{max} \sim 10^{-2}$ Hz, $v_{min} \sim 10^{-4}$ Hz. The spectrum of real tsunami waves lies precisely within these limits [Murty (1977); Pelinovsky (1996)].

The theoretical dependence (2.102) has been tested experimentally [Nosov, Shelkovnikov (1992)]. Use was made of the set-up shown in Fig. 2.16. The pneumatic generator, 30 cm long, simulated harmonic oscillations of an area of the ocean bottom. Practically, all the remaining parts of the hydrocanal were occupied by the wave-damping system representing a gentle slope covered with a plastic mesh. The wave was registered by the IR wavegraph at a distance of 10 cm from the boundary of the generation area. The results of experiments and of theoretical calculations for three different water depths in the hydrocanal are presented in Fig. 2.28. The experimental data are seen to comply with the theoretical dependence.

2.4 Generation of Tsunami Waves and Peculiarities of the Motion of Ocean Bottom at the Source

In this section we shall deal with the space (3D) problem of tsunami generation by ocean bottom displacements. Transition from plane (2D) models to the more realistic three-dimensional problem makes it possible to investigate the most important issue of the orientation of wave emission and of its relation to parameters of the source. The effect of oriented emission of tsunami waves from the source area can be due to various reasons, which are usually considered to comprise the geometrical shape of the deformation area of the ocean bottom, the transfer of horizontal momentum to masses of water, and the wave-guide properties of the bottom relief [Voight (1987); Dotsenko, Soloviev (1990)]. The last reason, generally speaking, is related to the tsunami propagation, and not to wave generation. The orientation of tsunami emission, due to source asymmetry, has been studied theoretically [Kajiura (1963, 1970); Dotsenko et al. (1993)], experimentally [Takahasi (1963)] and numerically [Marchuk, Titov (1993)].

2.4 Generation of Tsunami Waves in Space

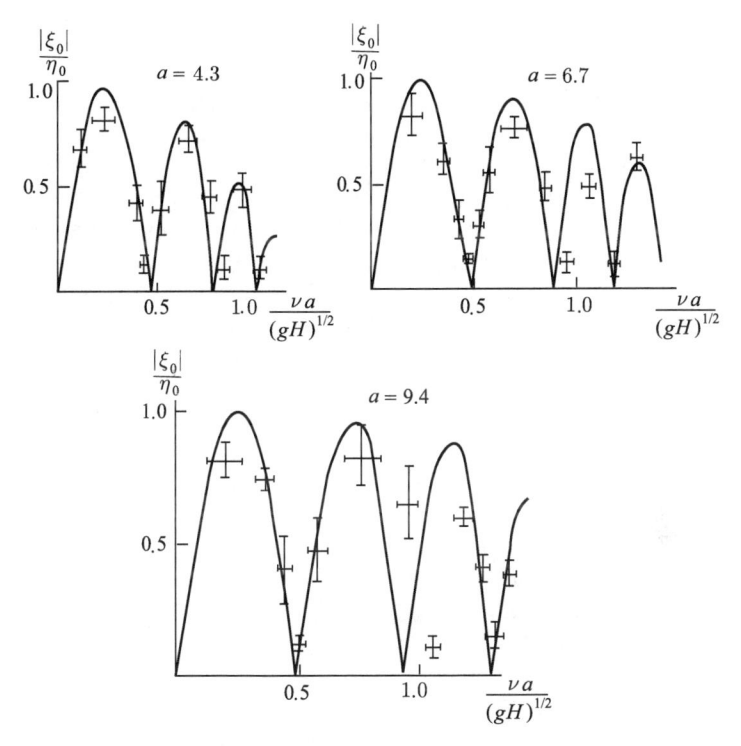

Fig. 2.28 Comparison of experimental and theoretical dependences of amplitudes of excited progressive wave versus frequency of bottom oscillations

The strikingly clear orientation of the Chilean tsunami of May 22, 1960, when the amplitude of the wave travelling in a direction, perpendicular to the South American coast, was several times larger than the amplitudes of waves propagating in other directions, initiated the appearance of a series of publications [Voit et al. (1980), (1981), (1982); Lebedev, Sebekin (1982)], in which the role was estimated of a horizontal motion of the ocean bottom in forming an oriented tsunami wave. In these works, the influence of a horizontal motion on the ocean was simulated by applying an effective mass force in the vicinity of the source, and, then, the properties of waves at a large distance from the generation area were studied. Thus, for example, it was established that a wave caused by a vertically directed mass force exhibits axial symmetry at long distances from the source, in spite of the perturbating force not being axially symmetric, while at the same time the wave front of a tsunami caused by a transfer of horizontal momentum remains anisotropic.

Numerical models of tsunamis that gained well-known popularity as powerful means for investigations [Chubarov et al. (1992); Kato, Tsuji (1995); Satake (1995); Satake, Imamura (1995); Tanioka, Satake (1996); Titov (1999); Myers, Baptista (1995); Suleimani et al. (2003); Zaitsev et al. (2005); Kowalik et al. (2005); Titov et al. (2005); Horrillo et al. (2006); Rivera (2006); Gisler (2008)], seem to have achieved a certain limit in their perfection, in the sense that for their further successful development it is necessary to introduce a number of essentially novel features, one of which consists in the following: wave excitation must be described realistically, i.e. it must be dealt with as a process extended in time. As a rule, in describing tsunami generation impulse displacements are considered, and, consequently, only geometric characteristics of the source, i.e. the distribution of residual bottom displacements in space, are taken into account. Here, the actual method (the time law followed by motions of the ocean bottom), by which the different residual displacements came about, is totally neglected. At the same time, the duration of processes at the source may amount to 100 s, and more [Satake (1995)]. For example, the process at the source of the Sumatran catastrophic tsunami of 2004 went on for about 1,000 s. In such a long period of time, a long wave is capable of covering a distance comparable to the size of a tsunami source, which means that a displacement cannot be assumed to exhibit an impulse character. Moreover, in [Dotsenko (1996); Nosov (1998)] it was established that the energy, amplitude and, even, orientation of tsunami waves is not only related to the geometric characteristics of the source, but also to the time law of motion of the ocean bottom.

In Sect. 2.2.1 the general solution was obtained, within the framework of potential theory, for the linear response of a layer of incompressible liquid of fixed depth to deformations of the ocean bottom, $\eta(x,y,t)$. We shall consider the following three model laws of ocean bottom deformation:

- Piston-like displacement

$$\eta_1(x,y,t) = \eta_S(x,y)\left(\theta(t)t - \theta(t-\tau)(t-\tau)\right)\tau^{-1}, \qquad (2.105)$$

- Membrane-like displacement

$$\eta_2(x,y,t) = \eta_S(x,y)\left(2\theta(t)t - 4\theta(t-\tau/2)(t-\tau/2)\right.$$
$$\left. + 2\theta(t-\tau)(t-\tau)\right)\tau^{-1}, \quad (2.106)$$

- Running displacement

$$\eta_3(x,y,t) = \eta_S(x-a,y)(1-\theta(x-vt)), \qquad (2.107)$$

where $\eta_S(x,y) = \eta_0\left(\theta(x+a) - \theta(x-a)\right)\left(\theta(y+b) - \theta(y-b)\right)$ is the space distribution of ocean bottom deformations, $\theta(z)$ is the Heaviside step-function. The active region has the shape of a rectangle of length $2a$ and width $2b$. The piston-like and membrane-like displacements are characterized by amplitude η_0 and duration τ, the running displacement by its amplitude η_0 and propagation velocity v. In the case of a running displacement the area of bottom deformations is shifted in the positive direction of axis $0x$ by the quantity a, so as to have motions of the ocean bottom start at the time moment $t = 0$.

2.4 Generation of Tsunami Waves in Space

We now introduce dimensionless variables (the sign '*' will be further dropped)

$$\{m^*, n^*\} = H\{m, n\}; \qquad \{x^*, y^*, z^*, a^*, b^*\} = H^{-1}\{x, y, z, a, b\};$$
$$\{t^*, \tau^*\} = \{t, \tau\} g^{1/2} H^{-1/2}; \qquad \{\xi^*, \zeta^*\} = \eta_0^{-1}\{\xi, \zeta\}. \tag{2.108}$$

Part of the integrals, present in formula (2.39), can be calculated analytically. Zipping intermediate calculations, we shall write out the formulae describing the perturbation of a free surface in the case of ocean bottom deformations of the form (2.105)–(2.107):

- Piston-like displacement

$$\xi_1(x, y, t) = \theta(t) \zeta_1(x, y, t) - \theta(t - \tau) \zeta_1(x, y, t - \tau), \tag{2.109}$$

- Membrane-like displacement

$$\xi_2(x, y, t) = 2\theta(t) \zeta_1(x, y, t)$$
$$- 4\theta(t - \tau/2) \zeta_1(x, y, t - \tau/2) + 2\theta(t - \tau) \zeta_1(x, y, t - \tau), \tag{2.110}$$

where

$$\zeta_1(x, y, t) = \frac{4}{\pi^2 \tau} \int_0^\infty \int_0^\infty dm\, dn$$

$$\times \frac{\sin(ma) \sin(nb) \cos(mx) \cos(ny) \sin\left((k \tanh k)^{1/2} t\right)}{mn \cosh k (k \tanh k)^{1/2}}, \tag{2.111}$$

$$k^2 = m^2 + n^2,$$

- Running displacement

$$\xi_3(x, y, t) = \frac{\eta_0}{2\pi^2 i} \int_0^\infty dn \int_{-\infty}^{+\infty} dm \frac{\exp\{imx\} \sin(nb) \cos(ny)}{\cosh(k) n}$$

$$\times \left(\frac{1 - \exp\left\{-i2a\left(m + \frac{(k \tanh(k))^{1/2}}{v}\right)\right\}}{m + \frac{(k \tanh(k))^{1/2}}{v}} \exp\left\{i(k \tanh(k))^{1/2} t\right\} \right.$$

$$\left. + \frac{1 - \exp\left\{-i2a\left(m - \frac{(k \tanh(k))^{1/2}}{v}\right)\right\}}{m - \frac{(k \tanh(k))^{1/2}}{v}} \exp\left\{-i(k \tanh(k))^{1/2} t\right\} \right).$$

$$\tag{2.112}$$

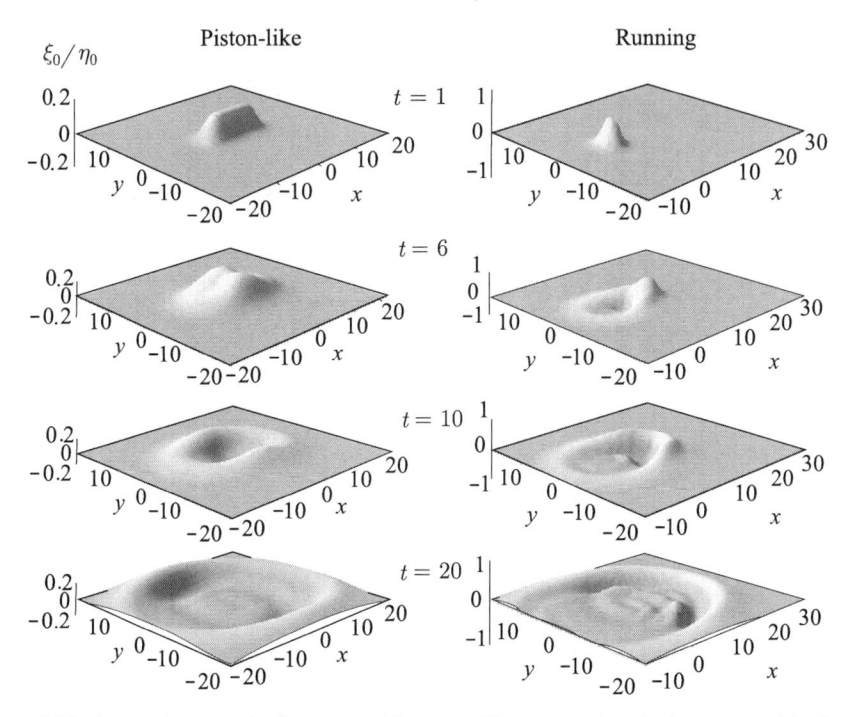

Fig. 2.29 Free-surface perturbations, caused by piston-like and running displacements of the bottom with parameters $a = 6, b = 2, v = 2$ ($\tau = 6$). Calculations are performed for the time moments t indicated in the figure

Formula (2.112) is valid, when $t \geqslant 2a/v$ (the reason that such a restriction exists is expounded in Sect. 2.3.1). The integrals in expressions (2.109)–(2.112) were calculated numerically.

Figure 2.29 presents the space structure of waves excited by piston-like and running displacements, which have ultimately resulted in identical residual deformations ($a = 6$, $b = 2$). The propagation velocity of a running displacement, $v = 2$, and the duration of a piston-like displacement, $\tau = 6$, satisfy the relationship $\tau v = 2a$. Calculations are performed in accordance with formulae (2.109) and (2.112). From the figure it is seen that in the case of a piston-like displacement the waves of maximum amplitude propagate in the negative and positive directions of axis 0y, i.e. in a direction perpendicular to the direction of maximum extension of the source. In the case of a running displacement the source emits waves of maximum amplitude at the Mach angle to the direction of propagation of the displacement (0x). Moreover, attention is immediately drawn to the fact that the amplitude of waves caused by a running displacement is significantly superior to the amplitude of waves in the case of a piston-like displacement.

For detailed investigation of the orientation of waves emitted from the source area wave time-bases were calculated at points lying on a circle of a certain radius ($r > \max[a, b]$), with its centre coinciding with the origin of the chosen reference

2.4 Generation of Tsunami Waves in Space

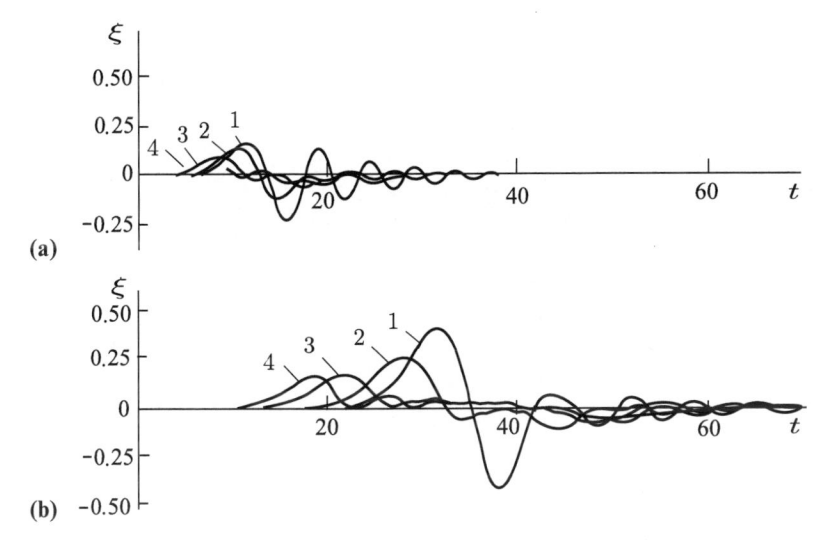

Fig. 2.30 Time-base of waves caused by piston-like displacement. Calculations are performed at points lying on a circle of radius $r = 10$ (**a**) and $r = 30$ (**b**) with its centre at the origin of the reference frame, for azimuthal angles $\alpha = 0, 30, 60, 90°$ (curves 1—4, respectively). The parameters of the bottom displacement: $a = 1, b = 5, \tau = 1$ (**a**) and $a = 3, b = 15, \tau = 1$ (**b**)

frame. Examples of such time-bases are presented in Fig. 2.30. The azimuthal angle was counted off from the positive direction of axis 0x. From the wave time-bases amplitude characteristics were determined, and the energy was estimated by the formula proposed in [Kajiura (1970)],

$$W = \rho g (gH)^{1/2} \int_0^T \int_\gamma \xi^2(t)\, dt\, d\gamma. \tag{2.113}$$

Formula (2.113) yields the energy that passed through the contour γ in time T. In our case, the contour γ was chosen to be the segment of a circle of radius **r**, given $\Delta\alpha = 10°$. Energy values were normalized to the quantity $W_0 = 2\rho g a b \eta_0^2$. The quantity W_0 corresponds to the potential energy of the initial free-surface elevation, exhibiting the shape of the residual bottom displacement.

Figure 2.31 presents, in the form of directional diagrams, the dependences of the amplitude of the first crest A_1 (a, b), of the 'maximum span' $A_{max} - A_{min}$ (c, d), and of the wave energy W_α (e, f) upon the azimuthal angle. Calculations are performed for piston-like and running displacements, the durations of which are chosen so as to satisfy the relationship $\tau = 2av^{-1}$. Thus, the process of wave excitation can be investigated by varying the parameter, common to both piston-like and running displacements, namely, the duration of the process in the active area. The dotted line in the figure shows the shape and orientation of the active area. From Fig. 2.31 the orientation is seen to be manifested most weakly for the amplitude of the first crest. The evolution of directional diagrams differs essentially for

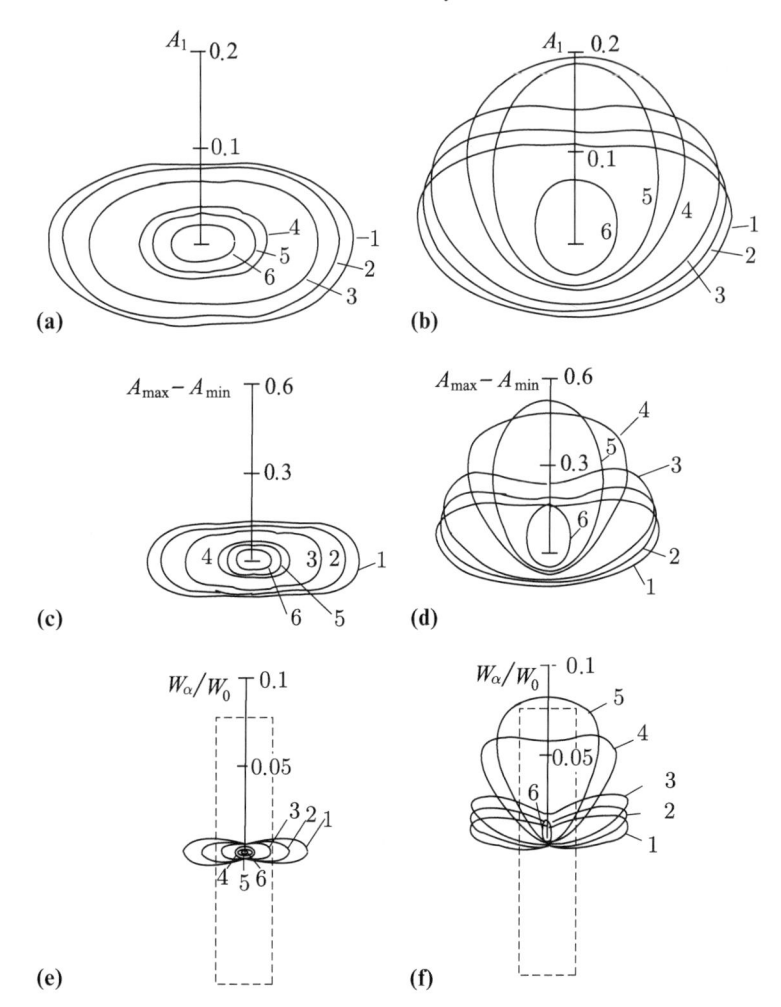

Fig. 2.31 Distribution of amplitude of the first crest (**a, b**), of the 'maximum span' (**c, d**) and of the energy (**e, f**) of a wave over the azimuthal angle. Curves 1–6 correspond to piston-like displacements (**a, c, e**) of durations $\tau = 2, 3.3(3), 5, 10, 12.5, 20$ and to running displacements (**b, d, f**) with propagation velocities $v = 5, 3, 2, 1, 0.8, 0.5$. The durations and propagation velocities are chosen so as to have $\tau = 2av^{-1}$. The dotted line shows the shape and orientation of the active area ($a = 5, b = 1$)

piston-like and running displacements, as the duration of the process at the source varies. When the duration of a piston-like displacement increases, the amplitude and energy of waves monotonously decreases, while their distribution over the azimuthal angle tends to be isotropic. Here, the wave of largest amplitudes and energy are always emitted in a direction, perpendicular to the largest extension of the source. In the case of a running displacement, as the duration of the process at the source increases (the propagation velocity of the displacement decreases), the directional diagrams gradually transform from two-pronged into single-pronged diagrams, and

2.4 Generation of Tsunami Waves in Space

Fig. 2.32 Total energy of waves generated by piston-like (1) and running (2) displacements versus duration of the process in the active area. In the case of a running displacement $\tau = 2av^{-1}$

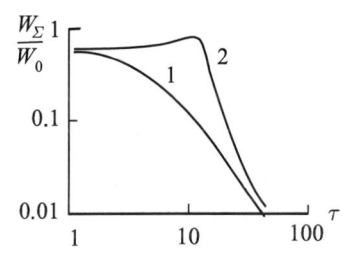

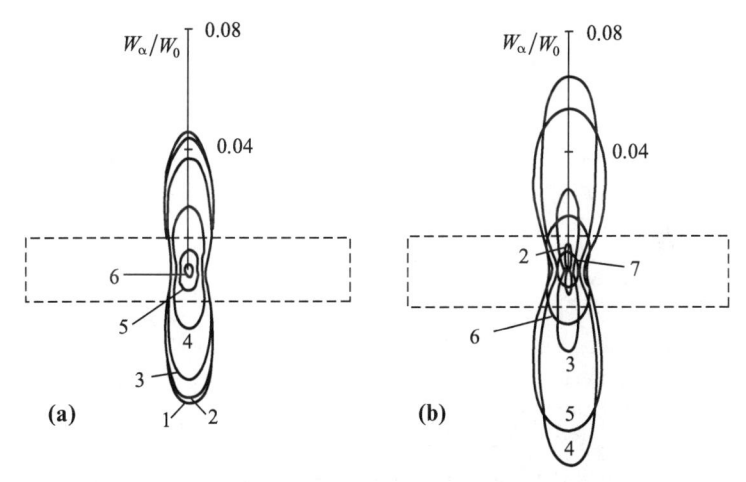

Fig. 2.33 Directional diagrams for emission energy of waves caused by piston-like (**a**) and membrane-like (**b**) displacements of ocean bottom. Curves 1–7 correspond to $\tau = 0.5, 1, 2, 4, 8, 16, 32$

the main part of energy starts to be emitted in the direction of propagation of the displacement. When the duration of the process at the source is small ($\tau = 2$), the directional diagrams for both cases investigated are practically identical. But in the case of large values of τ, the character of motion of the ocean bottom already exerts significant influence on the parameters of the excited wave.

Figure 2.32 presents the dependence of the total wave energy (integrated over all directions) upon the duration of the process at the source. In the case of a piston-like displacement (curve 1) the dependence is monotonous—the energy falls as the duration of the process increases. In the case of a running displacement (curve 2) the dependence reveals a maximum, which corresponds to coincidence of the propagation velocity of the displacement and the velocity of long waves. The running displacement is also seen to be noticeably more effective than the piston-like displacement within a wide range of τ values. When durations of the process at the source are small or large, the efficiencies of both mechanisms of wave generation are approximately identical.

Figure 2.33 presents energy directional diagrams for waves due to piston-like and membrane-like displacements. The dotted line shows the shape and orientation

Fig. 2.34 Directional coefficient versus duration of bottom displacement. Curves 1, 2 correspond to piston-like displacement, 3, 4 to membrane-like displacement. Curves 1, 3 are obtained for $a=1, b=5, r=10$, curves 2, 4 for $a=3, b=15, r=30$

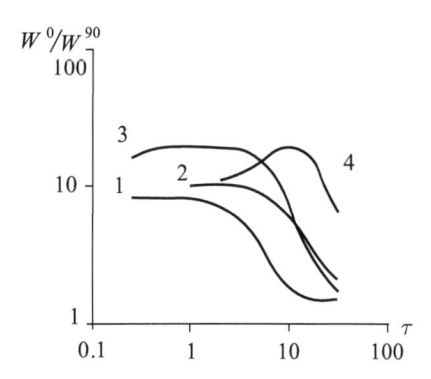

of the model source. Calculations have been performed for various deformation durations of the bottom. The main part of energy is seen to be emitted in a direction, perpendicular to the direction of maximum extension of the source, independently of the time law satisfied by motion of the bottom. But as the displacement duration increases, the diagram undergoes significant changes, the character of which does depend on the type of time law of motion of the bottom.

We shall now introduce the directional coefficient of emission as the ratio of energy fractions emitted in directions $\alpha = 0°$ and $90°$. Figure 2.34 shows the dependence of the directional coefficient upon the displacement duration. For a piston-like displacement the quantity $W^{0°}/W^{90°}$ decreases monotonously as the duration increases. But the corresponding dependence for a membrane-like displacement exhibits a non-monotonous character. It is interesting to note that in the case of a membrane-like displacement the directional coefficient may assume larger values than for a piston-like displacement. From the figure it can also be concluded that a decrease in the size of a source, if its shape (the ratio a/b) is conserved, results in a weakening of the directionality, especially when its size becomes comparable to the depth.

Why does the duration of the bottom deformation (or type of time law) influence the orientation of wave emission? The point is that an asymmetric source forms waves of differing wavelengths in different directions. In other words, there exists an effective horizontal size of a source, depending on the direction. Above, it was shown (Fig. 2.12) that, when the displacement duration increases, the wave amplitude falls more rapidly in the case of a source of smaller horizontal extension. Therefore, if the source is elongated, then the degree of emission orientation inevitably decreases, as the displacement duration rises. The behaviour of the directional coefficient in the case of a membrane-like displacement is explained in a similar manner.

Figure 2.35 presents the total wave energy (integrated over all directions) versus the duration of the bottom displacement. In the case of a piston-like displacement the energy monotonously decreases as the duration increases, while in the case of the membrane-like displacement there exists a certain 'optimal' duration, for which the energy is maximal. For this 'optimal' duration the membrane-like displacement turns out to be more effective, than even a piston-like displacement of small duration. Similar dependences were obtained in [Dotsenko, Soloviev (1990); Dotsenko (1996)] within the framework of linear theory of long waves.

2.4 Generation of Tsunami Waves in Space

Fig. 2.35 Total energy of waves due to piston-like and membrane-like displacements versus displacement duration. The curves are numbered like in Fig. 2.34

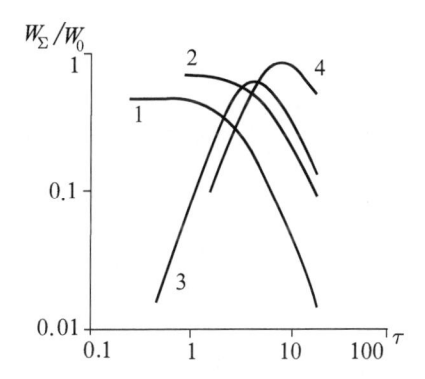

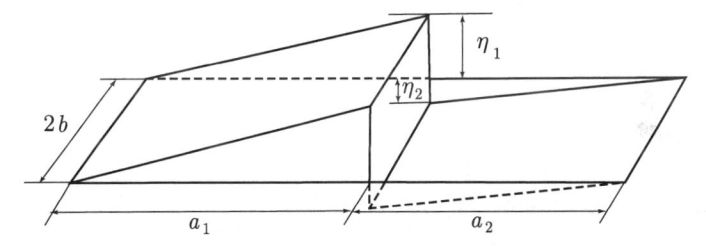

Fig. 2.36 Model of residual deformations of the bottom for a displacement alternating in sign

Analysis of space distribution of vertical residual displacements shows that at the tsunami source there usually exist two regions: an elevation and a depression of the ocean bottom [Van Dorn (1964); Dotsenko et al. (1986); Satake (1995); Kato, Tsuji (1995)]. We shall term such a displacement alternating in sign. We shall briefly dwell upon certain peculiarities of the directionality of wave emission, related to such bipolar deformation of the ocean bottom. Consider the displacement alternating in sign to be described by the formula [Nosov et al. (1999)]:

$$\eta_4(x,y,t) = \Big(\eta_1 a_1^{-1}(x+a_1)\left(\theta(x+a_1) - \theta(x)\right)$$
$$+ \eta_2 a_2^{-1}(x-a_2)\left(\theta(x) - \theta(x-a_2)\right)\Big)$$
$$\times \Big(\theta(y+b) - \theta(y-b)\Big)\Big(t\tau^{-1}\theta(t) - (t-\tau)\tau^{-1}\theta(t-\tau)\Big). \quad (2.114)$$

The space distribution of the deformation amplitude of the bottom, determined by formula (2.114), is shown in Fig. 2.36. As to the time law of bottom deformation, we consider a displacement with residual deformation.

The perturbation of a free surface in the case of a displacement alternating in sign is calculated in accordance with the following formula:

$$\xi_4(x,y,t) = \theta(t)\zeta_4(x,y,t) - \theta(t-\tau)\zeta_4(x,y,t-\tau),$$

where

$$\zeta_4(x,y,t) = \frac{2}{\pi^2 \tau} \int_0^\infty \int_0^\infty dm\, dn \, \frac{\sin\left[(k\tanh(k))^{1/2} t\right]}{\cosh(k)(k\tanh(k))^{1/2}} \frac{\cos(ny)\sin(nb)}{n}$$
$$\times \left[\cos(mx) \left(\frac{\eta_1}{a_1 m^2} [1-\cos(ma_1)] - \frac{\eta_2}{a_2 m^2} [1-\cos(ma_2)] \right) \right.$$
$$\left. + \sin(mx) \left(\frac{\eta_1}{a_1 m^2} \sin(ma_1) - \frac{\eta_1}{m} + \frac{\eta_2}{a_2 m^2} \sin(ma_2) - \frac{\eta_2}{m} \right) \right]. \quad (2.115)$$

Integration over the components m and n of the wave vector in formula (2.115) was performed numerically.

Figure 2.37 shows the example of the space structure of waves, excited by a bottom displacement alternating in sign with parameters $a_1 = 7$, $a_2 = 3$, $b = 10$, $\tau = 1$. The active area (its horizontal projection) exhibits the shape of a rectangle of size 10×20 (along axes $0x$ and $0y$, respectively). The 'fault' passes along the axis $0y$, the elevation and depression areas of the ocean bottom correspond, respectively, to negative and positive values of the x coordinate. From the figure it is seen that waves of maximum amplitude propagate in a direction, perpendicular to the 'fault'

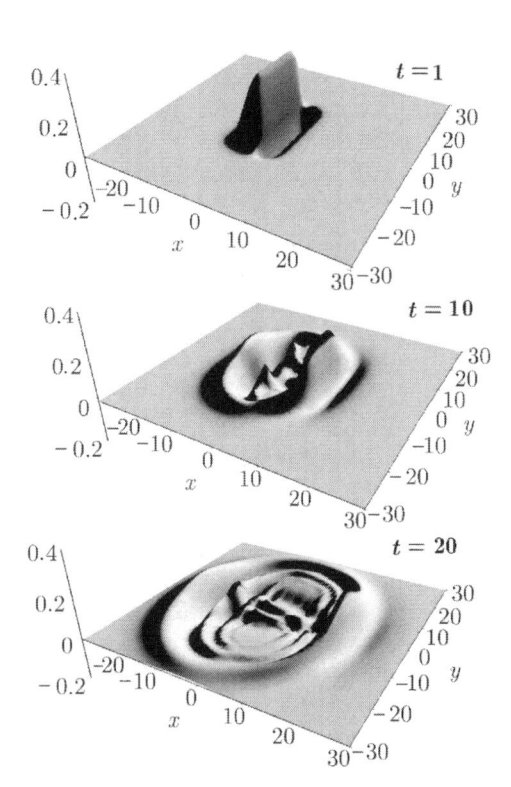

Fig. 2.37 Perturbation of free surface caused by displacement of bottom alternating in sign. Calculations are performed for a sequence of time moments (indicated in the figure) for $a_1 = 7$, $a_2 = 3$, $b = 10$, $\tau = 1$

2.4 Generation of Tsunami Waves in Space

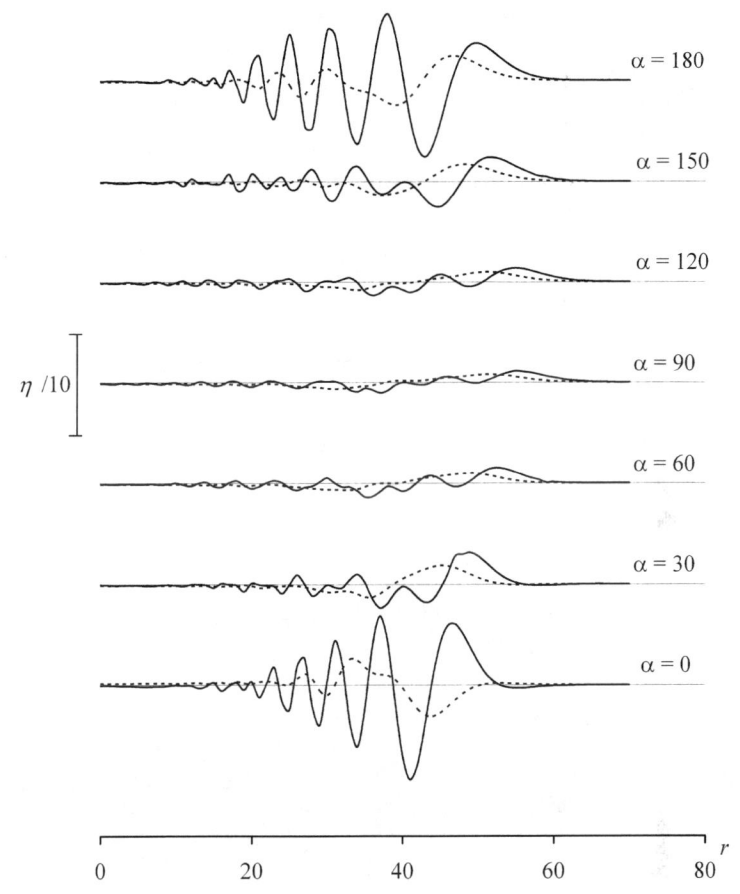

Fig. 2.38 Wave profiles (perturbation of free surface as function of radius r) for various azimuthal angles α at time moment $t = 50$. The parameters of the displacement are $a_1 = 7$, $a_2 = 3$, $b = 10$, $\tau = 1$ (solid line), $\tau = 10$ (dotted line)

direction. Besides the above, attention must be drawn to the following non-trivial peculiarity: the leading wave in the train exhibits negative polarity, not throughout the entire semiplane $x > 0$, but only in a certain sector, the angle of which is noticeably smaller than 180°. Note that in accordance with the theory of long waves the polarity of the leading tsunami wave is determined by the sign of deformation of the ocean bottom at the nearest point of the source. Potential theory provides a more precise result, demonstrating that the polarity of the leading wave is related to a greater number of parameters.

Figure 2.38 presents wave profiles, calculated for various azimuthal angles α and for two displacement durations $\tau = 1, 10$ (solid and dotted lines, respectively). The waves are characterized by significant dispersion, and their amplitudes depend essentially upon the direction of propagation. The energy directional diagrams are presented in Fig. 2.39. The energy values were normalized to the quantity

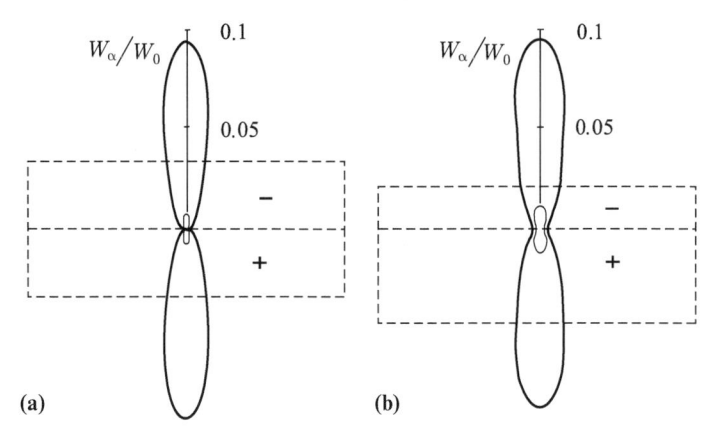

Fig. 2.39 Directional diagrams of energy emission for waves, caused by a bottom displacement alternating in sign. The shape and orientation of the source ($a_1 = a_2 = 5$ (**a**), $a_1 = 7$, $a_2 = 3$ (**b**)) are shown by the dotted line. Calculations are performed for $b = 10$, $\tau = 1$ (thick line), $\tau = 10$ (thin line)

$W_0 = \rho g b \left(\eta_1^2 a_1 + \eta_2^2 a_2 \right)/3$, the physical meaning of which consists in it representing the potential energy of the initial elevation of the water, similar in shape to the residual deformation of the ocean bottom. From Fig. 2.39 it is seen that in the case of symmetry ($a_1 = a_2 = 5$) no wave whatever is emitted in the 'fault' direction (90° and 270°). This effect is evident from general arguments: perturbations that happen to be identical in shape, but differ in their signs of bottom deformations, cancel each other out along the symmetry axis. Such a symmetry is, naturally, quite improbable in nature. Nevertheless, for the non-symmetric case ($a_1 = 7$, $a_2 = 3$), also, the prevalent part of wave energy is emitted perpendicularly to the 'fault', while the energy flux along the 'fault' direction is nearly 40 times smaller. An increase of the duration of the bottom deformation reduces the energy of the generated waves and weakens their directionality.

In conclusion, it must be stressed that the characteristics of tsunami waves may depend essentially not only on the shape of the source and the space distribution inside it of residual deformations of the ocean bottom, but also on how these deformations developed in time.

References

Abe K. (1978): A dislocation model of the 1933 Sanriku earthquake consistent with tsunami waves. J. Phys. Earth. **26**(4) 381–396
Basov B. I., Kaistrenko V. M., Levin B. W., et al. (1984): Some results of physical simulation of tsunami wave excitation and propagation. In: Tsunami generation and wave runup on shore. Radiosvyaz, Moscow (in Russian), pp. 68–72
Belokon' V. I., Goi A. A., Reznik B. L., Smal' N. A. (1986): Tsunami excitation by a seismic wave packet subject to dispersion. Tsunami researches (in Russian), (1) 28–36 Moscow

References

Bobrovich A. V. (1988): Tsunami wave excitation by fissure propagating on the bottom. In: Theoretical foundations, methods and technical means of tsunami prognosis. Theses of reports to symposium (in Russian). Obninsk. pp. 36–37

Chubarov L. B., Shokin Yu. I., Simonov K. V. (1992): Using Numerical Modelling to Evaluate Tsunami Hazard Near the Kuril Island. Nat. Hazard. (5) 293–318

Dotsenko S. F., Sergeevsky B. Yu., Cherkesov L. V. (1986): Spatial tsunami waves, caused by ocean surface displacements alternating in sign. Tsunami researches (in Russian), (1) 7–14, Moscow

Dotsenko S. F., Soloviev S. L. (1988): Mathematical simulation of tsunami excitation by dislocation of ocean bottom. Sci. Tsunami Hazards **6**(1) 31–36

Dotsenko S. F., Soloviev S. L. (1990a): Mathematical modelling of tsunami excitation processes by displacements of the ocean bottom. Tsunami researches (in Russian), (4) 8–20, Moscow

Dotsenko S. F., Soloviev S. L. (1990b): Comparative analysis of tsunami excitation by 'piston' and 'membrane' bottom displacements. Tsunami researches (in Russian), (4) 21–27

Dotsenko S. F., Sergeevsky B. Yu. (1993): Dispersion effects during directed tsunami wave generation and propagation. Tsunami researches (in Russian), (5) 21–32 Moscow

Dotsenko S. F., Soloviev S. L. (1995): On the role of residual displacements of the ocean bottom in tsunami generation by submarine earthquakes. Oceanology (in Russian), **35**(1) 25–31

Dotsenko S. F. (1996): The influence of ocean floor residual displacements on the efficiency of directed tsunami generation. Izvestiya – Atmos. Ocean Phys. **31**(4) 547–553

Garder O. I., Dolina I. S., Pelinovsky E. N., Poplavsky A. A., Fridman V. E. (1993): Tsunami wave generation by gravitational lithodynamic processes. Tusnami studies (in Russian), (5) 50–60

Gisler G. R. (2008): Tsunami simulations. Annu. Rev. Fluid Mech. **40** 71–90

Grilli S. T., Ioualalen J. M., Kirby J. T., Watts P., Asavant J. and Shi F. (2007): Source Constraints and Model Simulation of the December 26, 2004, Indian Ocean Tsunami. Journal of Ocean Engineering. **133**(6) 414–428

Handbook for Tsunami Forecast in the Japan Sea. (2001): Earthquake and Tsunami Observation Division, Seismological and Volcanological Department, Japan Meteorological Agency, 22

Hammack J. L. (1973): A note on tsunamis: their generation and propagation in an ocean of uniform depth. J. Fluid Mech. **60** 769–799

Hammack J. L. (1980): Baroclinic tsunami generation. J. Phys. Oceanogr. **10**(9) 1455–1467

Horrillo J., Kowalik Z., Shigihara Y. (2006): Wave dispersion study in the Indian Ocean tsunami of December 26, 2004. Science of Tsunami Hazards **25**(1) 42–63

Ji, C., D. J. Wald, D. V. Helmberger (2002): Source description of the 1999 Hector Mine, California earthquake; Part I: Wavelet domain inversion theory and resolution analysis, Bull. Seismol. Soc. Am. **92**(4) 1192–1207

Kajiura K. (1963): The leading wave of tsunami. Bull. Earthquake. Res. Inst. Tokyo Univ. **41**(3) 535–571

Kajiura K. (1970): Tsunami source, energy and directivity of wave radiation. Bull. Earthquake Res. Inst. Tokyo Univ. **48**(5) 835–869

Kanamori H. (1972): Mechanism of tsunami earthquakes. Phys. Earth Planet Int. **6** 346–359

Kanamori, H., Anderson D. L. (1975): Theoretical basis of some empirical relations in seismology. Bull. Seismol. Soc. Am. **65** 1073–1095 (1975)

Kanamori H., Brodsky E. E. (2004): The physics of earthquakes. Rep. Prog. Phys. **67** 1429–1496

Kato K., Tsuji Y. (1995): Tsunami of the Sumba Earthquake of August 19, 1977. J. Nat. Disaster Sci. **17**(2) 87–100

Kostitsyna O. V., Nosov M. A., Shelkovnikov N. K. (1992): A study of nonlinearity in the process of tsunami generation by sea floor motion. Moscow Univ. Phys. Bull. **47**(4) 83–86

Kowalik Z., Murty T. S. (1987): Influence of the size, shape and orientation of the earthquake source area in the Shumagin seismic gap on the resulting tsunami. Notes and Correspondens, (7) 1057–1062

Kowalik Z., Knight W., Logan T., Whitmore P. (2005): Numerical modelling of the global tsunami: Indonesian tsunami of 26 December 2004. Sci. Tsunami Hazard **23**(1) 40–56

Kulikov E. A., Rabinovich A. B., Fain I. V., Bornhold B. D., Thomson R. E. (1998): Tsunami generation by landslides at the Pacific coast of North America and the role of tides. Oceanology **38**(3) 323–328

Landau L. D., Lifshitz E. M. (1987): Fluid Mechanics, V.6 of Course of Theoretical Physics, 2nd English edition. Revised. Pergamon Press, Oxford-New York-Beijing-Frankfurt-San Paulo-Sydney-Tokyo-Toronto

Lebedev A. N., Sebekin B. I. (1982): Generation of a Directed Tsunami Wave in a Coastal Zone. Izvestia Akademii nauk SSSR. Fizika atmosfery i okeana **18**(4) 399–407

Levin B. W. (1978): Review of works on experimental modelling of the tsunami excitation process (in Russian). In: Methods for calculating tsunami rise and propagation. Nauka, Moscow, pp. 125–139

Levin B. W., Nosov M. A. (2008): On the possibility of tsunami formation as a result of water discharge into seismic bottom fractures. Izvestiya, Atmos. Oceanic Phys. **44**(1) 117–120

Levin B. W., Soloviev S. L. (1985): Variations of the field of mass velocities in the pleistoseist zone of an underwater earthquake (in Russian). DAN SSSR **285**(4) 849–852

Lobkovsky L. I., Baranov B. V. (1982): On tsunami excitation in subduction zones of lithospheric plates (in Russian). In: Processes of tsunami excitation and propagation. Publishing Department, RAS, pp. 7–17

Marchuk An. G., Chubarov L. B., Shokin Yu. I. (1983): Numerical simulation of tsunami waves (in Russian). Nauka, Siberian Branch, Novosibirsk

Marchuk An. G., Titov V. V. (1993): Influence of the source shape on tsunami wave formation (in Russian). Tsunami Res. (5) 7–21

Mirchina N. P., Pelinovsky E. N. (1987): Dispersive amplification of tsunami waves (in Russian). Oceanology **27**(1) 35–40

Miyoshi, H. (1954): Generation of the tsunami in compressible water (Part I), J. Oceanogr. Soc. Jpn. **10**(1–9)

Murty T. S. (1977): Seismic sea waves – tsunamis. Bull. Fish. Res. Board Canada **198**, Ottawa

Myers, E. P., Baptista A. M. (1995): Finite Element Modeling of the July 12, 1993 Hokkaido Nansei–Oki Tsunami, Pure Appl. Geophys. **144**(3/4) 769–802

Nikiforov A. f., Uvarov V. B. (1984): Special Functions of Mathematical Physics (in Russian). Nauka, Moscow

Nosov M. A.(1996): A comparative study of tsunami excited by piston-type and traveling-wave bottom motion. Volcanol. Seismol. (17) 693–698

Nosov M. A. (1999): Tsunami generation in compressible ocean. Phys. Chem. Earth (B) **24**(5) 437–441

Nosov M. A. (1992): Generation of tsunami by oscillations of a sea floor section. Moscow Univ. Phys. Bull. **47**(1) 110–112

Nosov M. A. (1998): On the directivity of dispersive tsunami waves excited by piston-type and traveling-wave sea-floor motion. Volcanol. Seismol. **19** 837–844

Nosov M. A., Shelkovnikov N. K. (1991): Method for measuring submillimeter waves on water surface. Moscow Univ. Phys. Bull. **46**(3) 106–108

Nosov M. A., Shelkovnikov N. K. (1992): Generation of surface waves in a fluid layer by periodic motions of the bottom. Izvestiya, Atmos. Ocean. Phys. **28**(10–11) 833–834

Nosov M. A., Shelkovnikov N. K. (1995): Tsunami generation by traveling sea-floor shoves. Moscow Univ. Phys. Bull. **50**(4) 88–92

Nosov M. A., Shelkovnikov N. K. (1997): The excitation of dispersive tsunami waves by piston and membrane floor motions. Izvestiya, Atmos. Ocean. Phys. **33**(1) 133–139

Nosov M. A., Mironyuk S. V., Shelkovnikov N. K. (1999): Bottom displacements of alternating signs and leading tsunami wave (in Russian). In: Collection 'Interaction in the lithosphere–hydrosphere–atmosphere system', **2**. Publishing Dept. of MSU Phys. Faculty, Moscow, pp. 193–200

Novikova L. E., Ostrovsky L. A. (1979): Excitation of tsunami waves by traveling displacement of the ocean bottom. Marine Geodesy **2**(4) 365–380

Ohmachi T., Tsukiyama H., Matsumoto H. (2001): Simulation of tsunami induced by dynamic displacement of seabed due to seismic faulting. Bull. Seismol. Soc. Am. **91**(6) 1898–1909

Okada (1985): Surface deformation due to shear and tensile faults in a half-space. Bull. Seismol. Soc. Am. **75**(4) 1135–1154

References

Okal E. A., Synolakis C. E. (2004): Source discriminants for near-field tsunamis. Geophys. J. Int. **158** 899–912

Pelinovsky E. N. (1996): Hydrodynamics of tsunami waves (in Russian). Institute of Applied Physics, RAS, Nizhnii Novgorod

Pelinovsky E., Talipova T., Kurkin A., Kharif C. (2001): Nonlinear mechanism of tsunami wave generation by atmospheric disturbances. Nat. Hazard. Earth Syst. Sci. **1** 243–250

Rivera P. C. (2006): Modeling the Asian tsunami evolution and propagation with a new generation mechanism and a non-linear dispersive wave model. Sci. Tsunami Hazard. **25**(1) 18–33

Satake K. (1995): Linear and nonlinear computations of the 1992 Nicaragua earthquake tsunami. PAGEOPH **144**(3/4) 455–470

Satake K., Imamura F. (1995): Tsunamis: seismological and disaster prevention studies. J. Phys. Earth **43**(3) 259–277

Satake K., Wang K., Atwater B. F. (2003): Fault slip and seismic moment of the 1700 Cascadia earthquake inferred from Japanese tsunami descriptions. J. Geophys. Res. **108**(B11) 2535, doi:10.1029/2003JB002521

Sretensky L. N. (1977): Theory of wave motions of liquids (in Russian). Nauka, Moscow

Suleimani E., Hansen R., Kowalik Z. (2003): Inundation modeling of the 1964 tsunami in Kodiak Island, Alaska. In: Submarine Landslides and Tsunamis (edited by Yalciner A. C., Pelinovsky E. N., Okal E. Synolakis C. E.) **21** 191–201. Kluwer, Dordrecht

Sveshnikov A. G., Tikhonov A. N. (1999): Theory of functions of a complex variable (in Russian). Nauka, Fizmatlit, Moscow

Takahasi R. (1934): A model experiment on the mechanism of seismic sea wave generation. Part 1. Bull. Earthquake Res. Inst. (12) 152–178

Takahasi R. (1963): On some model experiment on tsunami generation. In: Intern. Union Geodesy and Geophys. Monogr. **24** 235–248

Tanioka Y., Satake K. (1996): Fault Parameters of the 1896 Sanriku Tsunami Earthquake Estimated from Tsunami Numerical Modeling. Geophys. Res. Lett. **23**(13) 1549–1552

Titov V. V., Mofjeld H. O., Gonzalez F. I., Newman J. C. (1999): Offshore forecasting of Alaska-Aleutian subduction zone tsunamis in Hawaii. NOAA Technical Memorandum ERL PMEL-114

Titov V. V., Gonzalez F. I., Bernard E. N., et al. (2005): In: Real-Time Tsunami Forecasting: Challenges and Solutions. Nat. Hazard. **35**(1), Special Issue, pp. 41–58. U.S. National Tsunami Hazard Mitigation Program

Van Dorn W. G. (1964): Source mechanism of the tsunami of March 28, 1964, in Alaska. In: Proc. 9th Conf. Coastal Eng., Lisbon, pp. 166–190

Vasilieva G. V. (1981): On wave excitation in shallow water. In: Tsunami wave propagation and runup on shore (in Russian), pp. 67–69. Nauka, Moscow

Voight S. S. (1987): Tsunami waves. Tsunami researches (in Russian). (2) 8–26

Voight S. S., Lebedev A. N., Sebekin B. I. (1980): Certain tsunami wave peculiarities, related to characteristics of the perturbation source. In: Tsunami theory and effective prognosis (in Russian), pp. 5–11. Nauka, Moscow

Voit S. S., Lebedev A. N., Sebekin B. I. (1981): Formation of a Directed Tsunami Wave at an Excitation Focus. Izvestia Akademii nauk SSSR. Fizika atmosfery i okeana **17**(3) 296–304

Voight S. S., Lebedev A. N., Sebekin B. I. (1982): On the generation of a directed tsunami wave by a horizontal bottom displacement In: Processes of tsunami excitation and propagation (in Russian), pp. 18–23. IO RAN, Moscow

Watts P., Grilli S. T., Imamura F. (2001): Coupling of tsunami generation and propagation codes. In: ITS Proceedings, Session 7, Numbers 7–13, pp. 811–823

Yagi Y. (2004): Source rupture process of the 2003 Tokachi–oki earthquake determined by joint inversion of teleseismic body wave and strong ground motion data. Earth Planets Space **56** 311–316

Zaitsev A. I., Kurkin A. A., Levin B. W., et al. (2005): Numerical simulation of catastrophic tsunami propagation in the Indian Ocean (December 26, 2004). Doklady Earth Sci. **402**(4) 614–618

Chapter 3
Role of the Compressibility of Water and of Non-linear Effects in the Formation of Tsunami Waves

Abstract The necessity is substantiated for taking into account the compressibility of water in describing behaviour of water column in tsunami source. Within the framework of linear potential theory of a compressible liquid in a basin of fixed depth, the general analytical solution is constructed for 2D and quasi-3D (cylindrical symmetry) problems of the generation of acoustic–gravitational waves by bottom deformations of small amplitudes. Manifestations of compressibility of the water column in the problem of tsunami generation are studied, making use of the example of model bottom deformation laws (piston, membrane and running displacements). The main difference between the behaviour of a compressible water column as compared to an incompressible model medium is shown to consist in the formation of elastic oscillations exhibiting significant amplitudes and a discrete spectrum. Characteristic features of the dynamics of acoustic-gravitational waves in a basin of variable depth are described. Records of bottom pressure gauges are used for analysing manifestations of elastic oscillations of the water column at the source of the 2003 Tokachi-Oki tsunami. The mechanism is considered of tsunami formation, related to non-linear energy transfer from 'high-frequency' induced or elastic oscillations of the water column to 'low-frequency' gravitational waves.

Keywords Water compressibility · tsunami generation · T-phase · acoustic-gravity waves · linear potential theory · Laplace transformation · Fourier transformation · analytical solution · bottom displacement · quarter-wave resonator · elastic oscillations · normal modes · waveguide · numerical simulation · Tokachi-Oki 2003 · JAMSTEC · bottom pressure · frequency spectra · non-linear effects · Euler's equations · non-linear tsunami source

The issue of accounting for the compressibility of water in the problem of tsunami generation has been raised repeatedly in the literature [Sells (1965); Kajiura (1970); Pod'yapolsky (1978); Yanushauskas (1981); Boorymskaia et al. (1981); Ewing et al. (1950); Levin (1981); Selezov et al. (1982); Garber (1984); Zhmur (1987); Nosov (1999), (2000); Nosov, Sammer (1998); Panza et al. (2000); Ohmachi (2001); Nosov, Kolesov (2002), (2003); Nosov et al. (2005), (2007); Gisler (2008)]. However, in most of the tsunami models the ocean is considered to be an incompressible medium. Only one of the manifestations of water being compressible in the case of underwater earthquakes, namely the T-phase, has been studied relatively well [Soloviev et al. (1968), (1980); Brekhovskikh (1974); Kadykov (1986), (1999); Lysanov (1997); Okal (2003); Okal et al. (2003)]. The range of frequencies between 1 and 100 Hz is usual for the T-phase. We note, here, that in this chapter we will be interested in hydroacoustic phenomena that are related to another frequency range (~ 0.1 Hz) and are localized in the immediate vicinity of the tsunami source.

The necessity of taking into account non-linear effects during tsunami formation is related to the fact that in the case of seismic motions of the seabed exhibiting small amplitudes, the velocities of these motions may turn out to be quite significant. The non-linear mechanism of tsunami generation was first considered in [Novikova, Ostrovsky (1982)]. This line of research was further developed in [Nosov, Skachko (2001a, b); Nosov, Kolesov (2002), (2005) and Nosov et al. (2008)].

3.1 Excitation of Tsunami Waves with Account of the Compressibility of Water

3.1.1 Preliminary Estimates

If the process is treated from a formal physical point of view [Landau, Lifshits (1987)], then a liquid can be considered incompressible only when $\Delta \rho / \rho \ll 1$, where ρ is the density of the liquid. As it is known, the necessary condition for the above to be valid is that the motions of the liquid exhibit small velocities, as compared to the velocity of sound. In the case of stationary motion this condition is sufficient. The problem of tsunami generation is evidently non-stationary, so one more, additional, condition must be fulfilled. In the problem of tsunami generation both conditions are of the following form:

1. $v \ll c$.
2. $\tau \gg (H c^{-1}, L c^{-1})$

where v is the characteristic mass velocity of motion (of water or of particles of the ocean bottom), c is the velocity of sound in water, τ is the duration of bottom displacement, H is the ocean depth and L is the characteristic horizontal size of the source. Note that even in those rare cases, when the authors of one or another investigation substantiate application of the theory of incompressible liquids in the tsunami problem, the second condition is always forgotten. The characteristic values of the indicated parameters are the following: $v \sim 1$ m/s, $c \sim 1{,}500$ m/s, $H \sim 4{,}500$ m, $L \sim 10\text{–}100$ km, $\tau \sim 1\text{–}100$ s. The first condition is seen to be well satisfied, while the second can be violated in many cases. In the case of a running displacement (a fault ripped open or a surface seismic wave) the first condition will also be violated.

For problems concerning tsunami propagation in the open ocean and waves running up a coast, the first condition remains the same and is quite fulfilled. The second condition assumes the following form: $T \gg (H c^{-1}, \lambda c^{-1})$, where T is the tsunami period and λ is the wavelength. With account of the obvious relationship $T(gH)^{1/2} = \lambda$ and of the fact that the tsunami period is usually tens and even hundreds of minutes long, fulfilment of the second condition is also doubtless.

It is interesting to compare the energies of acoustic and gravitational waves, excited by one and the same mechanism: by vertical displacement of part of the ocean bottom. We shall consider a column of an ideal homogeneous compressible (or incompressible) liquid with a free upper surface of thickness H located on an absolutely hard bottom in the field of gravity exhibiting the free-fall acceleration g. At a certain moment of time an area S of the ocean bottom starts to move vertically with a constant velocity v. The motion will take place during a time interval τ, upon which the ocean bottom stops. Such a process results in a residual displacement of the ocean bottom of height $\eta_0 = v\tau$ over an area S. It is known that in an incompressible liquid, when $\tau \ll S^{1/2}(gH)^{-1/2}$, at the time moment $t = \tau$ the shape of the surface perturbation is close to the shape of the residual displacement of the bottom, so the energy transferred to the ocean by the moving bottom can be estimated (from above) as the potential energy of the initial elevation of area S and height $\xi_0 = \eta_0$:

$$W_1 = 0{,}5 \rho g S \xi_0^2. \tag{3.1}$$

Within the framework of the model of compressible liquids, the energy of acoustic waves [Landau, Lifshitz (1987)], excited by the motion of the ocean bottom, described above, has the following form:

$$W_2 = c \rho S \eta_0^2 \tau^{-1}. \tag{3.2}$$

It is readily verified that, within the range of τ values peculiar to real seismic events, the ratio $W_2/W_1 = 2c/(g\tau) > 1$. In other words, a significant part of the energy transferred from the moving bottom to the ocean exists in the form of acoustic waves. As time passes, this energy can be transferred to seismic waves or to other forms of motion of the water column. In any case, elastic oscillations represent an energetically significant effect, which must be taken into account.

The obtained estimates are expediently compared with natural data. Taking advantage of the empirical relationships (2.3)–(2.5) and of formulae (3.1) and (3.2), it is possible to calculate the energy of a gravitational tsunami wave and the energy of elastic waves in water depending on the earthquake magnitude. We shall compare these quantities with the earthquake energy E [J], estimated by the formula [Puzyrev (1997)]

$$\lg E = 1.8M + 4. \tag{3.3}$$

In calculations we assumed the duration of the ocean bottom to be half the duration of the process at the earthquake source, T_{hd}, while the area of the tsunami source is calculated via its radius by the elementary formula $S = \pi R_{TS}^2$.

Figure 3.1 presents the dependences of the earthquake energy (1), the energy of a gravitational tsunami wave (2) and of the energy of elastic waves (3) upon the earthquake magnitude. It can be seen that the formation of a gravitational tsunami wave requires <1% of the earthquake energy, which is in good agreement with the data of [Levin (1981)] and [Voight (1987)]. But tens of times more—up to

Fig. 3.1 Earthquake energy (1), energy of gravitational tsunami wave (2) and energy of elastic waves in water (3) versus earthquake magnitude

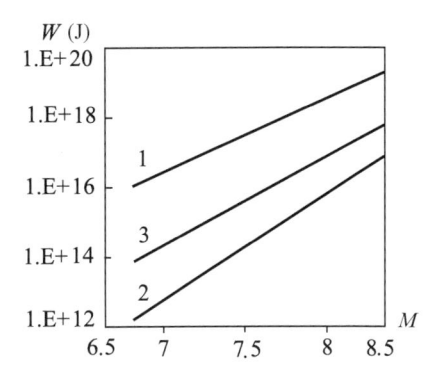

several percent of the earthquake energy can be transferred to elastic waves. Thus, submarine earthquakes are capable of exciting powerful acoustic waves, and noticeable manifestations are to be expected of the effect of water compressibility in the case of tsunami generation.

Further we shall consider the mathematical model of a compressible water column under the assumption of an absolutely rigid ocean bottom. Such an assumption simplifies the problem noticeably and permits to concentrate on manifestations of the compressibility of the water column. A realistic formulation of the problem should, naturally, take into account the elasticity properties of the ocean bottom. The problem, thus formulated, was first considered by G.S. Pod'yapolsky [Pod'yapolsky (1968a, b), (1978)]. This topic was further developed in analytical studies [Gusyakov (1972), (1974)], [Alexeev, Gusyakov (1973)], [Zvolinsky (1986)], [Zvolinsky et al. (1991), (1994)] and [Sekerzh-Zen'kovich et al. (1999)]. In recent years publications have appeared, in which attempts are made of numerical simulation of the dynamics of a compressible water column with account of the elasticity properties of the ocean bottom [Panza et al. (2000); Ohmachi (2001)]. In particular, these properties will be manifested in the time interval, during which elastic oscillations of a water column exist, being limited owing to 'leakage' of energy into the ocean bottom.

Consider elastic oscillations to be caused by a horizontally homogeneous vertical deformation of the ocean bottom, taking place within quite an extended area. Then the problem becomes one-dimensional along the vertical coordinate. We write the evolution equation for the energy W of elastic waves contained in a water column of thickness H as

$$\frac{dW}{dt} = -\frac{W}{\tau_s}.$$

The energy W obviously decreases exponentially with time. The quantity τ_s, characterizing the damping time, can be deduced from the following arguments. During the propagation of an elastic wave from the ocean bottom to the surface and back $(2H/c)$ its energy will be reduced by the quantity $D_0 W$, where $D_0 = 4\rho\rho_b c c_b (\rho c + \rho_b c_b)^{-2}$ is the transition coefficient at the 'water-bottom' boundary for a normally incident elastic wave, ρ, ρ_b and c, c_b are the densities of and elastic wave propagation velocities in water and the ocean bottom rock, respectively.

3.1 Excitation of Tsunami Waves with Account of the Compressibility of Water

In the case of the ocean bottom, longitudinal waves are intended, since in the one-dimensional case considered transverse and surface seismic waves are not excited. The maximum period of elastic oscillations of a water column with a free surface is known to be $T_0 = 4H/c$. Thus, we obtain the following formula for the damping time: $\tau_s = T_0/2D_0$.

We shall consider the density of water and the propagation velocity of sound in water to be $\rho = 1{,}000 \text{ kg/m}^3$ and $c = 1{,}500 \text{ m/s}$, respectively. In the case of rock, making up the ocean bottom, its density and the velocity of longitudinal waves in it vary within the respective limits $1{,}400 < \rho_b < 3{,}500 \text{ kg/m}^3$ and $1{,}700 < c_b < 8{,}000 \text{ m/s}$. Correspondingly, the transition coefficient varies within the limits $0.19 < D_0 < 0.95$. The lower limits of the indicated ranges correspond to friable sedimentary rock. Usually, the ocean bottom has a stratified structure. The effective reflection of elastic waves takes place from the acoustic base—a certain sufficiently dense and high-velocity column. Thus, for example, in the case of $\rho_b = 3{,}000 \text{ kg/m}^3$ and $c_b = 7{,}000 \text{ m/s}$, we obtain $\tau_s \approx 2T_0$. So, it takes two periods for the energy of elastic oscillations to be reduced by a factor of e, and four periods for the oscillation amplitude.

3.1.2 General Solution of the Problem of Small Deformations of the Ocean Bottom Exciting Waves in a Liquid

In this section the general solution is constructed for the problem of small deformations of the basin bottom exciting acoustic–gravity waves in a column of liquid. The problem is formulated exactly like in Sect. 2.2.1, with the sole exception consisting in that, here, the liquid is considered to be compressible.

We shall consider an infinite, in plane $0xy$, column of an ideal compressible homogeneous liquid of constant depth H in the field of gravity (Fig. 2.8). The origin of the Cartesian reference system $0xyz$ will be located on the free non-perturbed plane with axis $0z$ directed vertically upward. Motion of the liquid is caused by deformations of the bottom of small amplitude ($A \ll H$). Moreover, we shall assume the deformation velocity of the bottom, and consequently, the velocity of motion of particles of the liquid to be small quantities. This makes it possible to neglect the non-linear term $(\mathbf{v}\nabla)\mathbf{v}$ in the Euler equation. Dynamical variations of density ρ' and pressure p' in the liquid are also small as compared to the equilibrium quantities ($\rho' \ll \rho_0$, $p' \ll p_0$). Assuming the equilibrium flow velocity to be zero, we substitute the pressure and density, expressed as $p = p_0 + p'$, $\rho = \rho_0 + \rho'$, into the Euler and the continuity equations. Neglecting small quantities of the second order and introducing the potential of the flow velocity, F ($\mathbf{v} = \nabla F$), we arrive at the wave equation

$$\frac{\partial^2 F}{\partial t^2} - c^2 \Delta F = 0. \tag{3.4}$$

where $c = \sqrt{(\partial p/\partial \rho)_s}$ represents the velocity of sound.

The boundary conditions on the surface of the liquid and on the bottom remain the same as in the case of a noncompressible liquid [Yanushkauskas (1981)],

$$F_{tt} = -gF_z, \qquad z = 0, \qquad (3.5)$$
$$F_z = \eta_t, \qquad z = -H. \qquad (3.6)$$

If the depth of the basin, H, is a function of the horizontal coordinate, then the following formula can be applied as the boundary condition on the bottom:

$$\frac{\partial F}{\partial \mathbf{n}} = (\mathbf{U}, \mathbf{n}), \qquad (3.7)$$

where $\mathbf{n}$ is the normal to the basin bottom plane, $\mathbf{U}$ is the deformation velocity vector of the bottom. From a physical point of view, this condition signifies equality of the liquid flow velocity component, normal to the bottom, and the motion velocity of the bottom in the same direction (a no-flow condition).

Displacement of the free surface of the liquid from its equilibrium position, the dynamical pressure and the flow velocity are calculated via the potential by the following formulae:

$$\xi(x,y,t) = -g^{-1}F_t(x,y,0,t), \qquad (3.8)$$
$$p'(x,y,z,t) = -\rho_0 F_t(x,y,z,t), \qquad (3.9)$$
$$\mathbf{v}(x,y,z,t) = \nabla F(x,y,z,t). \qquad (3.10)$$

Note that the problem for an incompressible liquid, (2.29)–(2.31), is a limit case of the problems (3.4)–(3.6) in the case of $c \to \infty$. It is readily seen that equation (3.4) reduces, in the limit, to the Laplace equation (2.29). Therefore, the results, obtained for an incompressible liquid in Sects. 2.2–2.4, will serve as a reliable benchmark. Moreover, an answer will be obtained to the question concerning the difference between the behaviours of compressible and incompressible liquids, and as to when this difference may be neglected.

We shall restrict ourselves to resolving the two following two-dimensional problems: the plane problem (in a Cartesian reference frame) and the quasithree-dimensional problem (exhibiting cylindrical symmetry).

3.1.2.1 Cartesian coordinates

In a Cartesian reference frame, the solution of the problem (3.4)–(3.6) is sought via the respective Laplace and Fourier transformations relative to the time and space coordinates in the following form:

$$F(x,z,t) = \int_{s-i\infty}^{s+i\infty} dp \int_{-\infty}^{+\infty} dk\, \Phi(z,p,k) \exp\{pt - ikx\}. \qquad (3.11)$$

3.1 Excitation of Tsunami Waves with Account of the Compressibility of Water

Substituting formula (3.11) into expression (3.4), we obtain an equation for determining function $\Phi(z, p, k)$:

$$\Phi_{zz} - \alpha^2 \Phi = 0, \tag{3.12}$$

where $\alpha^2 = k^2 + p^2 c^{-2}$.

The solution of equation (3.12) is well known and can be written in the form

$$\Phi(z, p, k) = A \cosh(\alpha z) + B \sinh(\alpha z),$$

where A and B are arbitrary numerical coefficients.

Applying the boundary condition for a free surface, (3.5), we find the relationship between the coefficients:

$$B = -A p^2 (g\alpha)^{-1}.$$

With the aid of the boundary condition for the basin bottom, (3.6), we determine the coefficient A:

$$A = -\frac{p\Psi(p,k)}{\alpha \sinh(\alpha H) + p^2 g^{-1} \cosh(\alpha H)},$$

where function $\Psi(p, k)$ represents the Laplace and Fourier transforms of the space–time law of motion of the bottom $\eta(x, t)$:

$$\Psi(p,k) = \frac{1}{4\pi^2 i} \int_0^\infty dt \int_{-\infty}^{+\infty} dx\, \eta(x,t) \exp(-pt + ikx). \tag{3.13}$$

Thus, we have function $\Phi(z, p, k)$ in the following form:

$$\Phi(z,p,k) = \frac{p\Psi(p,k)}{\alpha \sinh(\alpha H) + p^2 g^{-1} \cosh(\alpha H)} \left(p^2 (g\alpha)^{-1} \sinh(\alpha z) - \cosh(\alpha z) \right). \tag{3.14}$$

Substitution of expression (3.14) into formula (3.11) yields the final expression for calculation of the potential corresponding to an arbitrary space–time law determining the displacement of the bottom, $\eta(x, t)$. In investigating analytical models, it is of interest to consider the behaviour of the free surface of the liquid (as the most illustrative characteristic), the displacement of which relative to its unperturbed level is expressed via the potential in accordance with formula (3.8). As a result we obtain the following expression:

$$\xi(x,t) = g^{-1} \int_{s-i\infty}^{s+i\infty} dp \int_{-\infty}^{\infty} dk \frac{p^2 \Psi(p,k)}{\alpha \sinh(\alpha H) + p^2 g^{-1} \cosh(\alpha H)} \exp\{pt - ikx\}. \tag{3.15}$$

In choosing the concrete function for describing the space–time law determining the motion of the bottom, part of the integrals in expression (3.15) can be calculated analytically.

3.1.2.2 Cylindrical Coordinates

The cylindrical reference frame will be introduced in a standard manner with respect to the Cartesian reference frame, described in Sect. 3.1.2. The origin of the cylindrical reference frame will be located on the free unperturbed surface, axis $0z$ will be directed vertically upward. As a source of elastic gravitational waves we shall consider axially symmetric movements of the bottom proceeding in accordance with the law $\eta(r, t)$. The wave equation in the cylindrical reference system exhibits the following form:

$$r^{-1}(rF_r)_r + F_{zz} = c^{-2}F_{tt}. \tag{3.16}$$

The boundary conditions on the surface, (3.5), and on the bottom, (3.6), retain their form in the cylindrical reference frame.

The solution of the problems (3.16), (3.5) and (3.6) is sought applying the separation of variables in the form of the inverse Laplace transformation:

$$F(r,z,t) = \int_{s-i\infty}^{s+i\infty} dp\, R(r,p) Z(z,p) \exp\{pt\}. \tag{3.17}$$

Substituting formula (3.17) into (3.16), we obtain equations for determining functions $R(r,p)$ and $Z(z,p)$:

$$r^{-1}(rR_r)_r + k^2 R = 0 \tag{3.18}$$

$$Z_{zz} - \alpha^2 Z = 0, \tag{3.19}$$

where $\alpha^2 = k^2 + p^2 c^{-2}$. The solutions of equations (3.18) and (3.19) are well known [Nikiforov, Uvarov (1984)]. Making use of this result, one can represent the general solution of equation (3.16) as follows:

$$F(r,z,t)$$
$$= \int_0^\infty dk \int_{s-i\infty}^{s+i\infty} dp \exp\{pt\}\, J_0(kr)\, (A(p,k)\cosh(\alpha z) + B(p,k)\sinh(\alpha z)), \tag{3.20}$$

where J_0 is the zeroth-order Bessel function of the first kind.

With the aid of the boundary condition on the free surface, (3.5), we find the relationship between the coefficients A and B:

$$B(p,k) = -A(p,k) p^2 (g\alpha)^{-1}.$$

The boundary condition on the bottom, (3.6), permits to determine the coefficient $A(p,k)$:

$$A(p,k) = -\frac{pk\Psi(p,k)}{\alpha\left(\sinh(\alpha H) + p^2 g^{-1}\alpha^{-1}\cosh(\alpha H)\right)}, \tag{3.21}$$

where function $\Psi(p,k)$ represents the Laplace and Fourier–Bessel transforms of the space–time law of movements of the bottom, $\eta(x,t)$:

$$\Psi(p,k) = \frac{1}{2\pi i}\int_0^\infty dr \int_0^\infty dt\, \eta(r,t)\, r J_0(kr) \exp\{-pt\}. \tag{3.22}$$

We shall further consider the behaviour of a free surface, the displacement of which from its unperturbed level is expressed through the potential as follows:

$$\xi(r,t) = -g^{-1} F_t(r,0,t). \tag{3.23}$$

With use of formula (3.20) expression (3.23) acquires the following form:

$$\xi(r,t) = -g^{-1}\int_0^\infty dk \int_{s-i\infty}^{s+i\infty} dp\, p \exp\{pt\} J_0(kr) A(p,k). \tag{3.24}$$

3.1.3 Piston and Membrane Displacements

We shall start exposition of the peculiarities of tsunami formation in a compressible ocean by considering the axially symmetric problem [Nosov (2000)]. As sources of acoustic–gravity waves we choose two model displacements of the bottom: the piston and membrane displacements,

$$\eta_1(r,t) = \eta_0\bigl(1-\theta(r-R)\bigr)\left(\frac{\theta(t)t - \theta(t-\tau)(t-\tau)}{\tau}\right), \tag{3.25}$$

$$\eta_2(r,t) = \eta_0\bigl(1-\theta(r-R)\bigr)$$
$$\times \left(\frac{2\theta(t)t - 4\theta(t-0{,}5\tau)(t-0{,}5\tau) + 2\theta(t-\tau)(t-\tau)}{\tau}\right), \tag{3.26}$$

The displacement amplitude η_0 is the same throughout the entire active zone, exhibiting a circular shape of radius R, and is zero outside this region. The duration of the displacement is τ.

We introduce the dimensionless variables (the asterisk '*' will be dropped):

$$\begin{aligned}
&k^* = kH; & &p^* = pHc^{-1}; & &\alpha^* = \alpha H; \\
&R^* = RH^{-1}; & &r^* = rH^{-1}; & &z^* = zH^{-1}; \\
&t^* = tcH^{-1}; & &\tau^* = \tau cH^{-1}; & &c^* = c\,(gH)^{-1/2}.
\end{aligned} \tag{3.27}$$

Making use of the general solution (3.24), we obtain expressions describing motion of the free surface of a compressible liquid in the case of piston-like (ξ_1) and membrane-like (ξ_2) displacements of the bottom:

$$\xi_1(r,t) = \theta(t)\zeta(r,t) - \theta(t-\tau)\zeta(r,t-\tau), \tag{3.28}$$

$$\xi_2(r,t) = 2\theta(t)\zeta(r,t) - 4\theta(t-0,5\tau)\zeta(r,t-0,5\tau)$$
$$+ 2\theta(t-\tau)\zeta(r,t-\tau), \tag{3.29}$$

where

$$\zeta(r,t) = \frac{\eta_0 c^2 R}{2\pi i \tau} \int_0^\infty dk \int_{s-i\infty}^{s+i\infty} dp \frac{\exp\{pt\} J_0(rk) J_1(Rk)}{\alpha \sinh(\alpha) + p^2 c^2 \cosh(\alpha)}. \tag{3.30}$$

As a function of the complex parameter p the integrand in (3.30) has two or an infinite number (depending on the sign of α^2) of poles located on the axis Im(p) = 0. An incompressible liquid ($c = \infty$) represents a special case of the problem dealt with. The solution for an incompressible liquid can be obtained by a formal substitution of $\alpha \to k$ in formula (3.30). The integrand, here, will have only two first-order poles $p_0^{1,2} = \pm i c^{-1} (k \tanh(k))^{1/2}$, which permits to perform integration over the parameter p analytically. In the case of an incompressible liquid, this results in the function $\zeta(r,t)$, entering into formulae (3.28) and (3.29), assuming the following form:

$$\zeta(r,t) = \frac{\eta_0 c R}{\tau} \int_0^\infty dk \frac{J_0(rk) J_1(Rk) \sin(tc^{-1}(k \tanh(k))^{1/2})}{\cosh(k)[k \tanh(k)]^{1/2}}. \tag{3.31}$$

The integrals in formulae (3.30) and (3.31) were calculated numerically for $c = 8$ and $R = 1, 5$ and 10.

Figure 3.2 presents the example of time evolvents, showing the displacement of a free surface at two fixed points (at the centre of the active zone and outside it) for compressible and incompressible liquids. The insets show the behaviour of the free surface of an incompressible liquid at long times. The theory of a compressible liquid is seen to provide a more reliable description of the movement of the surface from the point of view of moment of time the perturbation arrives at the given point. Before the long gravitational wave arrives at point $r = 20$, acoustic precursors of noticeable amplitude are observed. The main difference in behaviour between compressible and incompressible liquids consists in the existence of 'fast' oscillations of the surface with a prevalent period, equal to $4H/c$. Oscillations take place against the background of the development of a slower gravitational wave. The origin of surface oscillations is due to the excitation of standing acoustic waves in the natural quarter-wave resonator of a 'column of compressible liquid with a free surface on the rigid bottom'. The resonator exhibits a set of frequencies: $v_k = 0.25 c (1+2k) H^{-1}$, where $k = 0, 1, 2, 3, \ldots$. Precisely the lowest mode corresponds to the period observed.

It is quite probable that such a resonator plays an important part in the formation of seaquakes. In the case of depths of several kilometres, usual for oceans, the eigen

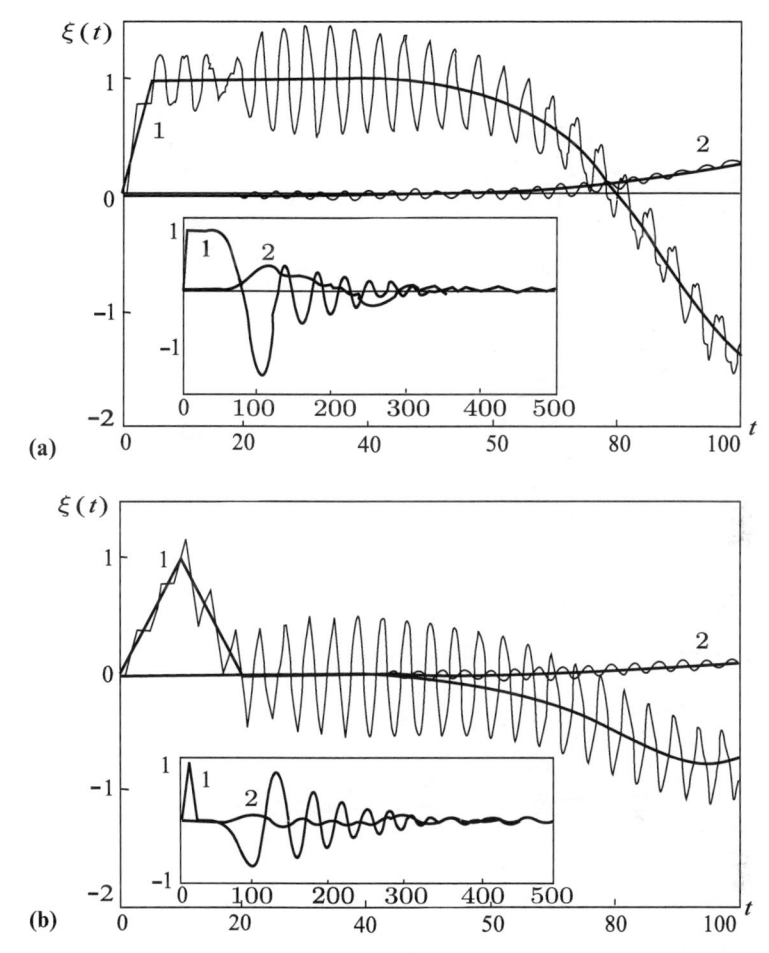

3.1 Excitation of Tsunami Waves with Account of the Compressibility of Water

Fig. 3.2 Examples of time evolvents of free-surface perturbations of compressible (thin line) and incompressible (thick line) liquids, caused by displacements of the bottom of duration $\tau = 5$ with a residual deformation (**a**) and of duration $\tau = 20$ without any residual displacement (**b**). Curves 1, 2 correspond to $r = 0, 20$, respectively. Calculation were performed for $R = 10$

frequencies of the resonator lie precisely within the range of frequencies of seismic processes. Therefore, extremely effective transmission of energy is possible from an oscillating ocean bottom to the thick volume of water.

Figure 3.3 presents the dependences of the maximun free-surface displacement amplitudes for compressible and incompressible liquids versus the durations of bottom displacements. Calculations were performed for three different sizes of the source ($R = 1, 5$ and 10). The maximum amplitudes were determined from the time evolvents in accordance with the following formulae:

- For model of compressible liquid:

$$A = \eta_0^{-1}\left(\max\left(\xi_{\text{comp}}(t) - \xi_{\text{incomp}}(t)\right) - \min\left(\xi_{\text{comp}}(t) - \xi_{\text{incomp}}(t)\right)\right);$$

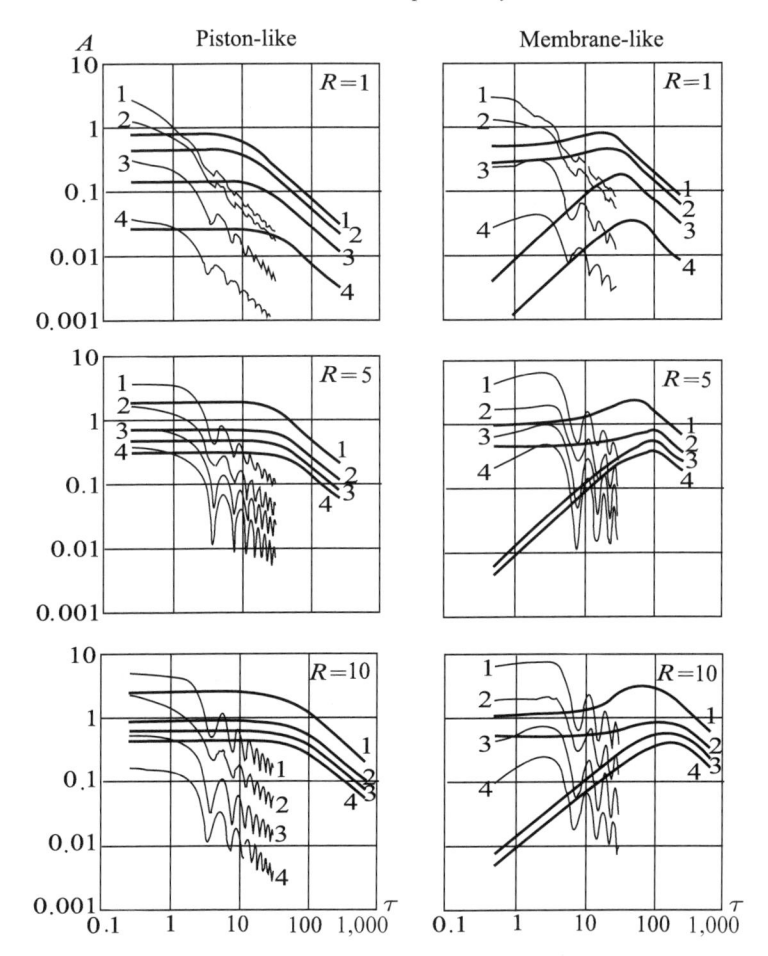

Fig. 3.3 Dependences of maximum amplitude of 'fast' surface oscillations (thin line) and of maximum amplitude of gravitational waves (thick line) upon the durations of bottom displacements with residual deformations (piston-like displacement) and without displacement (membrane-like displacement). Curves 1–4 correspond to distances $r = 0$, 10, 20 and 40 from the centre of the source. Calculations were performed for $R = 1$, 5 and 10

- For model of incompressible liquid:

$$A = \eta_0^{-1}\left(\max\left(\xi_{\text{incomp}}(t)\right) - \min\left(\xi_{\text{incomp}}(t)\right)\right).$$

In the case of incompressible liquids the dependences of amplitudes and energies of gravitational waves upon the durations of bottom displacements have been repeatedly investigated theoretically and experimentally by different authors [Hammack (1973); Dotsenko (1996a), (1996b); Nosov, Shelkovnikov (1997)]. The

3.1 Excitation of Tsunami Waves with Account of the Compressibility of Water 111

character of the curves presented in Fig. 3.3 for the case of incompressible liquids is in good agreement with the results of indicated publications and Sects. 2.3 and 2.4.

In the case of an *incompressible* liquid the characteristic features of the amplitude dependence on the displacement duration are the following. When a displacement of the ocean bottom is accompanied by a residual displacement, the dependence is characterized by the presence of a plateau at small values of the parameter τ and by a monotonous drop at large τ values. When no residual displacement accompanies the displacement, the dependence considered has a local maximum that shifts to the right as the source radius increases. Short displacements of the ocean bottom without residual displacement result in the formation of very weak surface perturbations outside the source.

Enhancement of the size of the active zone leads to an increase in both the amplitude of 'fast' surface oscillations and the amplitude of gravitational waves. It must be noted that an increase in the size of the active zone causes a noticeable change in the amplitude only for values of the parameter $R < 5$, after which the dependence reaches saturation.

The specific non-monotonous character of the curves, related to the model of compressible liquids, is due to the aforementioned resonance properties. As the displacement duration increases, the amplitude of 'fast' oscillations tends to decrease. It is interesting to note that in the case of large values of parameter τ all the dependences (both for compressible and incompressible liquids) behave like τ^{-1}. Given other conditions being equal, the amplitude of 'fast' surface oscillations above the source may be several times larger than the amplitude of surface displacements of an incompressible liquid.

The duration of real ocean bottom displacements lies within the range 0.4–40 (1–100 s). On the basis of data presented in Fig. 3.3 it is possible to conclude that the maximum amplitude of gravitational waves proper in the case of bottom displacements, involving residual deformation, is weakly sensitive to variations of the parameter τ. In the case of displacements of the ocean bottom without residual deformation the amplitudes of gravitational waves, going beyond the limits of the source, undergo quite significant changes within the range we are interested in. For any type of displacement the amplitude of 'fast' surface oscillations depends strongly on the displacement duration.

Figure 3.4 presents the dependences of amplitudes of gravitational waves and of 'fast' surface oscillations upon the distance from the source centre r. The data correspond to the duration of the ocean bottom displacement, $\tau = 1$, which does not violate the general nature of the conclusions, since the form obtained for the solution of (3.28) and (3.29) reveals that the parameter τ does not affect the decrease of amplitude with distance. In all cases, the amplitude varies weakly immediately above the source zone. Outside the source zone the amplitude of gravitational waves decreases approximately like $r^{-1/2}$ (corresponding to the known asymptotic estimates [Pelinovsky (1996)]), while the amplitude of oscillations drops like r^{-2}. Here, displacements of the ocean bottom with and without residual deformation lead practically to the same oscillation amplitude, while the amplitudes of gravitational waves differ noticeably. Figure 3.4 permits to conclude that 'fast' surface

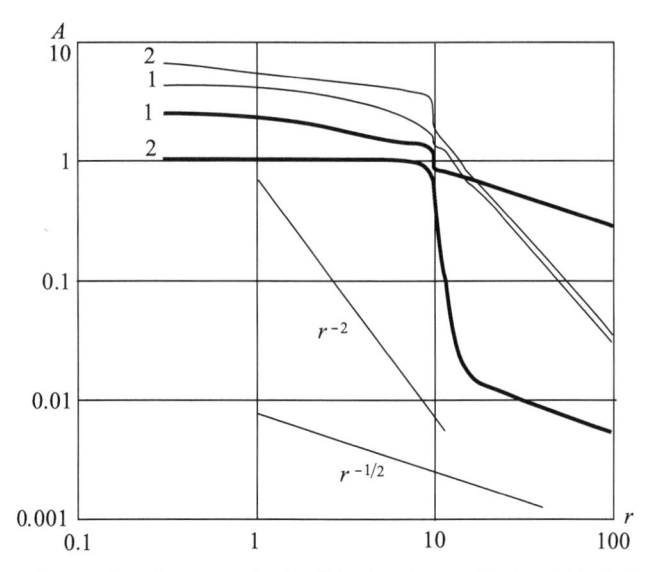

Fig. 3.4 Dependences of maximum amplitude of 'fast' surface oscillations (thin line) and of maximum amplitude of gravitational waves (thick line) upon the distance from the source centre. Curves 1, 2 correspond, respectively, to displacements of the bottom with residual deformation (piston-like displacement) and without residual deformation (membrane-like displacement). Calculations are performed for $\tau = 1, R = 10$

oscillations are to be considered local effects, the appearance of which should be noticeable either immediately at the tsunami source or at relatively small distances, not exceeding several sizes of the source.

The general picture of tsunami excitation in a compressible ocean can be represented as follows. When a vertical displacement of the ocean bottom occurs the water column is shifted correspondingly, and under the force of gravity it gradually starts to spread out, at the same time undergoing elastic oscillations. Therefore, the tsunami source not only serves as a source of gravitational tsunami waves, but also of low-frequency acoustic waves, the emission of which is possible at the characteristic frequencies $v_k = 0.25c\,(1+2k)\,H^{-1}$. Waves of low energy-carrying modes exhibit lengths that significantly exceed the width of the underwater acoustic channel and that, consequently, cannot be captured by it. In this case the entire thickness of the ocean must serve as the waveguide, while the elastic waves considered will effectively be scattered on irregularities of the ocean bottom and of the water surface and be absorbed by the elastic ocean bottom. Most likely, it is precisely for this reason that at large distances from the source only relatively weak components of the signal are observed at frequencies $v > 1$ Hz, and precisely they are termed the T-phase.

3.1.4 The Running Displacement

In dealing with the running-displacement problem in Sect. 2.3.3 we noted that movements of the ocean bottom of such types are characterized by high propagation velocities, at which the theory of incompressible liquids cannot be applied. The velocity, with which the fault ruptures at the earthquake source, the fissure propagating along the bottom and surface seismic waves are all phenomena characterized by velocities exceeding the speed of sound in water. And it is only in the case of underwater slumps (landslides) that the velocity of a running slide is significantly inferior to the speed of sound in water. Therefore, the aim of this section is the construction of a mathematical model for the excitation of waves by a running displacement of the ocean bottom in a compressible liquid.

Consider a plane problem, the general formulation of which corresponds to (3.4)–(3.6). We shall choose the model law of motion of the ocean bottom in the case of a running displacement to be of the form (Fig. 2.10)

$$\eta(x,t) = \eta_0 \left(\theta(x) - \theta(x-a) \right) \left(1 - \theta(x - vt) \right), \tag{3.32}$$

where $\theta(z)$ is the Heaviside step-function. The residual deformation of the ocean bottom, η_0, is the same over the entire active zone of length a and equals zero outside this zone. The horizontal propagation velocity of the displacement is v. A similar problem has been resolved in Sect. 2.3.1 for the case of an incompressible liquid.

We shall apply the general solution of the problem (3.15) and pass to dimensionless variables in accordance with formulae (3.27), which in the case of a running displacement must be complemented with the expression $v^* = v(gH)^{-1/2}$ (we drop the asterisk '*'). As a result we arrive at the following expression describing the surface perturbation of a compressible liquid, caused by a running displacement of the ocean bottom:

$$\xi(x,t) = \frac{\eta_0 c^2}{4\pi^2 i} \int_{-\infty}^{+\infty} dk \int_{s-i\infty}^{s+i\infty} dp \frac{p(\exp\{a\gamma\} - 1)\exp\{pt - ikx\}}{\gamma \cosh(\alpha)(\alpha \tanh(\alpha) + p^2 c^2)}, \tag{3.33}$$

where $\gamma = (ik - pcv^{-1})$, $\alpha^2 = k^2 + p^2$.

As a function of the complex parameter p the integrand expression has two or an infinite number (depending on the sign of α^2) of poles located on the axis $\mathrm{Im}(p) = 0$. Since the positions of the poles are determined from the solution of a transcendental equation, and, besides, they depend on the parameter k, over which external integration is performed, further analysis of expression (3.33) was carried out numerically. Formula (2.72), obtained in Sect. 2.3.1, is the analogue of expression (3.33) for the case of incompressible liquids.

The following parameter values were chosen for calculations: $c = 8$, $a = 10$, which at ocean depths of 4,000 m approximately corresponds to the velocity of sound in water, 1,500 m/s, and to a horizontal size of the source equal to 40 km. The propagation velocity of the displacement, v, was varied within limits from 0.125 up to 32 (from 23 up to 6,000 m/s).

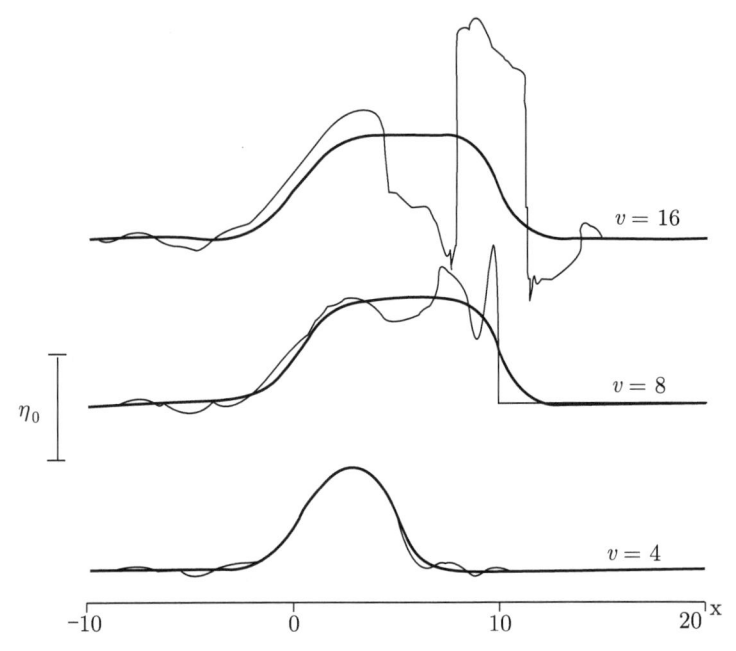

Fig. 3.5 Profiles of surface displacement of a liquid at moment of time $t = 10$ for different propagation velocities of the running displacement, v. The thick and thin lines correspond to incompressible and compressible liquids, respectively

Figure 3.5 presents displacement profiles of the surface of a liquid, $\xi(x)$, for the time moment $t = 10$, calculated within the framework of models for compressible and incompressible liquids for three displacement propagation velocities, $v = 4$, 8 and 16. In all cases, an account of the compressibility led to a significantly more subtly structurized perturbation of the surface, differing from zero only at those points, at which the elastic wave, formed by the running displacement, had time to arrive. As it is seen from the figure, when $v = 4$, the differences between free-surface perturbations for compressible and incompressible liquids is not so large, but in the case of high velocities the difference becomes quite significant. When $v \geqslant c$, the profile is characterized by the presence of steep fronts and of an original periodic structure, which is a consequence of multiple reflections from the surface and from the ocean bottom of the front of the elastic wave, formed by the front edge of the running displacement. From mathematical physics it is known that, when an elastic wave is reflected from a free surface, it changes polarity. Thus, for this reason positive and negative fronts alternate.

The results of calculations of time evolvents $\xi(t)$ for the centre of the active zone ($x = 5$) are presented in Fig. 3.6. The main feature, distinguishing the behaviour of a compressible liquid, consists in the rise in the source area of surface oscillations with a prevalent period equal to four. Oscillations take place against the background of a developing slower gravitational wave. The rise of surface oscillations is due

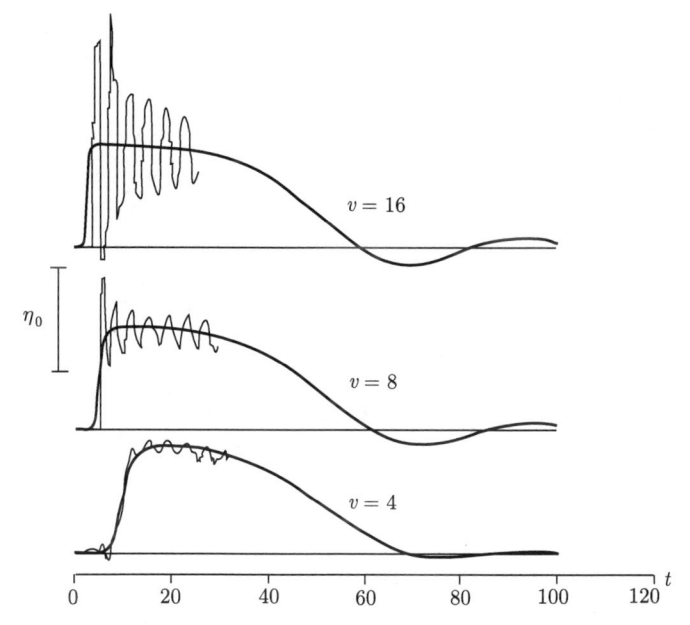

3.1 Excitation of Tsunami Waves with Account of the Compressibility of Water

Fig. 3.6 Time evolvents of surface displacements of liquids at the centre of the active zone for different displacement propagation velocities v. The thick and thin lines correspond to incompressible and compressible liquids, respectively

to the excitation of standing acoustic waves in the natural resonator of a 'column of compressible liquid with a free surface on the rigid bottom'. Similar oscillations arise in the case of piston-like and membrane-like displacements of the ocean bottom (Fig. 3.2).

Within the framework of the model applied, the damping of oscillations is due to the outflow of elastic wave energy from the generation area. The oscillation damping process proceeds, in this case, faster, than in the case of vertical displacements of the ocean bottom, which is related to the existence of a large number of elastic wave rays, deviated from the vertical direction. In real natural conditions the damping will proceed even more rapidly owing to losses occurring, when the elastic waves are reflected and scattered from the boundaries 'water–bottom' and 'water–atmosphere'.

In Fig. 3.7, the dependence of the maximum amplitude of surface displacement at the centre of the active zone ($x = 5$) is presented in a semilogarithmic scale as a function of the velocity v. From the figure it is seen that for propagation velocities of the displacement inferior to $v = 4$ ($v = c/2 \sim 750$ m/s) practically no difference exists between the models for compressible and incompressible liquids. Both theories reveal the presence of a local maximum at $v = 1$, corresponding to resonance excitation of gravitational waves. At high velocities the model for incompressible liquids more than twice underestimates the free-surface displacement.

Fig. 3.7 Maximum amplitudes of surface displacements for incompressible (1) and compressible (2) liquids at the centre of the active zone versus the propagation velocity of the bottom displacement, v

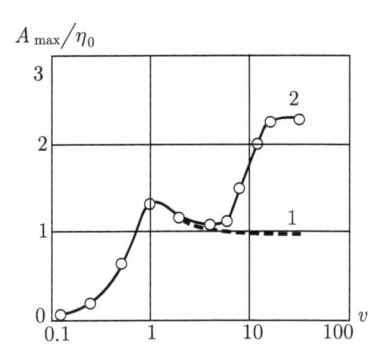

3.1.5 Peculiarities of Wave Excitation in a Basin of Variable Depth

Analytical resolution of the problem of movements of a compressible liquid in a basin with an irregular bottom encounters significant complications, while in the general case it is not even possible. Therefore, in studying peculiarities of the excitation of elastic gravitational waves in a basin of *variable depth* it is expedient to apply numerical simulation [Nosov, Kolesov (2003)]. It must be noted that numerical methods have also to be applied in dealing with analytical solutions (for calculating integrals). Given all the obvious advantages of analytical solutions, direct numerical simulation often turns out to be much more efficient.

We shall consider the plane problem (3.4)–(3.6).

Numerical resolution implies using a region of finite dimensions for calculations. Thus, besides the boundary conditions on the bottom and on the surface, conditions must be formulated for the left and right boundaries of the calculation region. As such conditions, the conditions for free second-order transition (of elastic waves) were chosen [Marchuk et al. (1983)]:

$$c\frac{\partial^2 F}{\partial x \partial t} - \frac{\partial^2 F}{\partial t^2} + \frac{c^2}{2}\frac{\partial^2 F}{\partial z^2} = 0, \qquad x = x_{\min}, x_{\max}. \tag{3.34}$$

Equation (3.4) and the boundary conditions (3.5), (3.7) and (3.34) were reduced to a dimensionless form in accordance with formulae $(x^*, z^*) = (x, z)H_{\max}^{-1}$, $t^* = tH_{\max}^{-1}c$, where $H_{\max}$ is the maximum depth of the basin.

The distribution of depths chosen for calculations imitated transition from the shelf zone through the continental slope towards the abyssal plain (Fig. 3.8a). The parameter $L = 80$ km was not varied. The depths H_1 and H_2 were varied between 0.5 and 8.5 km. The maximum steepness of the slope amounted to 0.1. The tsunami source was located on the slope and represented a displacement involving residual deformation. The form of the space–time law of motion of the bottom deformation, $\eta(x,t) = X(x)T(t)$, is shown in Fig. 3.8b. Movement of the bottom occurred in a direction normal to the surface (of the bottom). The displacement duration varied between 1 and 100 s.

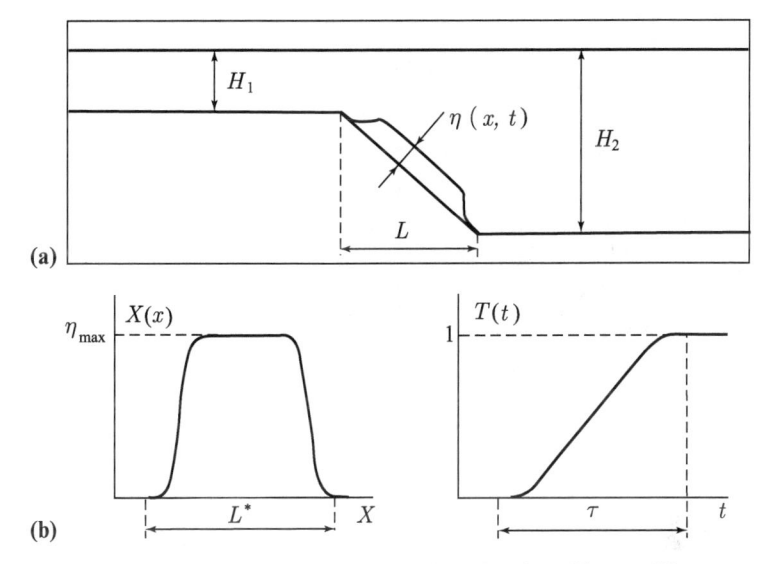

3.1 Excitation of Tsunami Waves with Account of the Compressibility of Water

Fig. 3.8 Shape of calculation region (**a**). Space–time law of motion of bottom (**b**)

Resolution of the problem (3.4), (3.5), (3.7) and (3.34) was performed by the explicit finite-difference method on a rectangular mesh with fixed (but not identical) horizontal and vertical steps. As the stability condition we adopted the Courant criterion $\Delta t < \min(\Delta x, \Delta z)/c$, where Δt is the time step, Δx and Δz are the space steps.

Figures 3.9 and 3.10 show the calculated free-surface disturbance at time moment $t = 1{,}000$ s for different depth values H_1 and H_2. The displacement duration was 10 s. Figure 3.9 corresponds to a fixed depth of the shallow-water area, $H_1 = 0.5$ km; here, the depth H_2 is varied. Figure 3.10 demonstrates the results of calculations for the case of a fixed average depth of the calculation region: the slope 'rotates' about its central point $x = 0, z = -4.5$ km. Practically, the wave perturbation of the surface in all the cases consists of a slow gravitational component and of a fast acoustic (or elastic) component.

In the case of a horizontal ocean bottom ($H_1 = H_2 = 4.5$ km) elastic oscillations of the water column above the source area continue to be present for a long time. This is due to the wave vectors retaining directions close to vertical, and the energy of elastic oscillations slowly leaves the source area. The appearance of even a very insignificant slope of the bottom (1:160 for $H_1 = 4.25$ km, $H_2 = 4.75$ km) alters the picture drastically. In the deep part of the basin an acoustic precursor is observed of significant amplitude, the amplitude of elastic surface oscillations in the shallow region remains practically intact. Oscillations immediately above the source are already close to conclusion by the time moment $t = 1{,}000$ s. Further enhancement of the slope's steepness, first, leads to a decrease in the amplitude of the acoustic precursor in the shallow-water part of the basin, and, subsequently, to its total disappearance. Hence follows the important conclusion that it is impossible to register an acoustic precursor in shallow water, for example, by variations of the sea level.

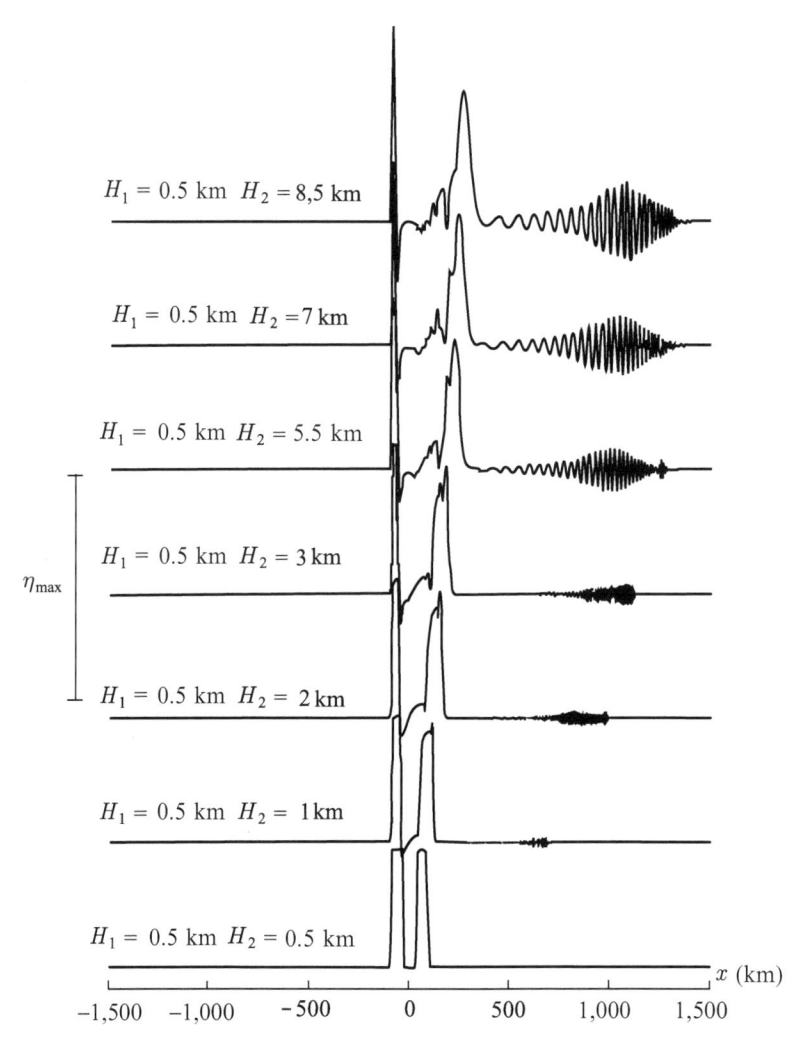

Fig. 3.9 Perturbation of free surface at time moment $t = 1,000$ s

The absence is also to be noted of any manifestation of the compressibility effect on the surface in the case of insignificant depths $H_1 = H_2 = 0.5$ km (Fig. 3.9). Enhancement of the slope's steepness is accompanied by an increase in the propagation velocity of the acoustic precursor towards the deep-water area. Here, the region of maximum amplitudes is intended, the front of acoustic perturbation, naturally, travels with the velocity of a sonic wave.

In the deep-water region, the wavelength of the acoustic precursor and its amplitude may reach values comparable to the length and amplitude of a tsunami gravitational wave. The wavelength of the acoustic precursor in the deep-water area grows starting from the front towards the 'tail'. This effect is a direct consequence

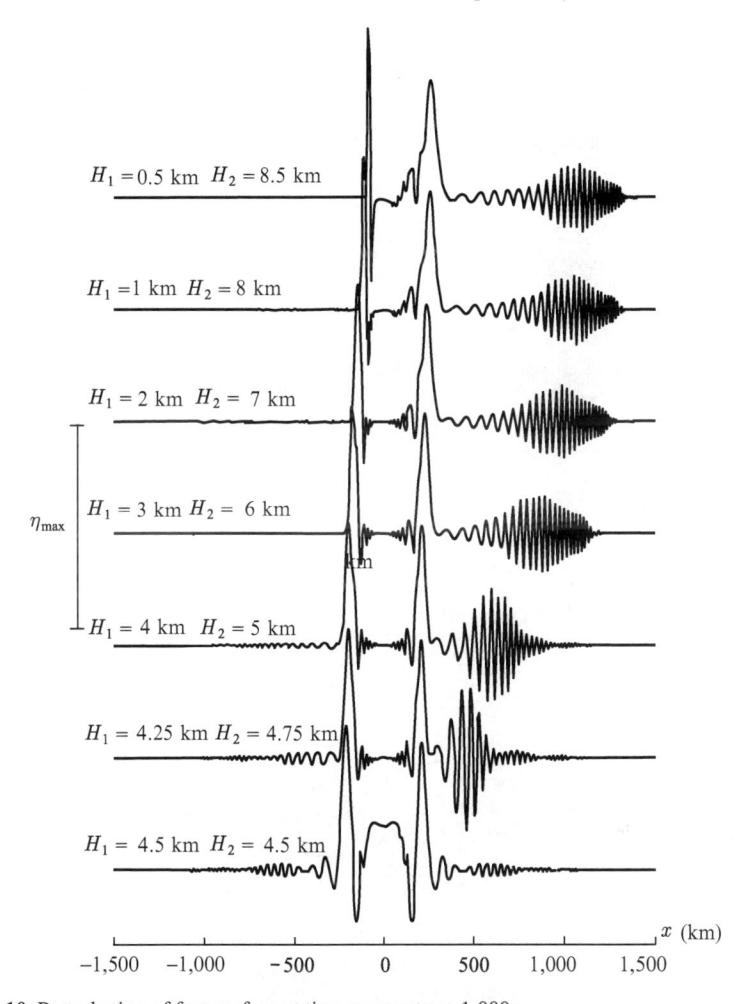

3.1 Excitation of Tsunami Waves with Account of the Compressibility of Water   119

Fig. 3.10 Perturbation of free surface at time moment $t = 1{,}000$ s

of the conditions for wave formation on an inclined ocean bottom: the source of high-frequency oscillations (the shallow-water area) is 'switched on' faster than its low-frequency analogue (the deep-water area).

As compared to the case of elastic surface perturbation, the amplitude characteristics of gravitational waves are not so sensitive to variations in the ocean bottom profile. Nevertheless, the wave of higher amplitude can be seen to propagate into the shallow-water area. Anyhow, the wave entering the deep-water area exhibits greater energy.

The dynamic pressure happens to be the most illustrative characteristic for describing the wave field throughout the thickness of a water column. In natural conditions precisely the dynamic pressure can be measured in a most simple

$p/\rho c v_{max}$

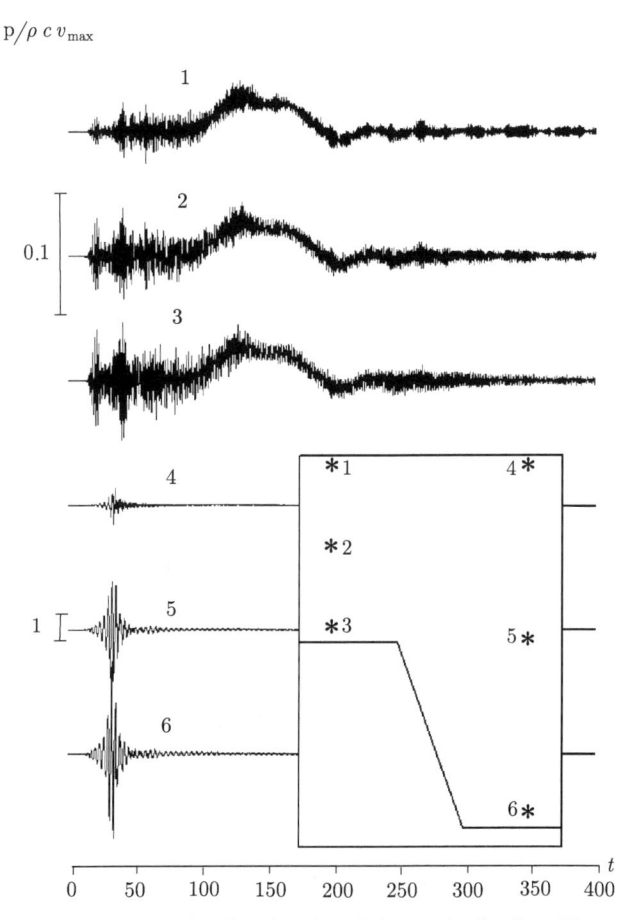

Fig. 3.11 Dynamic pressure, calculated as function of time at fixed points. The pressure is normalized to the quantity $\rho c v_{max}$, where v_{max} is the maximum velocity of movement of the ocean bottom. The scale units for realizing 1–3 and 4–6 are different (indicated in the figure for curves 2 and 5, respectively)

manner (with the aid of hydrophones). Figure 3.11 presents an example of characteristic dependences of the dynamic pressure versus time, calculated for six fixed points, the location of which is shown in the inset. From the figure it is seen that the amplitude of dynamic pressure related to elastic waves (the short-period component) increases as it approaches the ocean bottom. The contribution given by the gravitational surface wave (long-period component at points 1–3) is noticeable only in the shallow-water region against the background of quite weak elastic oscillations. In the deep-water area the amplitude of dynamic pressure reaches a significantly higher value, and the main contribution to the perturbation is precisely due to the acoustic, but not gravitational, component.

3.1 Excitation of Tsunami Waves with Account of the Compressibility of Water

We shall further analyse the peculiarities of the space distribution of the dynamic pressure amplitude for various shapes of the relief of the ocean bottom and conditions for wave generation. We define the amplitude dynamic pressure in accordance with formula

$$p_{\max}(x,z) = \max_{0<t<\Theta} [p(t,x,z)],$$

where Θ is the moment of time, when the acoustic and gravitational waves have already had time to leave the point considered.

The influence the ocean bottom slope in the source area has on the space distribution of the dynamic pressure amplitude is illustrated by Fig. 3.12 (see colour

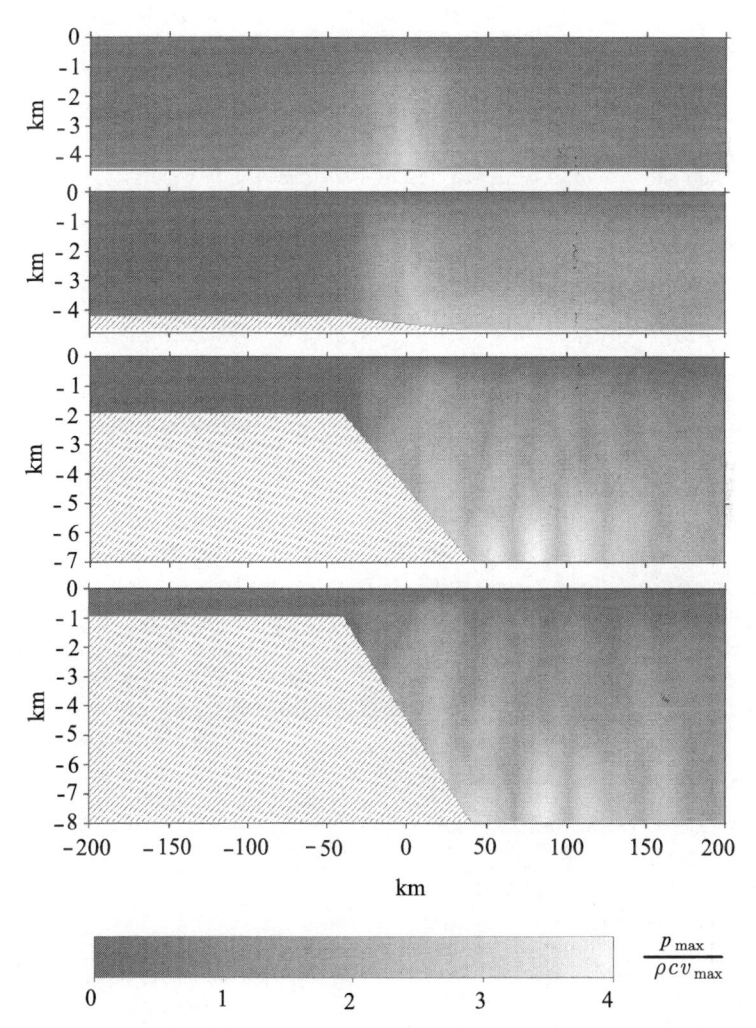

Fig. 3.12 Space distribution of maximum dynamic pressure. The calculation is performed at $\tau = 10\,\text{s}$ for various profiles of the ocean bottom (see also Plate 5 in the Colour Plate Section on page 313)

section). From the figure it is seen that the amplitude of dynamic pressure reaches its maximum values near the ocean bottom, while the presurface region is characterized by minimum values of the dynamic pressure. This property is a direct consequence of the boundary condition at the free water surface. When the bottom is flat, the region of maximum pressures is localized immediately above the source, and the amplitude of the signal reaches noticeable values near the ocean bottom ($p_{max} \sim 0.5\rho c v_{max}$) at significant distances from the source, ~ 200 km, also. In Fig. 3.12, the amplitude of the dynamic pressure is normalized to the quantity $\rho c v_{max}$, where v_{max} is the maximum velocity of motion of the ocean bottom. The appearance of even a very insignificant slope angle in the vicinity of the source leads to a shift of the region of maximum pressures towards large depths. In this case the amplitude of the signal in the shallow-water region is noticeably reduced. Further enhancement of the ocean bottom slope angle results in the maximum pressure values being achieved already outside the source area (in the deep-water region), while propagation of the acoustic signal into the shallow-water region is strongly suppressed. Thus, for example, if $H_1 = 1$ km and $H_2 = 8$ km in the shallow-water region, then the dynamic pressure remains at a level $\sim 0.02 \rho c v_{max}$, and the main contribution to this quantity is not due to the acoustic, but to the surface gravitational wave. At the same time, in the deep-water area the pressure amounts to $3\rho c v_{max}$ and more.

In the case of short displacements the region of maximum dynamic pressure may be observed not only near the ocean bottom, but also inside the thick water column, which is due to the excitation of higher modes of elastic oscillations. In the case of long-duration displacements the dynamic pressure becomes homogeneous in the vertical direction, because effects of water compressibility lose their first-priority significance, and the pressure related to gravitational waves starts to prevail.

In the case of a basin of variable depth, for instance, when the source of waves is located on the sloping ocean bottom, a most important result consists in the shallow-water region turning out to be practically closed to the penetration of elastic waves. The acoustic signal in the shallow-water region being suppressed depends on *two reasons*: the *first* is trivial. The underwater slope is so oriented that the source emits elastic waves into the deep-water region. But this reason is not the sole and even less the principal one. The *second* reason is related to the wave-guide properties of a column of compressible liquid, limited by a free surface and by an absolutely rigid bottom. It is known [Brekhovskikh, Goncharov (1982)] or [Tolstoy, Kley (1987)] that the dispersion relation for normal modes in such a waveguide has the form:

$$k_x^n = \pi \left(\frac{4}{T^2 c^2} - \frac{1}{H^2} \left(n - \frac{1}{2} \right)^2 \right)^{1/2}, \qquad (3.35)$$

where k_x is the x-component of the wave vector, T is the period of elastic waves and n is the mode number ($n = 1, 2, 3 \ldots$). It is seen that for fixed period T and depth H the horizontal wave number will be real only for a finite number of modes. These modes will be propagating modes. For modes of higher numbers k_x becomes a purely imaginary quantity, consequently, the perturbation in the wave decreases exponentially in the x direction. The situation is possible, when in the deep-water

3.1 Excitation of Tsunami Waves with Account of the Compressibility of Water

part of the basin there exist several (or one) propagating modes of period T, while in the shallow-water part no mode exists for such a period. Assuming in expression (3.35) $n = 1$, one can readily find the critical period for the given depth H:

$$T_c = \frac{4H}{c}. \qquad (3.36)$$

The frequency corresponding to the critical period is called the cut-off frequency. Modes of periods inferior to T_c do not propagate in the considered waveguide. Formula (3.36) also permits to calculate the critical depth H_c for a given period of elastic waves, T. An elastic wave will not penetrate the region, where the depth is smaller than the critical depth, $H < H_c = cT/4$. A displacement of the ocean bottom with residual deformation of duration τ forms elastic waves of period $T \sim \tau$, which are capable of penetrating down to depths $H_c \sim c\tau/4$. The examples of calculations, presented in Figs. 3.9, 3.10 and 3.12 correspond to $\tau = 10$ s, i.e. the critical depth amounts to $H_c = 3.75$ km. From the figures it is seen that manifestations of compressibility of the water column correspond to those cases, when the depths exceed the critical value H_c.

Note one more interesting effect, related to the shape of the ocean bottom relief. We intend the possibility for the lowest mode of elastic oscillations to be captured by regions of local depressions of the ocean bottom (deep-water trenches or hollows). Indeed, if the lowest mode originates in the region of a local maximum depth, H_{max}, then it exhibits the period $T_{max} = 4H_{max}/c$. This mode cannot leave the region, where it originated, since to do so it would have to propagate up the slope.

In conclusion of this section, we note that in tsunami catalogues [Soloviev et al. (1997), Soloviev, Go (1974), (1975)] cases are repeatedly mentioned, when tsunami waves throw out onto the coast deep-water fish (unknown species, 'sea monsters'). Moreover, cases have been described, when deep-water fish rised up to the surface before an earthquake. We shall present two quotations from the catalogue of tsunamis in the Mediterranean sea.

1783, February, 5, 12 h ±30min. Calabrian Arc. $38°25'$ N, $15°50'$ E.

Catastrophic Calabrian earthquake, which initiated a long period of seismic activity in the south-west of Italy continued for several years.

Unusual events at sea are described, which can be considered short-time precursors of the earthquake. At the beginning of February close to Messina and at other sites deep-water fish Chichirella started appearing in large numbers, although it usually does not leave the seabed and digs into the seabed silt.

1887, February (March), 23, 6 h 20 min. Ligurian sea, Italy, France. $43°42'$N, $08°03'$ E.

Strong earthquake that occupied an area of 570,000 km^2 Deep-water fish or fish rarely seen in winter were found thrown out onto the beaches of Nizza, San-Remo, Savona.

Such a behaviour of the marine inhabitants is readily explained by the distribution of the dynamic pressure amplitude. In attempts at avoiding the influence of uncomfortable changes in pressure, caused by the underwater earthquake, the fish goes to those regions, where the variations in pressure are minimal, i.e. to shallow-water regions or to the surface.

3.1.6 Elastic Oscillations of the Water Column at the Source of the Tokachi-Oki Tsunami, 2003

Till recently all the information on tsunami sources was obtained exclusively by remote measurements using mareographs (coastal or deep-water), hydroacoustic systems or seismographs. The absence of direct measurements at the tsunami sources, in part, explains why processes at the epicentral zones of underwater earthquakes have been studied relatively weakly.

The possibility, in principle, of investigating the formation of a tsunami at its source arose in 1996, when a set of registering devices, comprising sensors of ocean bottom pressure (JAMSTEC, Japan Agency for Marine-Earth Science and Technology) was established on the continental slope close to the islands of Japan. The sensors are connected to the registration site on the coast by cable lines. Variations of the pressure at the ocean bottom are recorded continually with a sampling frequency of 1 Hz.

The Tokachi-Oki earthquake of 2003 was the first strong seismic event, the epicentre of which was located in the immediate vicinity of the JAMSTEC sensors. According to the seismic catalogue NEIC, this event took place on September 25 at 19:50:06 UTC; the coordinates of its epicentre were 41.78 N, 143.86 E; its hypocentre depth was 27 km, its magnitude was 8.3 M_w (HRV). The earthquake gave rise to a tsunami wave, the height of which amounted to 4 m along the southeast coast of Hokkaido island.

In this section data are analysed concerning variations of the bottom pressure registered by JAMSTEC sensors at the source of the Tokachi-Oki tsunami of 2003 [Nosov et al. (2005), (2007); Nosov, Kolesov (2007)]. We consider the water column at the source to behave as a compressible medium—this is essential, here. A theoretical analysis of the role, played by water compressibility in the tsunami problem, carried out in Sect. 3.1.1 permits to assert that elasticity effects turn out to be essential only at the stage of tsunami generation by an earthquake, while wave propagation or the runup of a wave onto the coast can be described as the motion of an incompressible liquid. Simple estimation reveals that the energy of elastic oscillations of a water column in the tsunami source area may exceed the energy of the gravitational tsunami wave by more than an order of magnitude.

If the case of a horizontal absolutely rigid ocean bottom is considered, then the main difference in the behaviour of a compressible ocean as compared to an incompressible model medium consists in the formation of elastic oscillations of the water column, which are characterized by a discrete set of normal frequencies $v_k = c(1+2k)/4H$, where $k = 0, 1, 2 \ldots$, H is the ocean depth and c is the velocity of sound in water. For typical conditions of a tsunami source the minimal normal frequency $v_0 = c/4H \sim 0.1$ Hz is excited most effectively.

Real tsunami sources are, naturally, located not on a horizontal ocean bottom, but in a region of complex bathymetry. But the slope of the oceanic bottom usually does not exceed the value of 0.1. Therefore, the surface of the bottom can arbitrarily be represented as a set of quasihorizontal segments, each of which is characterized by

3.1 Excitation of Tsunami Waves with Account of the Compressibility of Water   125

its own depth and set of normal frequencies, corresponding to this depth. Thus, at a certain fixed point of the source there, first, takes place formation of elastic oscillations with normal frequencies, determined by the ocean depth at this point. Then, the spectrum of elastic oscillations can be enriched by high frequencies at the cost of waves arriving from neighbouring shallow-water regions. In Sect. 3.1.5 it was shown that owing to the existence of a cut-off frequency low-frequency oscillations, formed in adjacent deep-water regions do not propagate up the slope.

Note two features peculiar to compressibility effects, that explain why they have been studied weakly. First, elastic low-frequency oscillations of a water column can be revealed only at sufficiently large depths (in the open ocean), which hinders their direct registration. Second, the compressibility effects were not quite within the line of research of tsunamis, since, owing to the significant difference in frequency ranges, elastic oscillations were not considered capable of giving any contribution to a tsunami wave. Actually, such an assertion is erroneous, and the contribution of elastic oscillations to a tsunami wave can be provided for by non-linear mechanisms (Sect. 3.2).

Till recently the existence of elastic low-frequency oscillations of the water column at a tsunami source had not been confirmed by measurements in natural waters, and, therefore, the effect remained only theoretically predicted. To avoid confusion, the difference must be stressed between such a well-known phenomenon as the T-phase and the effects dealt with here. Not only the T-phase is related to a range of higher frequencies (1–100 Hz), but it is also registered at significant distances from the source.

Figure 3.13 shows the epicentre location of the 25.09.2003 earthquake and the circular region (dotted line) giving an idea of the size of the tsunami source. The radius of the source, R_{TS} (km), was estimated by the empirical formula (2.3). In the case considered $R_{TS} \approx 112$ km. At the tsunami source there happened to be two sensors of ocean bottom pressure, PG1 (41°42.076′ N, 144°26.486′ E) and PG2 (42°14.030′ N, 144°51.149′ E). The distances of the sensors from the earthquake epicentre were $R_{PG1} \approx 49$ km and $R_{PG2} \approx 96$ km.

Figures 3.14a and 3.15a present the change in time of the pressure near the ocean bottom, registered by sensors PG1 and PG2, respectively. The range of pressure variations, calculated as $p_{max} - p_{min}$, amounted to ≈ 398 kP at sensor PG1 and ≈ 348 kP at sensor PG2. If pressure variations are considered manifestations of elastic oscillations of the water column, resulting from a displacement, then it is possible to estimate the upper limit of the velocity of vertical motion of the bottom (strictly speaking, in the direction normal to the ocean bottom),

$$U \sim \frac{p_{max} - p_{min}}{\rho c},$$

where $\rho = 1{,}000$ kg/m^3 is the density of water and $c = 1{,}500$ m/s is the velocity of sound in water. Estimation yields quite reasonable values: $U_{PG1} \sim 0.27$ m/s, $U_{PG2} \sim 0.23$ m/s. Experience of numerical simulation of the process of tsunami generation in a compressible ocean (Sect. 3.1.5) permits to assert that the range

126 3 Role of the Compressibility of Water and of Non-linear Effects

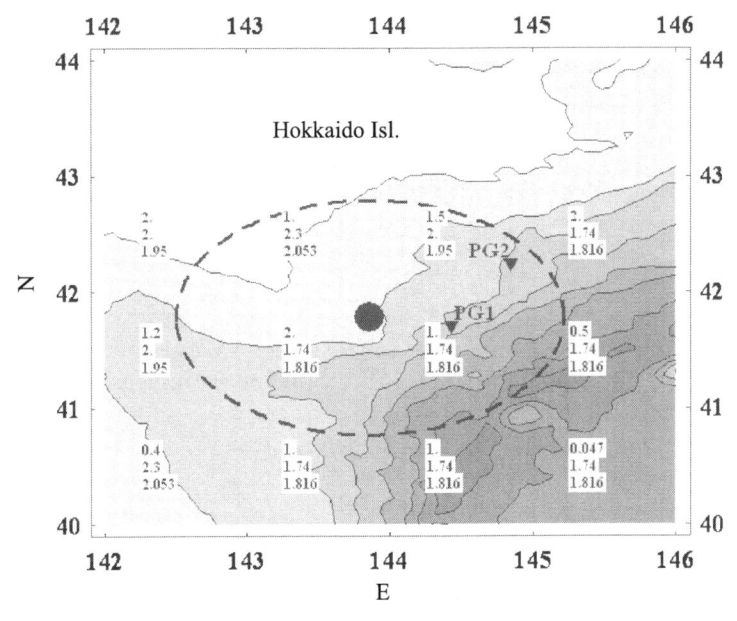

Fig. 3.13 Relative arrangement of the earthquake epicentre (full black circle) and sensors of ocean bottom pressure (triangles). The dotted line shows the region of ocean bottom deformation (estimation). The isobaths are drawn with an interval of 1 km. The numbers in the figure are characteristics of the sedimentary column (from top to bottom: thickness in km, velocity of longitudinal waves in km/s, density g/cm^3)

of pressures may be several times larger than the value of $\rho c U$. Therefore, the value of $U \sim 0.1$ m/s is a good estimate for the velocity of vertical motion of the ocean bottom. We recall that seismologists consider the destruction of buildings and civil structures to start when the ground moves with a velocity on the order of 0.1 m/s.

In Figures 3.14b and 3.15b, the variations of near-bottom pressure are presented in a magnified scale, permitting to see the formation of residual displacements of the ocean bottom, caused by the earthquake. To discard the high-frequency components (>0.02 Hz) the intial time series were subjected to numerical filtration. The result of filtration is shown in the figures by dotted lines. The smooth decrease of pressure down to the starting point of the earthquake is related to tidal variations of the ocean level. The behaviour of the dotted curve clearly shows that the earthquake resulted in the average pressure at sensor PG1 decreasing by $\Delta p_{PG1} \approx 4$ kP, and at sensor PG2 by the quantity $\Delta p_{PG2} \approx 1.5$ kP, which corresponds to a reduction of the water level (elevation of the bottom) by $\Delta H_{PG1} \approx 0.4$ m and $\Delta H_{PG2} \approx 0.15$ m, respectively ($\Delta H = \Delta p / \rho g$, where g is the acceleration of gravity). Note that residual deformations of the ocean bottom were first revealed from the data considered by the authors of [Watanabe et al. (2004)].

According to the Harvard CMT Catalog, the half duration of the process at the source amounted to $\tau_{EQ} = 33.5$ s. But from Figs. 3.14b and 3.15b it is seen that the pressure decreases during an essentially longer period of time (~ 900 s).

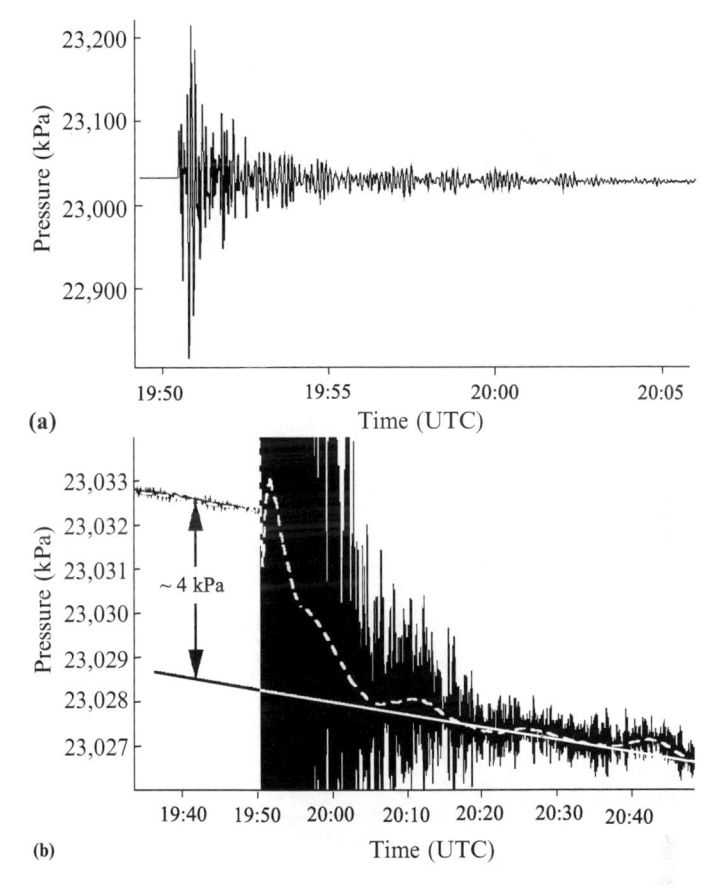

3.1 Excitation of Tsunami Waves with Account of the Compressibility of Water   127

Fig. 3.14 Variations of the pressure near the ocean bottom, registered by sensor PG1 at the tsunami source: '(**a**)' — initial data, '(**b**)' — variations of pressure in magnified scale. The arrow indicates variation of the hydrostatic pressure, caused by elevation of the ocean bottom. The dotted line represents the result of filtration of the initial signal

Evidently, the sensor registers not only the deformation process of the ocean bottom, but also the tsunami wave formation, which is observed as a relaxation of the water column 'elevated' by the displacement. We shall estimate the relaxation time as the propagation time of a long wave over a distance equal to the 'radius' of the tsunami source, $\tau_{TS} \sim R_{TS}(gH)^{-1/2}$. If the quantity H is assumed to be the ocean depth at the earthquake epicentre ($\sim 1{,}900$ m), then one obtains $\tau_{TS} \approx 821$ s, which is in good agreement with the observed value.

Comparison of data on the residual deformation and on the vertical velocity of motion of the ocean bottom, yields an estimate (lower) for the duration of the bottom displacement at the points, where the sensors are located, $\tau_{PG1} \sim \Delta H_{PG1}/U_{PG1} \approx 1.5$ s and $\tau_{PG2} \sim \Delta H_{PG2}/U_{PG2} \approx 0.65$ s. The values obtained are essentially smaller than the duration of the process at the earthquake source. This is not surprising, since the time τ_{EQ} is associated with the formation time

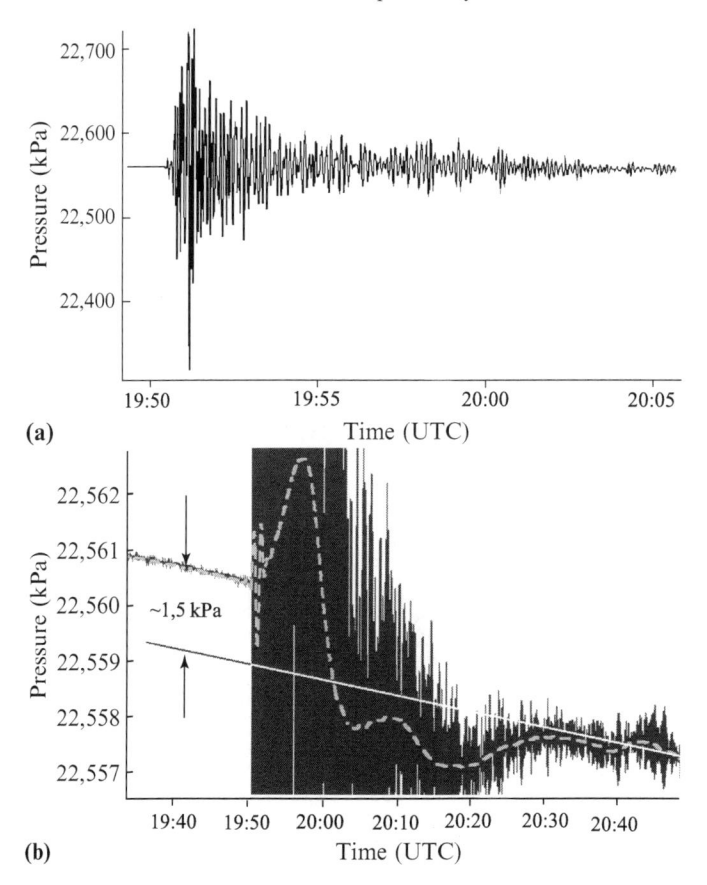

Fig. 3.15 Variations of pressure at ocean bottom, registered by sensor PG2. The notation is similar to the one in Fig. 3.14

of the fault, while the time τ_{PGi} demonstrates a totally different characteristic—the time the deformation of the ocean bottom lasts at the given point. In the event considered the fault propagated in the north-west direction during 50–60 s [Yagi (2004)], consequently, deformation of the ocean bottom did not occur simultaneously along the entire active region, but exhibited the character of a running displacement.

In Sect. 3.1.1 the ratio of the energy of elastic oscillations of a water column, W_2, and the tsunami energy W_1 (the potential energy of the intial elevation) was shown to be determined by the following simple formula:

$$\frac{W_2}{W_1} = \frac{2c}{g\tau}.$$

In the case of the observed duration of the ocean bottom deformation, $\tau \sim 1$ s, the energy of elastic oscillations should be about 300 times greater than the energy of the tsunami waves.

3.1 Excitation of Tsunami Waves with Account of the Compressibility of Water 129

From Figures 3.14a and 3.15a it is well seen that the most significant variations of pressure are observed during 10–15 min after the beginning of the earthquake. Therefore, in performing spectral analysis, we only dealt with lengths of time series amounting to 1,000 s. Before calculating a spectrum, the time series was reduced to the zeroth level by subtraction of the linear trend. It is readily shown that only the minimal normal frequency of elastic oscillations of the water column, v_0, can be observed in the spectra; all the other normal frequencies lie above the Nyquist frequency, equal to 0.5 Hz.

The frequency spectra of pressure variations, normalized to the maximum value, are presented in Fig. 3.16. The non-standard position of the upper plot permits to compare the locations of the principal maxima of the spectra, without overloading the figure. The 0.1–0.2 Hz frequency range is shown in detail in the inset.

From Fig. 3.16 it is seen that the energy of elastic oscillations is mostly concentrated within the 0.05–0.4 Hz interval. Both spectra exhibit clear main maxima. The spectrum corresponding to sensor PG1 has a maximum (resolved as several approximately equal peaks) in the 0.14–0.15 Hz range. The main maximum for sensor PG2 lies in the 0.15–0.16 Hz.

We stress that the positions of the maxima are different along the frequency scale. This is in favour of the assumption that the main maxima are not related to the spectral characteristics of the seismic source, but to the resonance response of the compressible water column at the minimal normal frequency. Sensor PG1 is situated in a deeper place, so a lower normal frequency corresponds to it.

For performing accurate calculation of the normal frequency v_0, information is necessary on the speed of sound (in the vertical profile) and the ocean depth at the given point. The depths of the ocean at the points, where the sensors were established, were determined by linear interpolation from the 2-min global database of the Earth's relief (ETOPO2, http://www.ngdc.noaa.gov/). The information on the location of the sensors, presented above, was taken from the official network page

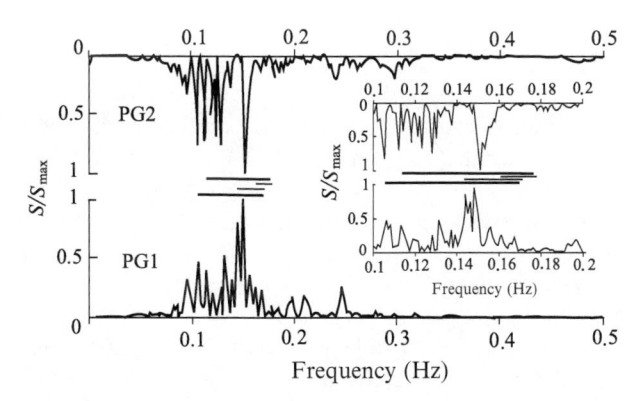

Fig. 3.16 Normalized spectra of near-bottom pressure variations, registered by sensors PG1 (lower) and PG2 (upper). Horizontal lines show the ranges of positions of the minimal normal frequency of elastic oscillations, calculated without account (thin lines) and with account (thick lines) of the sedimentary layer. The inset shows the 0.1–0.2 Hz frequency interval in detail

JAMSTEC (http://www.jamstec.go.jp/). Since the pressure sensors are located on the continental slope, the depths of the ocean at the sites, where they are established, undergo noticeable changes within a single step of the ETOPO2 net. To estimate the possible error in determining depths, the minimum and maximum depths were calculated inside the square 2×2 angular minutes with the centre at the point, where the sensor was established. The horizontal size of the square approximately corresponds to the depth of the ocean in the area considered, which can be considered an additional physical substantiation of the expedience of the choice of dimensions of the square. As a result, the following values were obtained: $H_{\min}^{PG1} = 2{,}256$ m, $H_{\max}^{PG1} = 2{,}578$ m and $H_{\min}^{PG2} = 2{,}170$ m, $H_{\max}^{PG2} = 2{,}300$ m. To estimate the error, related to values of the speed of sound, we assumed that it can vary within limits, known in marine acoustics, from $c_{\min} = 1{,}480$ m/s up to $c_{\max} = 1{,}545$ m/s.

From the values of depths and velocities, determined by the aforementioned method, calculations were performed of the ranges limited by the values $v_{0\min}^{PGi} = c_{\min}/4H_{\max}^{PGi}$ and $v_{0\max}^{PGi} = c_{\max}/4H_{\min}^{PGi}$. The maxima of the spectra, related to elastic oscillations of the water column, must lie precisely within these ranges. In Fig. 3.16 the ranges are shown by thin horizontal lines. We immediately note that variations of the near-bottom pressure occur at frequencies that correspond quite well to theoretically calculated ranges. This fact is evident that the sensors recorded elastic oscillations of the water column. But if we mention the exact positions of the main maxima of the observed spectra, they are seen to lie somewhat more to the left, than predicted by the theory. The shift of maxima towards low frequencies is explained by the acoustic base in the region dealt with being located under a powerful sedimentary layer. Therefore, for theoretical calculation of the position of the maxima of the spectra, it is expedient to consider not simply elastic oscillations of the water column on an absolutely rigid bottom, but the related oscillations of two columns: the water column with its free surface and the underlying sedimentary layer with its rigid lower boundary. For such a two-column system the set of normal frequencies γ is determined from the following transcendental equation:

$$\tan\left(\frac{2\pi\gamma H}{c}\right) \tan\left(\frac{2\pi\gamma H_s}{c_s}\right) = \frac{\rho_s c_s}{\rho c},$$

where H_s is the thickness of the sedimentary layer, c_s is the velocity of elastic longitudinal waves in the sedimentary layer, ρ_s is the density of the sedimentary layer.

According to the database of properties of sedimentary rock (http://mahi.ucsd.edu/Gabi/sediment.html), the thickness of the sedimentary layer in the region considered, the velocity of longitudinal waves and the density of sediments vary within the following respective limits: 47 m–2 km, 1.74–2.3 km/s and 1.816–2.053 g/cm^3. The space distributions of these characteristics are indicated by numbers in Fig. 3.13. It is seen that the possible variations in the velocity of longitudinal waves and in the density are insignificant, while the thickness of the sedimentary layer can change by more than an order of magnitude. Therefore, in calculating the value of the minimal normal frequency γ_0 we took into account the uncertainty, due to the thickness of the sedimentary layer, while the velocity of longitudinal

3.1 Excitation of Tsunami Waves with Account of the Compressibility of Water

waves and the density were set equal to $c_s = 1.74$ km/s and $\rho_s = 1.816$ g/cm^3, respectively. Besides, like in the previous case, account was taken of the range of variations of the speed of sound in water and of the ocean depths at the points, where the sensors were established.

In Fig. 3.16 the frequency ranges, calculated with account of the sedimentary layer, are shown by thick horizontal lines. Like it was to be expected, the ranges have become wider and shifted towards low frequencies. From the figure it is seen that with account of the sedimentary layer the positions of the main maxima are in good agreement with theoretical notions.

Note that quite a significant part of the energy of the spectra, depicted in Fig. 3.16, is to be attributed to frequencies, lying below the principal maxima. This effect cannot be explained within the framework of the absolutely-rigid-ocean-bottom model. The point is that the formation of low frequencies takes place at large depths, and owing to the existence of a cut-off frequency they cannot reach the sensors. But, in reality, the rock of the ocean bottom is not absolutely rigid. Therefore, from our point of view, low-frequency oscillations reach the sensors like seismic waves. We note that the low-frequency limits of the spectra $v_{min} \sim 0.05$ Hz comply well with the maximum depth of the ocean in the region dealt with, $v_{min} \sim c/4H_{max}$ ($H_{max} \approx 7,500$ m).

As to the high-frequency limits of the spectra (0.3–0.4 Hz), in this case it is not correct to relate them to any minimum depths. The point is that the deformation area also embraces part of the the island of Hokkaido, consequently, the ocean depth can decrease down to zero, and the minimum normal frequency can increase indefinitely. In such a situation one must turn to the possibility of the source (ocean bottom displacement) to cause perturbations of high frequencies. It is seen that the lower estimate, obtained above for the duration of the ocean bottom deformation, τ_{PGi}, complies quite well with the right-hand limit of the spectrum, $v_{max} \sim 1/\tau_{PGi}$

Knowledge of the high-frequency limit of the spectrum, v_{max}, permits to make one more interesting conclusion. The boundary, separating the two regions of the tsunami source, goes along the isobath $H_0 \sim c/4v_{max} \approx 1,000$ m. In the first of these two regions, where $H < H_0$, the ocean behaves like an incompressible liquid. In the second region, which is deeper, the water compressibility effects play an important role.

Analysis of direct measurements of the near-bottom pressure at the source of the Tokachi-Oki, 2003 tsunami (taken advantage of as an example) results in the revelation of general-physics regularities determining the behaviour of water as a compressible liquid:

- Low-frequency elastic oscillations of the water column have been revealed. These oscillations were shown to be one of the main dynamic processes at the tsunami source.
- Estimation has been performed of the velocity, amplitude and duration of the ocean bottom deformation at the tsunami source.
- The relationship has been established between the low-frequency limit of the spectrum of near-bottom pressure variations and the ocean depth in the tsunami source area. The high-frequency limit of the spectrum has been shown to depend upon the minimal duration of the ocean bottom deformation.

3.2 Non-linear Mechanism of Tsunami Generation

This section deals with the formation mechanism of tsunami waves due to the non-linear transfer of energy from 'high-frequency' induced or elastic oscillations of the water column to 'low-frequency' surface gravitational waves. Seismic movements of the ocean bottom are considered as the source of 'fast' oscillations of the water column. The 'traditional' tsunami generation mechanism, related to residual displacements substituting the water, naturally, remains in force, and in most cases precisely it plays the leading part. The non-linear mechanism provides additional contributions to the tsunami amplitude and energy. It is not excluded that in individual cases non-linear effects can also provide a determinative contribution to a tsunami wave.

At a first glance formulation of the problem, assuming the presence of periodic oscillations, may seem restricted. Actually, such a restriction is important only under the condition that the water column responds to movements of the ocean bottom like an incompressible liquid. In this case one must indeed consider *periodic* oscillations of a part of the bottom, which lead to corresponding induced oscillations of the incompressible water column. But, if the water column reacts to seismic movements of the ocean bottom like a compressible liquid, the necessity of periodic movements of the bottom vanishes, since any vertical displacements will be accompanied by elastic oscillations of the water column at normal frequencies.

In substantiating the application of linear theory in the tsunami generation problem one usually quotes the condition that the amplitude of the ocean bottom deformation be small as compared to the depth of the basin, $\eta \ll H$. Indeed, this condition is quite fulfilled in reality. But even when the amplitude of the ocean bottom displacement is small, the velocity of its movement may turn out to be sufficiently high for the manifestation of non-linear effects.

Before our studies were published [Nosov, Skachko (2001); Nosov, Kolesov (2002), (2005); Nosov et al. (2008)] there existed only a single work [Novikova, Ostrovsky (1982)], in which the possibility was investigated of tsunami formation resulting from a non-linear effect—the 'detection' of acoustic oscillations of the water column.

3.2.1 Base Mathematical Model

Before proceeding with the construction of a model describing non-linear effects, it is useful to present a description of the character of the linear response of a compressible water column to movements of the ocean bottom without residual displacement. The character of the response varies depending on the position of the spectrum of ocean bottom movements with respect to the two characteristic frequencies $(g/H)^{1/2}$ and $c/4H$, where c is the velocity of sound in water and g is the acceleration of gravity. Further, without losing generality, we shall not speak of a spectrum, but of a certain frequency of bottom oscillations, ν. Thus, if the

3.2 Non-linear Mechanism of Tsunami Generation

frequency of bottom oscillations, $v < (g/H)^{1/2}$, then the linear response of the water column represents a superposition of induced oscillations (in the source area) and of gravitational waves, emitted towards the distant zone. If the frequency lies within the range $(g/H)^{1/2} < v < c/4H$, then no gravitational waves arise, and movements of the water column exist only in the immediate vicinity of the source in the form of induced oscillations. As the frequency increases up to values $v > c/4H$, a qualitative change occurs in the dynamics of the linear response, the water column starts to behave like a compressible medium.

The three frequency ranges identified above are shown in the 'ocean depth–frequency of bottom oscillations' plane in Fig. 3.17. It is interesting to note that the dependences intersect at the hypothetical ocean depth of $H = 14{,}350$ m, consequently, in the conditions of the planet Earth the three indicated ranges exist at any point of the World Ocean. Non-linear effects are manifested only in the case of sufficiently high velocities of bottom movements, which can be characteristic of the frequency ranges 'II' and 'III'. Clearly, the low-frequency range 'I' is of no special interest.

The mathematical model of tsunami generation due to non-linear effects will be constructed on the basis of Euler's equations, assuming the liquid to be compressible,

$$\frac{\partial \mathbf{v}}{\partial t} + (\mathbf{v}, \nabla)\mathbf{v} = -\frac{\nabla p}{\rho} + \mathbf{g}, \tag{3.37}$$

$$\frac{\partial \rho}{\partial t} + \text{div}(\rho \mathbf{v}) = 0. \tag{3.38}$$

In the case of an incompressible liquid the density ρ will be assumed constant, while for the compressible liquid we shall consider the variations in pressure to be proportional to the variations in density, $p' = c^2 \rho'$.

Applying a device used, for instance, in turbulence theory or non-linear acoustics, we represent movement of the liquid as the sum of a slow (average) movement and of a fast (oscillatory) movement,

$$\mathbf{v} = \langle \mathbf{v} \rangle + \mathbf{v}', \quad p = \langle p \rangle + p', \quad \rho = \langle \rho \rangle + \rho'. \tag{3.39}$$

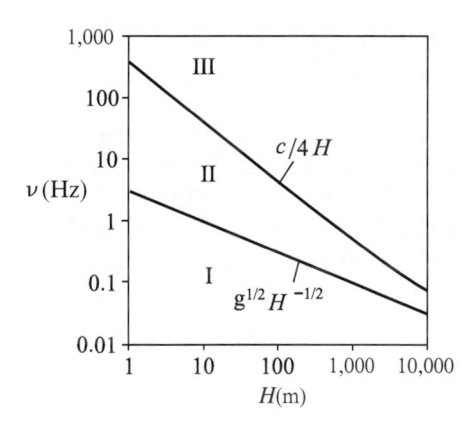

Fig. 3.17 Character of linear response (frequency ranges) of a water column of depth H to oscillations of the ocean bottom of frequency v

Substituting formulae (3.39) into equations (3.37) and (3.38) and averaging over the period of 'fast' oscillations, we obtain a set of equations for describing the average movement of the liquid,

$$\frac{\partial \langle \mathbf{v} \rangle}{\partial t} + (\langle \mathbf{v} \rangle, \nabla) \langle \mathbf{v} \rangle = -\frac{\nabla \langle p \rangle}{\langle \rho \rangle} + g - \langle (\mathbf{v}', \nabla) \mathbf{v}' \rangle + \frac{\langle \rho' \nabla p' \rangle}{\langle \rho \rangle^2}, \quad (3.40)$$

$$\frac{\partial \langle \rho \rangle}{\partial t} + \mathrm{div}(\langle \rho \rangle \langle \mathbf{v} \rangle) = -\mathrm{div}\langle \rho' \mathbf{v}' \rangle. \quad (3.41)$$

In performing the averaging we applied rules, similar to the Reynolds rules, applied in turbulence theory.

In the case of an incompressible liquid (range 'II') the mean is calculated from the period of the ocean bottom oscillations, and the average motion can, obviously, be described as the flow of an incompressible liquid. If the liquid is compressible (range 'III'), then as the period for averaging one should take the quantity $4H_{max}/c$, where H_{max} is the maximum depth of the basin. It is known that acoustic modes with periods superior to $4H_{max}/c$ do not exist, consequently, in this case, also, the mean movement can be described as the flow of an incompressible liquid. Taking into account that $\langle \rho \rangle = \mathrm{const}$ and neglecting the term quadratic in the average velocity, $(\langle \mathbf{v} \rangle, \nabla) \langle \mathbf{v} \rangle$, one arrives at the following system:

$$\frac{\partial \langle \mathbf{v} \rangle}{\partial t} = -\frac{\nabla \langle p \rangle}{\langle \rho \rangle} + g - \langle (\mathbf{v}', \nabla) \mathbf{v}' \rangle + \frac{\langle \rho' \nabla p' \rangle}{\langle \rho \rangle^2}, \quad (3.42)$$

$$\mathrm{div}(\langle \mathbf{v} \rangle) = -\frac{1}{\langle \rho \rangle} \mathrm{div}\langle \rho' \mathbf{v}' \rangle. \quad (3.43)$$

The expressions obtained differ from the usual linearized Euler equations for an incompressible liquid by the presence of the following new terms:

$$\Phi = -\langle (\mathbf{v}', \nabla) \mathbf{v}' \rangle + \frac{\langle \rho' \nabla p' \rangle}{\langle \rho \rangle^2} \equiv -\langle (\mathbf{v}' \nabla) \mathbf{v}' \rangle + \frac{\langle \nabla p'^2 \rangle}{2c^2 \langle \rho \rangle^2}, \quad (3.44)$$

$$s = -\frac{1}{\langle \rho \rangle} \mathrm{div}\langle \rho' \mathbf{v}' \rangle \equiv -\frac{1}{c^2 \langle \rho \rangle} \mathrm{div}\langle p' \mathbf{v}' \rangle, \quad (3.45)$$

which can be interpreted as a force field Φ and a distributed source of mass, s. The origin of the new terms is due to the non-linearity of the initial equations. The combined action of the force field and the distributed source of mass under certain conditions is capable of causing long gravitational waves. We shall speak of this action as a 'non-linear tsunami source'.

For calculating the waves caused by the action of the force field and the distributed source of mass, we shall apply the linear theory of long waves. The expedience of choosing this theory is, first of all, explained by the fact that we are interested in large-scale motions correlated in space (i.e. long waves), and, moreover, this way seems the most simple one.

3.2 Non-linear Mechanism of Tsunami Generation

We shall further restrict ourselves to dealing with the plane problem. We write equations (3.42) and (3.43) for the separate components:

$$\frac{\partial u}{\partial t} = -\frac{1}{\rho}\frac{\partial p}{\partial x} + \Phi_x, \tag{3.46}$$

$$\frac{\partial w}{\partial t} = -\frac{1}{\rho}\frac{\partial p}{\partial z} + \Phi_z - g, \tag{3.47}$$

$$\frac{\partial u}{\partial x} + \frac{\partial w}{\partial z} = s. \tag{3.48}$$

Neglecting vertical acceleration $\partial w/\partial t$, we integrate equation (3.47) over the vertical coordinate within limits from z to ξ. The result for the pressure is the following:

$$p(z) = p_{\text{atm}} + \rho g \xi - \rho g z - \rho \int_z^\xi \Phi_z \, dz^*, \tag{3.49}$$

where ξ is the displacement of the free surface and z is the running vertical coordinate, varying within the limits $-H \leqslant z \leqslant \xi$. Considering the free surface to deviate insignificantly from its equilibrium position ($\xi \ll H$), it is correct to perform integration over the vertical coordinate not up to $z = \xi$, but to $z = 0$. Substituting expression (3.49) into equation (3.46), we find:

$$\frac{\partial u}{\partial t} = -g\frac{\partial \xi}{\partial x} + \int_z^0 \frac{\partial \Phi_z}{\partial x} dz^* + \Phi_x. \tag{3.50}$$

Integration of formula (3.50) over dz within limits from $-H$ to 0 yields the following equation:

$$H\frac{\partial U}{\partial t} = -gH\frac{\partial \xi}{\partial x} + \int_{-H}^0 dz \int_z^0 \frac{\partial \Phi_z}{\partial x} dz^* + \int_{-H}^0 \Phi_x \, dz, \tag{3.51}$$

where U is the horizontal velocity value averaged along the vertical direction. We further integrate the continuity equation (3.48) over dz within the same limits:

$$H\frac{\partial U}{\partial x} + \frac{\partial \xi}{\partial t} = \int_{-H}^0 s \, dz. \tag{3.52}$$

In obtaining expression (3.52) account was taken of the no-flow condition on the ocean bottom, $w(x,-H,t) = 0$ (the ocean bottom is considered motionless for the mean movement), while the vertical velocity at the surface is expressed as the partial time derivative of the displacement ξ.

Further, calculating the partial derivatives with respect to x and t of equations (3.51) and (3.52), respectively, and excluding the mixed derivative $\partial^2 U/\partial x \partial t$ we arrive at the inhomogeneous wave equation

$$\frac{\partial^2 \xi}{\partial x^2} - \frac{1}{gH}\frac{\partial^2 \xi}{\partial t^2} = \frac{1}{gH}Q(x,t). \quad (3.53)$$

Thus, to calculate a long wave caused by the combined action of a force field and distributed sources of mass it is necessary to calculate the following function:

$$Q(x,t) = \int_{-H}^{0} dz \left(\frac{\partial \Phi_x}{\partial x} + \int_{z}^{0} \frac{\partial^2 \Phi_z}{\partial x^2} dz^* - \frac{\partial s}{\partial t} \right). \quad (3.54)$$

For a constant depth of the basin the solution of equation (3.53) is given by the well-known integral formula [Tikhonov, Samarsky (1999)]. In the general case, when the depth is a function of the horizontal coordinate, the equation is readily resolved numerically by the finite difference method.

3.2.2 Non-linear Mechanism of Tsunami Generation by Bottom Oscillations in an Incompressible Ocean

Suppose that in the process of an underwater earthquake a section of the ocean bottom oscillates with a frequency corresponding to range II. In this case the ocean behaves like an incompressible liquid, undergoing induced oscillations following movements of the bottom. From formulae (3.44) and (3.45) the non-linear tsunami source is seen to be manifested only as a force field,

$$\Phi_{\text{incompr}} = -\langle (\mathbf{v}', \nabla) \mathbf{v}' \rangle. \quad (3.55)$$

In this case the linear mechanism is not capable of leading to the formation of gravitational waves, but they may arise as a result of the action of the force field.

Calculation of the quantity Φ_{incompr} requires knowledge of the velocity field in the induced oscillations of the water column. The velocity field can be calculated from the solution of the problem within the framework of linear potential theory, (2.58)–(2.59). Let the law of motion of the ocean bottom, $\eta(x,t)$, have the form

$$\eta(x,t) = \eta_i(x)\big(\theta(t) - \theta(t-\tau)\big)\sin(\omega t), \quad i = 1, 2,$$
$$\eta_1(x) = \eta_0 \exp\{-x^2 a^{-2}\},$$
$$\eta_2(x) = \begin{cases} \eta_0, & |x| \leqslant b, \\ \eta_0[c^{-1}(b-|x|)+1], & b < |x| \leqslant b+c, \\ 0, & |x| > b+c, \end{cases} \quad (3.56)$$

where η_0 and ω are the amplitude and cyclic frequency, respectively, of ocean bottom oscillations, a,b,c are parameters characterizing the horizontal extension and shape of the space distribution of the amplitudes of bottom oscillations and θ is the Heaviside function. The model law of motion of the ocean bottom is

3.2 Non-linear Mechanism of Tsunami Generation

Fig. 3.18 Model law of motion of the ocean bottom: time part (**a**), space distribution of $\eta(x)$ (**b**) and (**c**)

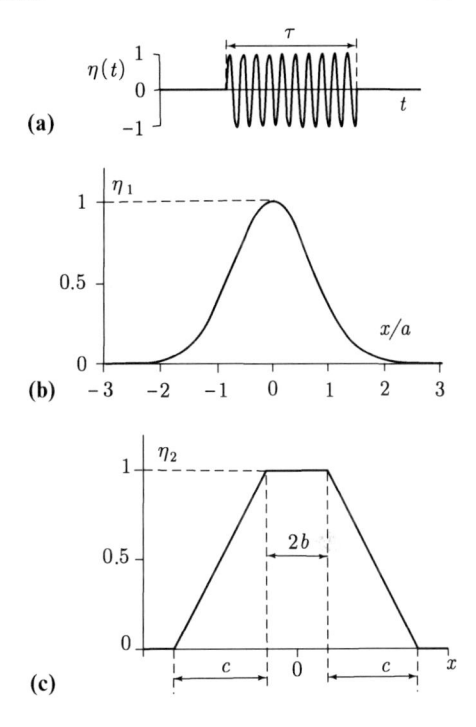

shown in Fig. 3.18. We shall consider ocean bottom oscillations to always terminate at the same phase, as when they started, otherwise the residual displacements of the ocean bottom will certainly excite a gravitational wave via the ordinary piston mechanism.

Dropping intermediate calculations, we present formulae for components of the flow velocity and displacement of the free surface in the case of ocean bottom oscillations described by expression (3.56) (for $\tau = \infty$),

$$u(x,z,t) = \frac{\eta_0 \omega}{\pi} \int_0^\infty dk \frac{\sin(kx)\cosh(kz)X_i(k)}{\cosh(k)(p_0^2 - \omega^2)}$$
$$\times \left(\cos(\omega t)\left(k + \omega^2 \tanh(kz)\right) - \cos(p_0 t)\left(k + p_0^2 \tanh(kz)\right)\right) \quad (3.57)$$

$$w(x,z,t) = -\frac{\eta_0 \omega}{\pi} \int_0^\infty dk \frac{\cos(kx)\cosh(kz)X_i(k)}{\cosh(k)(p_0^2 - \omega^2)}$$
$$\times \left(\cos(\omega t)\left(k\tanh(kz) + \omega^2\right) - \cos(p_0 t)\left(k\tanh(kz) + p_0^2\right)\right), \quad (3.58)$$

$$\xi(x,t) = \frac{\eta_0}{\pi} \int_0^\infty dk \frac{\omega \cos(kx)(\omega \sin(\omega t) - p_0 \sin(p_0 t))X_i(k)}{\cosh(k)(\omega^2 - p_0^2)}, \quad (3.59)$$

where

$$p_0^2 = k\tanh(k), \quad X_i(k) = \int_{-\infty}^{+\infty} dx \exp\{ikx\}\eta_i(x).$$

Expressions (3.57)–(3.59) contain under the integral sign dimensionless variables (the asterisk '*' is omitted)

$$k^* = Hk, \quad t^* = t\left(\frac{g}{H}\right)^{1/2},$$
$$\omega^* = \omega \left(\frac{H}{g}\right)^{1/2}, \quad (x^*, z^*, a^*, b^*, c^*) = \frac{1}{H}(x, z, a, b, c), \quad (3.60)$$

but the coefficients in front of the integrals are dimensional.

Numerical calculation of the flow velocity components has shown that in the frequency range considered, immediately after oscillations of the ocean bottom are 'switched on', each point of the liquid starts performing harmonic oscillations with an amplitude depending only on its coordinates,

$$u'(x,z,t) = u'(x,z)\cos(\omega t), \quad w'(x,z,t) = w'(x,z)\cos(\omega t). \quad (3.61)$$

Substituting formulae (3.61) into expression (3.55) and subsequently averaging over the period of oscillations, we obtain formulae for calculating the horizontal and vertical components of the force field, Φ_x and Φ_z, respectively,

$$\Phi_x(x,z) = -\frac{1}{2}\left(u'(x,z)\frac{\partial u'(x,z)}{\partial x} + w'(x,z)\frac{\partial u'(x,z)}{\partial z}\right), \quad (3.62)$$

$$\Phi_z(x,z) = -\frac{1}{2}\left(u'(x,z)\frac{\partial w'(x,z)}{\partial x} + w'(x,z)\frac{\partial w'(x,z)}{\partial z}\right). \quad (3.63)$$

Functions $u'(x,z)$ and $w'(x,z)$ can be calculated from formulae (3.57) and (3.58) at $t = 0$:

$$u'(x,z) = -\frac{\eta_0 \omega}{\pi} \int_0^\infty dk \frac{\sin(kx)\sinh(kz)X_i(k)}{\cosh(k)}, \quad (3.64)$$

$$w'(x,z) = \frac{\eta_0 \omega}{\pi} \int_0^\infty dk \frac{\cos(kx)\cosh(kz)X_i(k)}{\cosh(k)}. \quad (3.65)$$

As a result we arrive at the following expressions for the components of the force field:

3.2 Non-linear Mechanism of Tsunami Generation

$$\Phi_x(x,z) = -\frac{(\eta_0\omega)^2}{4\pi^2 H}\int_0^\infty dk_1 \int_0^\infty dk_2 \frac{X_i(k_1)X_i(k_2)k_2}{\cosh(k_1)\cosh(k_2)}$$
$$\times (\sin((k_1-k_2)x)\cosh((k_1+k_2)z) - \sin((k_1+k_2)x)\cosh((k_1-k_2)z)), \quad (3.66)$$

$$\Phi_z(x,z) = -\frac{(\eta_0\omega)^2}{4\pi^2 H}\int_0^\infty dk_1 \int_0^\infty dk_2 \frac{X_i(k_1)X_i(k_2)k_2}{\cosh(k_1)\cosh(k_2)}$$
$$\times (\cos((k_1-k_2)x)\sinh((k_1+k_2)z) - \cos((k_1+k_2)x)\sinh((k_1-k_2)z)). \quad (3.67)$$

Figure 3.19 presents a typical form of the force field in the case of the space distribution of the oscillation amplitude calculated for different sizes of the source, a. It is seen that, as parameter a increases, the vertical component of the force, Φ_z, decreases, while the dependence of the horizontal component Φ_x upon the vertical coordinate z becomes weaker and weaker.

For estimation of the relative contributions of the horizontal and vertical components of the force field to the amplitude of the long gravitational wave we take advantage of formula (3.54) for function $Q(x,t)$, entering into the right-hand part of the wave equation (3.53). With account of the set of dimensionless variables (3.60), adopted above, we have:

$$Q(x,t) = \int_{-1}^0 dz \left(\frac{\partial \Phi_x}{\partial x} + \int_z^0 \frac{\partial^2 \Phi_z}{\partial x^2} dz^* \right). \quad (3.68)$$

Substituting formulae (3.66) and (3.67) into (3.68) and performing the required differentiation and integration, we obtain

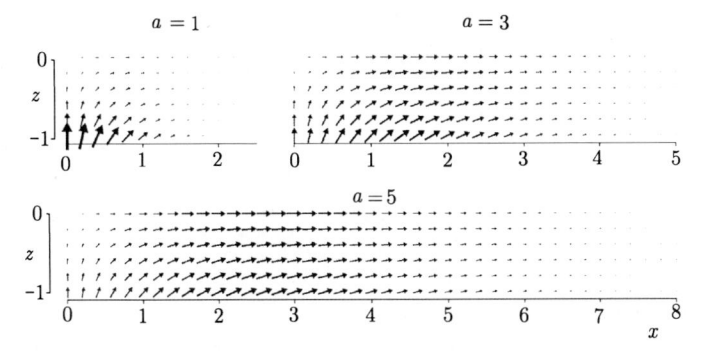

$$Q(x,t) = \text{'X'} + \text{'Z'}, \quad (3.69)$$

Fig. 3.19 Typical form of force field. The calculation is performed for the space distribution of the ocean bottom oscillation amplitude η_1 for $a = 1, 3$ and 5. The direction and length of the arrow corresponds to the vector $a\Phi_{\text{incompr}}$

where

$$`X` = \frac{(\eta_0\omega)^2}{4\pi^2 H}\int_0^\infty dk_1 \int_0^\infty dk_2 \frac{X_i(k_1)X_i(k_2)k_2}{\cosh(k_1)\cosh(k_2)}$$
$$\times \left(\frac{(k_1+k_2)}{(k_1-k_2)}\cos((k_1+k_2)x)\sinh((k_1-k_2))\right.$$
$$\left.-\frac{(k_1-k_2)}{(k_1+k_2)}\cos((k_1-k_2)x)\sinh((k_1+k_2))\right), \quad (3.70)$$

$$`Z` = \frac{(\eta_0\omega)^2}{4\pi^2 H}\int_0^\infty dk_1 \int_0^\infty dk_2 \frac{X_i(k_1)X_i(k_2)k_2}{\cosh(k_1)\cosh(k_2)}$$
$$\times \left(\frac{(k_1-k_2)^2}{(k_1+k_2)^2}\cos((k_1-k_2)x)\left((k_1+k_2)-\sinh(k_1+k_2)\right)\right.$$
$$\left.-\frac{(k_1+k_2)^2}{(k_1-k_2)^2}\cos((k_1+k_2)x)\left\{(k_1-k_2)-\sinh(k_1-k_2)\right\}\right). \quad (3.71)$$

The quantity $X(x)$ determines the contribution of the horizontal component of the force field to the formation of long gravitational (tsunami) waves, and the quantity $Z(x)$ determines the contribution of the vertical component.

Figure 3.20 presents functions $Q(x)$, $X(x)$ and $Z(x)$, which were calculated in accordance with formulae (3.69)–(3.71) for the space distribution of the ocean

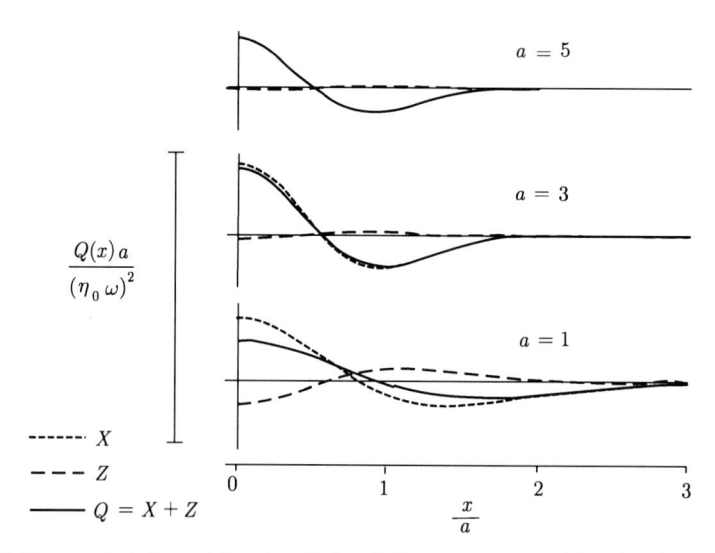

Fig. 3.20 Characteristic form of function $Q(x)$ and of its components $X(x)$ and $Z(x)$. The calculation is performed for the space distribution of the ocean bottom oscillation amplitude η_1 for $a = 1$, 3 and 5

3.2 Non-linear Mechanism of Tsunami Generation

bottom oscillation amplitude $\eta_1(x)$. From the figure it is seen that the terms $X(x)$ and $Z(x)$, as a rule, exhibit differing signs. This means the structure of the force field is such that the contribution of the horizontal force component to the gravitational wave formation is always partly compensated by the vertical component. In the case of a source of small size ($a \sim H$) this effect is capable of significantly reducing the wave amplitude. However, in the case of a large horizontal extension of the source ($a \gg H$) the action of the horizontal component turns out to prevail ($|X| \gg |Z|$). The dimensions of real tsunami sources are always significantly greater than the ocean depth, therefore, the contribution of the vertical component of the force field can be neglected.

Neglecting the contribution of the vertical component of the force field, $Z(x)$, we write equation (3.53) in a dimensionless form (in accordance with formulae (3.60)):

$$\frac{\partial^2 \xi}{\partial x^2} - \frac{\partial^2 \xi}{\partial t^2} = \frac{H}{g} \frac{\partial \overline{\Phi}_x}{\partial x}, \tag{3.72}$$

where $\overline{\Phi}_x = \int\limits_{-1}^{0} \Phi_x dz$ is the horizontal component of the force field, averaged along the vertical direction, ξ is the displacement of the free surface of the liquid from its equilibrium position, corresponding to the mean movement. We recall that there are, also, present above the oscillating ocean bottom fast oscillations of the surface, which are related to induced oscillations.

The solution of equation (3.72) is well known [Tikhonov, Samarsky (1999)]:

$$\xi(x,t) = \frac{H}{2g} \int\limits_0^t d\hat{t} \int\limits_{x-(t-\hat{t})}^{x+(t-\hat{t})} \frac{\partial \overline{\Phi}_x}{\partial \hat{x}} d\hat{x}. \tag{3.73}$$

Oscillations of the ocean bottom (3.56) take place during a finite period of time τ and exhibit fixed amplitude and frequency. Therefore we can write

$$\overline{\Phi}_x(x,t) = \overline{\Phi}_x(x)\big(\theta(t) - \theta(t-\tau)\big). \tag{3.74}$$

Substituting expression (3.74) into formula (3.73) and performing integration over the space variable, we obtain

$$\xi(x,t) = -\frac{H}{2g} \int\limits_0^t \big(\theta(\hat{t}) - \theta(\hat{t}-\tau)\big) \big(\overline{\Phi}_x(x+(t-\hat{t})) - \overline{\Phi}_x(x-(t-\hat{t}))\big) d\hat{t}. \tag{3.75}$$

The process of tsunami formation by the non-linear mechanism is shown in Fig. 3.21. Shifts of the surface of the liquid, ξ, were calculated by formula (3.75) as functions of the horizontal coordinate x for consecutive moments of time. A completely formed wave is sure to consist of a hump and depression, which have a zero total volume. The perturbation always starts with a positive phase. The wave length approximately corresponds to the size of the source.

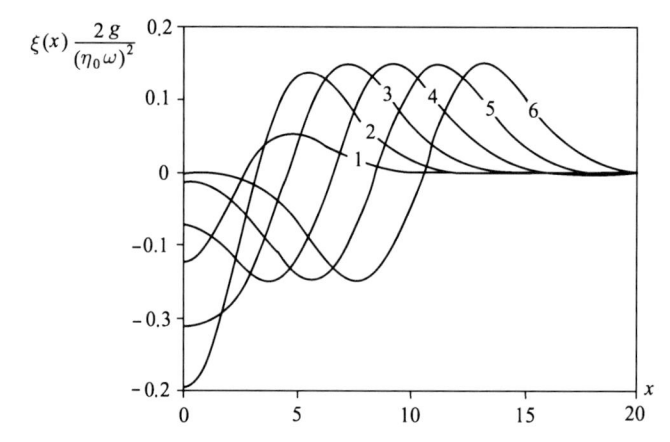

Fig. 3.21 Profile of wave formed by the non-linear mechanism in an incompressible ocean. The calculation is performed for consecutive moments of time $t = 2, 4, 6, 8, 10, 12$ (curves 1–6) for the case of η_1 and $a = 5$, $\tau = 3$

Fig. 3.22 Amplitude of long wave versus the duration of the source action. Curves 1–3 are calculated for the space distribution of η_1 for $a = 5$, 10 and 20, curves 4 and 5 for η_2 and $b = 2$, $c = 3$ (4) and $b = 1$, $c = 9$ (5)

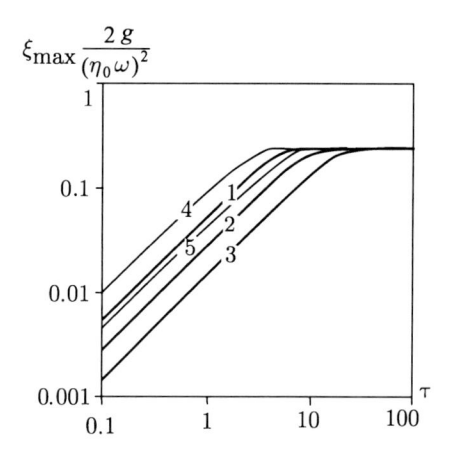

Figure 3.22 presents the wave amplitude ξ_{max} (the height of the hump) as a function of the duration of ocean bottom oscillations for different shapes of the space distribution of the oscillation amplitude. The quantity ξ_{max} increases monotonously with the duration of oscillations, but this increase is not without limit: the amplitude cannot exceed a certain value, which is practically independent of the shape of the space distribution of $\eta_i(x)$. The horizontal extension of the oscillating area of the ocean bottom noticeably affects the value of τ, at which the maximum amplitude is achieved: when the extension in space of the source is greater, the formation of a wave of maximum amplitude will require prolonged action of the source.

The non-linear effect considered can be briefly presented as follows. When oscillations of the basin bottom occur, the liquid is 'pushed out' of the region of most intense movements (the source), which is precisely what causes the

3.2 Non-linear Mechanism of Tsunami Generation

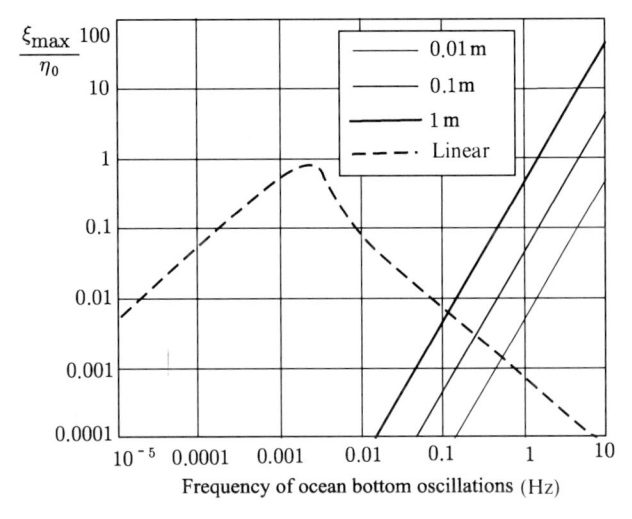

Fig. 3.23 Amplitude of gravitational waves, excited by oscillations of ocean bottom, versus oscillation frequency: linear and non-linear responses. Calculations are performed for exponential distribution of amplitude of bottom oscillations for $a = 10$ km and ocean depth of 1 km

formation of a gravitational wave. The amplitude of such a wave does not depend on the space law, governing variations in the amplitude of the ocean bottom oscillations (providing the law is sufficiently smooth), but depends on the velocity amplitude of oscillations, $\eta_0 \omega$, their duration τ and the horizontal size of the oscillating area.

The data presented in Fig. 3.22 permit to estimate the amplitude of a tsunami wave caused by the non-linear mechanism considered. Thus, for example, when the ocean depth is 1 km, oscillations of an area of the ocean bottom of the characteristic size of 20 km (the space distribution of η_1, $a = 10$), amplitude of oscillatory velocity of 10 m/s, lasting for 60 s, gives rise to a wave of amplitude 0.8 m.

For illustrative estimation of the contribution of the non-linear effect to the tsunami amplitude, Fig. 3.23 presents the dependence of the gravitational wave amplitude upon the frequency of ocean bottom oscillations. The wave amplitude is normalized to the amplitude of bottom oscillations. Calculation of the dependence is performed for the case of $\eta_1(x)$ for $a = 10$ km and $H = 1$ km. The oscillations of the ocean bottom, having started at a certain moment of time, are assumed to continue sufficiently long for the amplitude of the wave, formed by the non-linear mechanism, to reach the maximum value. The linear response (dotted line) is calculated using formula (3.59). Owing to the auxiliary problem being linear, this dependence is the same for any amplitude of bottom oscillations. The contribution of the non-linear effect is proportional to the square velocity of bottom oscillations; therefore, it depends on both the amplitude and the frequency of oscillations. Within the range 0.1–1 Hz this contribution is already capable of competing with the linear response and even of exceeding it.

3.2.3 Non-linear Tsunami Generation Mechanism with Account of the Compressibility of Water

This section deals with the tsunami generation mechanism related to the non-linear transfer of energy from 'high-frequency' elastic oscillations of the water column to 'low-frequency' surface gravitational waves. Elastic oscillations are the reaction of the water column to movements of the bottom of seismic origin. In this case movements of the ocean bottom may not be periodic, it is only important for their frequency spectrum to correspond to frequency range III ($v > c/4H$). Our aim is to find the relationship between characteristics of the 'low-frequency' gravitational wave and the parameters determining the ocean bottom displacement, and, also, a comparative analysis of the efficiencies of the piston and membrane mechanisms in tsunami generation.

In the case of a compressible liquid, a non-linear tsunami source is manifested as the action of a force field Φ and of a distributed source of mass, s. For calculation of these quantities knowledge is required of the fields of velocity $\mathbf{v}'$ and of dynamic pressure p', which we shall find by resolving the plane problem of the linear response of an ideal compressible liquid to small deformations of the ocean bottom, (3.4)–(3.6).

The problem was resolved numerically by the explicit finite difference method, using dimensionless variables ($x^* = x/H$, $t^* = tc/H$). The velocity of ocean bottom deformation was given by the following model laws:

$$U_{\text{pist}}(x,t) = v_{\max}\, \eta\left(\frac{x}{L}\right) \eta\left(\frac{t}{\tau}\right) \quad \text{(piston-like displacement)},$$

$$U_{\text{osc}}(x,t) = v_{\max}\, \eta\left(\frac{x}{L}\right) \sin\left(\frac{2\pi N t}{\tau}\right) \left(\theta(t) - \theta(t-\tau)\right)$$

(oscillations of ocean bottom),

where $\eta(\alpha) = 0.5\left(\tanh(20(\alpha - 0.15)) - \tanh(20(\alpha - 0{,}85))\right)$, $v_{\max}$ is the maximum deformation velocity value, $\theta(t)$ is the Heaviside function, L is the horizontal extension of the deformation area, τ is the duration of the deformation process and N is the number of oscillation periods (an integer number). The form of function $\eta(\alpha)$ is shown in Fig. 3.24. The piston-like displacement resulted in residual deformations of the ocean bottom, oscillations of the ocean bottom finished without residual deformations.

Fig. 3.24 Form of function determining the space-time law of ocean bottom deformation

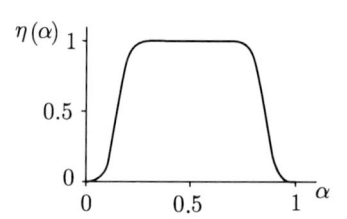

3.2 Non-linear Mechanism of Tsunami Generation

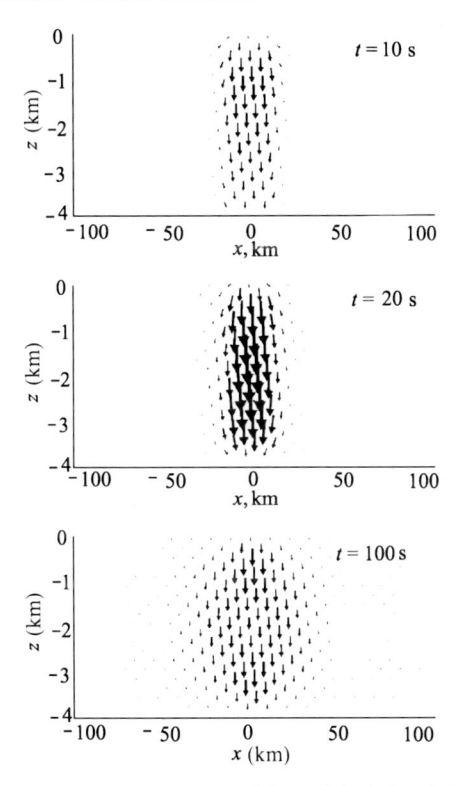

Fig. 3.25 Force field Φ at consecutive moments of time. Calculation for ocean bottom displacement with residual deformation, $\tau = 8$ s, $L = 40$ km, $H = 4$ km

Numerical calculations and theoretical estimates [Nosov, Kolesov (2002), (2005)] reveal the contribution of the force field to tsunami formation to be essentially greater than the contribution of distributed sources of mass. The characteristic form of the field Φ, calculated at consecutive moments of time, is presented in Fig. 3.25. The model parameters chosen for calculations are typical for a real tsunami source. At most of the points the field is directed nearly vertically downwards, which leads to the water being pushed out of the area of intense elastic oscillations. Thus, the leading wave, in this case also, is positive. The quantity $|\Phi|$ develops in time as follows. Being equal to zero at the initial moment of time, it reaches its maximum during the first tens of seconds, then, as the elastic waves leave the source area, it tends monotonously towards zero.

For the calculation of gravitational waves, due to the non-linear mechanism, we applied equation (3.53) written in dimensionless variables ($x^* = x/H$, $t^{**} = t\sqrt{g/H}$, $\xi^* = \xi g/v_{max}^2$)

$$\frac{\partial^2 \xi^*}{\partial x^{*2}} - \frac{\partial^2 \xi^*}{\partial t^{**2}} = Q^*(x^*, t^{**}). \tag{3.76}$$

Equation (3.76) was approximated by the explicit finite difference scheme. At the boundaries of the calculation region the condition of free passage was realized,

$$\frac{\partial \xi^*}{\partial t^{**}} = \mp \frac{\partial \xi^*}{\partial x^*}. \tag{3.77}$$

Since the fields Φ and s, determining the function Q^*, are a result of averaging over the time interval $\Delta t^* = 4$, the output of the model (3.4)–(3.6) was the discrete set: $Q^*(x^*, n\Delta t^*)$, where $n = 1, 2, 3, \ldots$. In passing to resolve the problems (3.76) and (3.77) the step in space Δx^* remained the same, while the time steps Δt^* and Δt^{**} were made to comply with each other as follows: within the time interval t^{**} from 0 to $4\sqrt{gH}/c$ function $Q^*(x^*, 4)$ was in force, within the time interval from $4\sqrt{gH}/c$ to $8\sqrt{gH}/c$ it was $Q^*(x^*, 8)$ and so on.

The main part of numerical experiments was carried out for values of the dimensionless parameters, corresponding to $H = 4$ km, $L = 20$, 40 and 80 km, $0.26 < \tau < 26$ s. The vertical step amounted to $\Delta z = 20$ m. The horizontal step was chosen to be such that 100 nodes could occupy the length L of the source ($\Delta x = 200$, 400 and 800 m). The time step was determined by the Courant condition $\Delta t < \Delta z/c$. In calculations the step $\Delta t = 0.009$ s was applied.

Figure 3.26 presents the typical time behaviour of function $Q^*(x^*)$, reflecting the action of a 'non-linear tsunami source'. The highest absolute values of $Q^*(x^*)$ are not achieved immediately, but only after the passage of a certain time ($t^* = 8$ in the example considered), upon which the intensity of the 'non-linear source' decreases monotonously, which is explained by elastic waves leaving the region where deformation of the ocean bottom occurred. It is important to note that the time the non-linear source is in action noticeably exceeds the duration of the ocean bottom displacement.

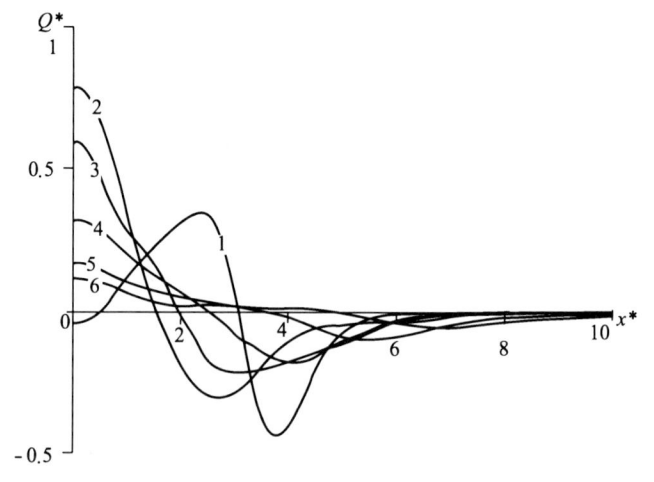

Fig. 3.26 Characteristic form of function $Q^*(x^*, t^*)$. Curves 1–6 correspond to $t^* = 4, 8, 12, 16, 20, 24$. The source parameters: $\tau = 8$ s, $L = 40$ km, $H = 4$ km

3.2 Non-linear Mechanism of Tsunami Generation

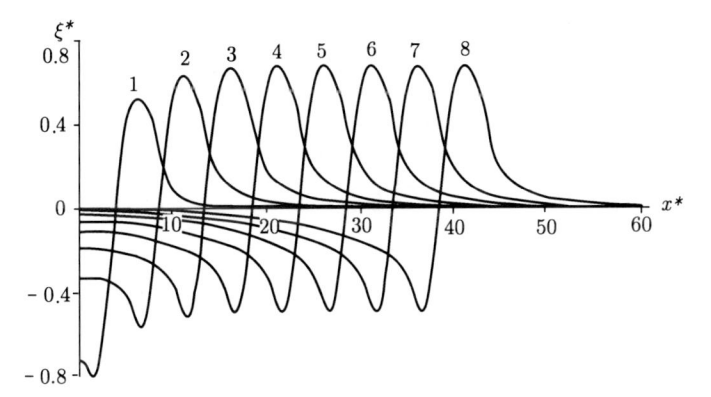

Fig. 3.27 Profiles of gravitational waves formed by a 'non-linear source'. Curves 1–8 are calculated for consecutive moments of time separated by intervals of 100 s. The source parameters: $\tau = 8\,\text{s}, L = 40\,\text{km}, H = 4\,\text{km}$

Figure 3.27 presents typical profiles of surface waves, formed by a 'non-linear source'. The action of this source leads to water being 'pushed out' of the source area, therefore, the waves always originate with a positive phase and finish with a negative phase.

From the profiles of the formed waves calculation was performed of the amplitude

$$A_N = \frac{v_{\max}^2}{g}\left(\text{Max}_{x^*}(\xi^*) - \text{Min}_{x^*}(\xi^*)\right) \equiv \frac{v_{\max}^2}{g} A^*(\tau^*, L^*), \qquad (3.78)$$

and of the energy.

$$W_N = \rho H g^{-1} v_{\max}^4 \int_{-\infty}^{\infty} \xi^2 dx^* \equiv \rho H g^{-1} v_{\max}^4 W^*(\tau^*, L^*). \qquad (3.79)$$

The result of calculations carried out for various durations of piston-like displacements, τ^*, and source sizes L^* ($\tau^* = \tau c/H$, $L^* = L/H$), were dimensionless functions of the dimensionless arguments $A^*(\tau^*, L^*)$ and $W^*(\tau^*, L^*)$.

Non-linear effects can, obviously, provide a noticeable contribution to a tsunami wave only in the case of sufficiently high velocities of the ocean bottom deformation, which is equivalent to displacements of small durations. Therefore, in calculations we only dealt with the range of $\tau < 8H/c$. From the point of view of traditional ideas, such displacements can be considered instantaneous ($\tau = 8H/c \ll L/\sqrt{gH}$); in the case of an instantaneous displacement, on the water surface an initial elevation is formed, which repeats the shape of residual deformations of the ocean bottom. Precisely, the evolution of this elevation generates tsunami waves in their classical sense. We shall term such a tsunami generation mechanism linear. The tsunami amplitude formed by the linear mechanism can be estimated as the amplitude of residual deformations of the ocean bottom,

$$A_L \approx \eta_0 = v_{max}\tau C_1, \quad C_1 = \int_0^1 \eta(\alpha)d\alpha \approx 0.7, \tag{3.80}$$

and the energy as the potential energy of the initial elevation

$$W_L \approx \frac{\rho g}{2}\int_{-\infty}^{+\infty}\xi^2(x,\tau)dx = \rho g v_{max}^2 \tau^2 L\frac{C_1^2 C_2}{2},$$

$$C_2 = \int_0^1 \eta^2(\alpha)d\alpha \approx 0.65. \tag{3.81}$$

Applying formulae (3.80) and (3.81), we obtain relationships permitting to calculate the relative contributions of the non-linear and the linear mechanisms to the amplitude and energy of tsunami waves:

$$\frac{A_N}{A_L} = \left(\frac{\eta_0 c^2}{gH^2}\right)\frac{A^*(\tau^*,L^*)}{C_1^2 \tau^{*2}}, \tag{3.82}$$

$$\frac{W_N}{W_L} = \left(\frac{\eta_0 c^2}{gH^2}\right)^2 \frac{2W^*(\tau^*,L^*)}{C_1^4 C_2 \tau^{*4} L^*}, \tag{3.83}$$

where η_0 is the amplitude of the vertical ocean bottom deformation. From formulae (3.82) and (3.83) the quantities A_N/A_L and W_N/W_L are seen to be determined to a large extent by the dimensionless combination $\eta_0 c^2 g^{-1} H^{-2}$.

Figures 3.28 and 3.29 present the dependences of quantities A_N/A_L and W_N/W_L upon the piston-like displacement duration. The calculation is performed for three different relationships between the source size and the ocean depth. The curves being non-monotonous for $\tau^* > 1$ is due to the modal structure of elastic oscillations of the water column (the minimum normal frequency corresponds to $\tau^* = 4$). When $\tau^* < 1$, the dependences investigated behave approximately like the power functions τ^{*-1} and τ^{*-2}. An increase in the horizontal size of the source leads to an insignificant enhancement of the role of the non-linear mechanism.

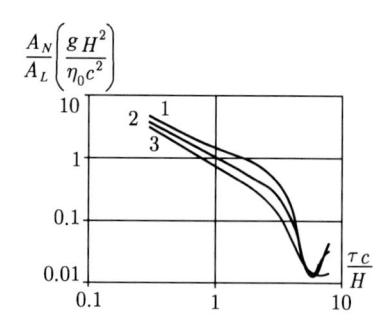

Fig. 3.28 Ratio between amplitudes of tsunami waves formed by the non-linear (A_N) and the linear (A_L) mechanisms versus the displacement duration. Curves 1–3 are drawn for $L/H = 20$, 10 and 5

3.2 Non-linear Mechanism of Tsunami Generation

Fig. 3.29 Ratio between energies of tsunami waves formed by the non-linear (A_N) and the linear (A_L) mechanisms versus the displacement duration. Curves 1–3 are drawn for $L/H = 20$, 10 and 5

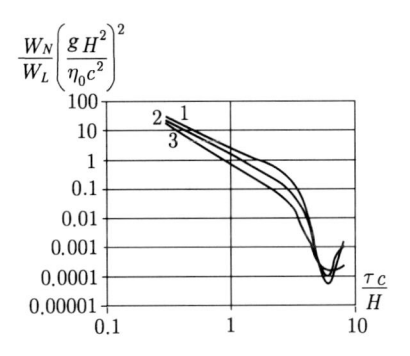

Taking advantage of the data presented in Figs. 3.28 and 3.29 one can readily perform the following estimations. For an ocean depth of 1.5 km, displacement duration and amplitude of 1 s and 1 m, respectively, the contribution of the non-linear mechanism to the tsunami amplitude will be at a level of 10%, and to its energy of 1%. The contribution of the non-linear mechanism may increase as the amplitude of the ocean bottom displacement increases or the displacement duration decreases, but, most likely, the linear mechanism will continue to prevail in the case of a piston-like displacement.

The non-linear mechanism may provide for an essential contribution to the amplitude of a tsunami wave in the case of ocean bottom oscillations at one of the normal frequencies, $v_k = c(1+2k)/4H$, $k = 0, 1, 2, \ldots$ (resonance pumping of energy). According to linear theory, ocean bottom oscillations without residual displacements at frequencies $v > \sqrt{g/H}$ do not produce gravitational waves (see Sect. 2.3.4). In conditions of the planet Earth $v_k > \sqrt{g/H}$, consequently, in the case of ocean bottom oscillations with frequencies v_k only the non-linear mechanism can give rise to tsunamis.

Calculations carried out for $U(x,t) = U_{\text{osc}}(x,t)$ have revealed the following. If an area of the ocean bottom of dimension $L = 40$ km at a depth of $H = 4$ km undergoes $N = 10$ oscillations of frequency $v_0 = c/4H \approx 0.094$ Hz and amplitude 0.3 m, then the non-linear mechanism produces a tsunami of amplitude ~ 0.5 m. In similar conditions, but at a higher frequency $v_3 = 7c/4H \approx 0.65$ Hz, the tsunami amplitude will already amount to ~ 1.2 m. If the frequency of ocean bottom oscillations differs noticeably from the normal frequency, then the efficiency of the non-linear mechanism decreases significantly. Thus, for example, if $v = 0.55$ Hz ($v_2 < v < v_3$), the tsunami amplitude will only be of the order of 6 cm.

In conclusion, we note that the frequencies of seismic oscillations of the ocean bottom lie within the range of several first normal frequencies of the water column, v_k, which creates favourable conditions for realization of the non-linear mechanism of tsunami generation.

References

Alexeev A. S., Gusyakov V. K. (1973): Numerical simulation of the process of tsunami wave and seismoacoustic wave excitation during earthquakes in the ocean. In: Works of IV All-Union symposium on wave diffraction and propagation (in Russian). **2** 194–197

Brekhovskikh L. M. (ed.) (1974): Acoustics of the ocean (in Russian). Nauka, Moscow

Brekhovskikh L. M., Goncharov V. V. (1982): Introduction to the mechanics of continuous media (as applied to wave theory) (in Russian). Nauka, Moscow

Boorymskaia R. N., Levin B. W., Soloviev S. L. (1981): Kinematical criterion for a submarine earthquake to be tsunamigenic (in Russian). DAN SSSR. **261**(6) 1325–1329

Dotsenko S. F. (1996a): The influence of ocean floor residual displacements on the efficiency of directed tsunami generation. Izvestiya – Atmos. Ocean Phys. **31**(4) 547–553

Dotsenko S. F. (1996b): Excitation of tsunamis due to oscillations in a section of the floor. Izvestiya Atmos. Ocean Phys. **32**(2) 244–249

Ewing W. M., Tolstoy I., Press F., (1950): Proposed use of the T phase in tsunami warning systems, Bull. Seism. Soc. Am., 40, 53–58

Garber M. R. (1984): Improved model for long-period wave excitation in ocean and atmosphere by underwater earthquakes (in Russian). DVNII Transactions No 103, pp. 14–18. Gidrometeoizdat, Leningrad

Gisler G. R. (2008): Tsunami simulations. Annu. Rev. Fluid Mech. **40** 71–90

Gusyakov B. K. (1972): Excitation of tsunami waves and of oceanic Rayleigh waves during a submarine earthquake. In: Mathematical problems of geophysics (in Russian). Publ. Dept. of Computation Center, No 3, pp. 250–272 SB RAS USSR, Novosibirsk

Gusyakov B. K. (1974): On the relationship between a tsunami wave and the source parameters of the underwater earthquake. In: Mathematical problems of geophysics (in Russian), No 5, Part I, pp. 118–140. Publ. Dept. of Computation Center, SB RAS USSR, Novosibirsk

Hammack J. L. (1973): A note on tsunamis: their generation and propagation in an ocean of uniform depth. J. Fluid Mech. 60 769–799

Kadykov I. F. (1986): The acoustics of submarine earthquakes (in Russian). Nauka, Moscow

Kadykov I. F. (1999): Submarine low-frequency acoustic noise of the ocean (in Russian). Editorial URSS, Moscow

Kajiura K. (1970): Tsunami source, energy and directivity of wave radiation. Bull. Earthquake Res. Inst. Tokyo Univ. **48**(5) 835–869

Landau L. D., Lifshitz E. M. (1987): Fluid Mechanics, V.6 of Course of Theoretical Physics, 2nd English edition. Revised. Pergamon Press, Oxford-New York-Beijing-Frankfurt-San Paulo-Sydney-Tokyo-Toronto

Levin B. W. (1981): On the source and hydromechanics of an underwater earthquake (in Russian). In: Tsunami wave propagation and runup on shore, pp. 5–10 Nauka, Moscow

Lysanov Yu. P. (1997): Capture by underwater acoustic channel of hydroacoustic waves, generated during submarine earthquakes in the deep ocean (in Russian). Acoust. J. **43**(1) 92–97

Marchuk An. G., Chubarov L. B., Shokin Yu. I. (1983): Numerical simulation of tsunami waves (in Russian). Nauka, Siberian Branch, Novosibirsk

Nikiforov A. F., Uvarov V. B. (1984): Special functions of mathematical physics (in Russian). Nauka, Moscow

Nosov M. A. (1999): A model for tsunami generation by bottom movements incorporating water compressibility. Volcanol. Seismol. **20** 731–741

Nosov M. A. (2000): On the tsunami generation in the compressible ocean by vertical bottom displacements. Izvestiya, Atmos. Ocean. Phys. **36** 5 718–726

Nosov M. A., Kolesov S. V. (2002): Non-linear mechanism of tsunami generation in a compressible ocean. In: Proceedings of the International Workshop 'Local Tsunami Warning and Mitigation', Moscow, pp. 107–114

References

Nosov M. A., Kolesov S. V. (2003): Tsunami generation in compressible ocean of variable depth. In: Submarine landslides and tsunamis, Edited by Yalciner A. C. Pelinovsky E., et al. pp. 129–137 Kluwer, Dordrecht, Boston, London

Nosov, M. A., Kolesov, S. V. (2005): Nonlinear tsunami generation mechanism in compressible ocean. Vestnik Moskovskogo Universita. Ser. 3 Fizika Astronomiya (3) 51–54

Nosov M. A., Kolesov S. V. (2007): Elastic oscillations of water column in the 2003 Tokachi-oki tsunami source: in-situ measurements and 3-D numerical modelling. Nat. Hazards Earth Syst. Sci. **7** 243–249

Nosov M. A., Sammer K. (1998): Tsunami excitation by a moving bottom displacement in compressible water. Moscow Univ. Phys. Bull. **53**(6) 67–70

Nosov M. A., Shelkovnikov N. K. (1997): The excitation of dispersive tsunami waves by piston and membrane floor motions. Izvestiya, Atmos. Ocean. Phys. **33**(1) 133–139

Nosov M. A., Skachko S. N. (2001a): Nonlinear tsunami generation mechanism. Nat. Hazards Earth Syst. Sci. **1** 251–253

Nosov M. A., Skachko S. N. (2001b): Non-linear mechanism of tsunami generation by bottom oscillations (in Russian). Moscow University Bulletin, Series 3, Phys. Astronom. (5) 57–60

Nosov M. A., Kolesov S. V., Ostroukhova A. V., Alekseev A. B., Levin B. W. (2005): Elastic oscillations of the water layer in a tsunami source. Doklady Earth Sciences **404**(7) 1097–1100

Nosov M. A., Kolesov S. V., Denisova A. V., Alekseev A. B., Levin B. W. (2007): On the near-bottom pressure variations in the region of the 2003 Tokachi-Oki tsunami source. Oceanology **47**(1) 26–32

Nosov M. A., Kolesov S. V., Denisova A. V. (2008): Contribution of nonlinearity in tsunami generated by submarine earthquake, Adv. Geosci. **14** 141–146

Novikova L. E., Ostrovsky L. A. (1982): On the acoustic mechanism of tsunami wave excitation (in Russian). Oceanology **22**(5) 693–697

Ohmachi T. (2001): Tsunami simulation taking into account seismically induced dynamic seabed displacement and acoustic effects of water. In: Book of Abstracts, NATO ADVANCED RESEARCH WORKSHOP 'Underwater Ground Failures on Tsunami Generation, Modeling, Risk and Mitigation', May 23–26, pp. 45–47, Turkey, Istanbul

Okal E. A. (2003): T Waves from the 1998 Papua New Guinea earthquake and its aftershocks: timing the tsunamigenic slump. Pure Appl. Geophys. **160** 1843–1863

Okal E. A., Alasset P. J., Hyvernaud O., Schindele F. (2003): The deficient T waves of tsunami earthquakes. Geophys. J. Int. **152** 416–432

Panza F. G., Romanelli F., Yanovskaya T. B. (2000): Synthetic tsunami mareograms for realistic oceanic models. Geophys. J. Int. **141** 498–508

Pelinovsky E. N. (1996): Hydrodynamics of tsunami waves (in Russian). Institute of Applied Physics, RAS, Nizhnii Novgorod

Pod'yapolsky G. S. (1968a): Excitation of a long gravitational wave in the ocean by a seismic source inside the crust (in Russian). Izv. RAS. Earth Phys. (1)

Pod'yapolsky G. S. (1968b): On the relationship between a tsunami wave and the underground source, that generated it (in Russian), In: The tsunami problem. Nauka, Moscow

Pod'yapolsky G. S. (1978): Tsunami excitation by an earthquake. In: Methods for calculating tsunami rise and propagation (in Russian), pp. 30–87 Nauka, Moscow

Puzyrev N. N. (1977): Methods and objects of seismic studies (in Russian). Publishing House of RAS Siberian Branch, NITs OIGGM, Novosibirsk

Sekerzh-Zen'kovich S. Ya., Zakharov D. D., Timokhina A. O., Shingareva I. K. (1999): Tsunami wave excitation in an inhomogeneous ocean by seismic-type sources inside the Earth's crust (in Russian). In: Collection 'Interaction in the lithosphere–hydrosphere–atmosphere system', **2**, pp. 233–240. Publishing Dept. of MSU Phys. Faculty, Moscow

Selezov I. T., Tkachenko V. A., Yakovlev V. V. (1982): On the influence of water compressibility on tsunami wave generation (in Russian). In: Processes of tsunami excitation and propagation, pp. 36–40. Publishing house of USSR AS, Moscow

Sells C. C. H. (1965): The effect of a sudden change of shape of the bottom of a slightly compressed ocean. Philos. Trans. Roy. Soc. Lond. (A), (1092) 495–528

Soloviev S. L., Voronin P. S., Voronina S. I. (1968): Seismic hydroacoustic data on the T wave (review of the literature) (in Russian). In: The tsunami problem, pp. 142–173. Nauka, Moscow

Soloviev S. L., Go C. N. (1974): Catalogue of tsunamis on the western coast of the Pacific Ocean (173–1968) (in Russian). Nauka, Moscow

Soloviev S. L., Go C. N. (1975): Catalogue of tsunamis on the eastern coast of the Pacific Ocean (1513–1968) (in Russian). Nauka, Moscow

Soloviev S. L., Belavin Yu. S., Kadykov I. F., U Ton Il' (1980): Registration of T phases in earthquake signals in the north-western part of the Pacific Ocean (in Russian). Volcanol. Seismol. (1) 60–69

Soloviev S. L., Go C. N., Kim Kh. S., et al. (1997): Tsunamis in the Mediterranean Sea, 2000 BC–1991 AD (in Russian), Nauchnyi mir, Moscow

Tikhonov A. N., Samarsky A. A. (1999): Equations of mathematical physics (in Russian). Publishing House of Moscow University

Tolstoy, I. and Clay, C. S. (1987): Ocean acoustics – theory and experiment in underwater sound, 2nd ed. American Institute of Physics, New York

Voight S. S. (1987): Tsunami waves. Tsunami Res. (in Russian) (2) 8–26

Watanabe T., Matsumoto H., Sugioka H., et al. (2004): Offshore monitoring system records recent earthquake off Japan's northernmost island. Eos **85**(2) 13 January

Yagi Y. (2004): Source rupture process of the 2003 Tokachi–oki earthquake determined by joint inversion of teleseismic body wave and strong ground motion data. Earth Planets Space **56** 311–316

Yanushkauskas A. I. (1981): Cauchy-Poisson theory for a compressible liquid (in Russian). In: Tsunami wave propagation and runup on shore. pp. 41–55 Nauka, Moscow

Zhmur V. V. (1987): Surface phenomena above the sources of strong underwater earthquakes. Tusnami Stud. (in Russian) (2) 62–71

Zvolinsky N. V. (1986): On the seismic mechanism of tsunami wave generation. Izv. AN SSSR, Ser. Earth Phys. (in Russian) (3) 3–15

Zvolinsky N. V., Nikitin I. S., Sekerzh-Zen'kovich S. Ya. (1991): Generation of tsunami and Rayleigh waves by a harmonic expansion center. Izv. AN SSSR, Ser. Earth Phys. (in Russian). (2) 34–44

Zvolinsky N. V., Karpov I. I., Nikitin I. S., Sekerzh-Zen'kovich S. Ya. (1994): Generation of tsunami and Rayleigh waves by a harmonic two-dimensional rotation center. Izv. AN SSSR, Ser. Earth Phys. (in Russian) (9) 29–33

Chapter 4
The Physics of Tsunami Formation by Sources of Nonseismic Origin

Abstract The physics is described of tsunami formation by sources of nonseismic origin: landslides, volcanic eruptions, meteorological causes and cosmic bodies falling into the ocean. Short descriptions are given of certain remarkable historical events (with the exception of cosmogenic tsunamis). Approaches to the mathematical description of tsunami generation by these sources are expounded. Basic regularities, relating parameters of a source and of the tsunami wave generated by it are presented.

Keywords Tsunami generation · gravitational surface wave · earthquake · landslide · slump · mud flow · river tsunami · erosion · sedimentary layer · viscous fluid · long waves · Froude number · volcano · volcanic eruption · caldera collapse · explosive eruption · underwater volcano · pyroclastic flow · equivalent source · stationary-phase method · meteotsunami · anemobaric waves · resonance · Proudman resonance · internal waves · storm surges · tension of friction · atmospheric pressure · long-wave theory · meteorite · cosmogenic tsunami · asteroid · kinetic energy · parametrization · numerical simulation · dispersion

Tsunami generation is mainly caused by sharp vertical displacements of separate areas of the ocean bottom, taking place during strong underwater earthquakes. Details of this process are described in Chapters 2 and 3. But seismotectonic movements are not the only possible mechanisms of tsunami formation. A significant number of events are caused by landslides (slumps), processes related to volcanic eruptions and meteorological causes. In accordance with the historical database of tsunamis in the Pacific Ocean (Institute of Computational Mathematics and Mathematical Geophysics of the RAS Siberian Branch [SB], Novosibirsk), 79% of events were due to earthquakes, 6% to landslides, 5% to volcanic eruptions and 3% to meteorological causes. The sources of the remaining 7% of events are still unknown.

Recently, tsunami generation by meteorites falling into the ocean has been the issue of active discussions. Such events are extremely rare. Such an event may even never have occurred during the whole history of our civilization. But, bearing in mind the scale of such a catastrophe, the authors considered it necessary to present certain results of studies of this tsunami generation mechanism.

4.1 Tsunami Generation by Landslides

After tsunamis of seismotectonic origin, most often encountered are so-called *landslide tsunamis*. This term stands for gravitational surface waves caused by underwater landslides and mud flows, fragments of steep coasts, rock and icebergs, and sometimes, even buildings in harbours, collapsing into the water. At present, only in the Pacific region over 80 tsunamis are known to have been caused by the mechanisms indicated. As a rule, landslide tsunamis are considered local events. But studies, performed in recent years, reveal that landslides can give essential additional contributions to tsunamis generated by strong earthquakes [Gusiakov (2001)].

As compared to the horizontal dimensions of seismic sources (10^4–10^5 m), coastal and underwater landslides usually exhibit smaller scales (10^2–10^3 m). The largest known in geological history Storegga landslide took place in the late quaterny period in the region of the steep continental slope off the coast of Norway [Jansen et al. (1987); Harbitz (1992)]. Its horizontal extension is estimated to have amounted to tens of kilometers.

In spite of their local character, the destructive force of landslide tsunamis is no less than that of waves of seismotectonic origin. Such tsunamis are particularly dangerous in narrow straits, fjords and closed gulfs and bays [Murty (1977); Jiang, LeBlond (1992)]. Among the best known events one must mention the catastrophic tsunamis in Lituya Bay (Lituya Bay, South-East Alaska, 1958) and in Vaiont Valley (Vaiont Valley, Northern Italy, 1963). The tsunami in Lituya Bay was caused by the fall of rock matter at the bay apex into the water, which led to the formation of a huge wave, the run-up height of which amounted to 524 m [Miller (1960); Murty (1977); Lander (1996)]. The catastrophe in Vaiont Valley resulted in the destruction of an entire city, and about 2,000 people died [Wiegel et al. (1970); Murty (1977)].

Landslide tsunamis are characterized by a high repetition rate at certain parts of the coast. For example, in situ studies at Lituya Bay, carried out after the catastrophic event of 1958, revealed that gigantic waves, caused by landslides, had also occurred there previously—in 1853–1854 (120 m), in 1874 (24 m), in 1899 (60 m) and in 1936 (150 m) [Miller (1960)]. Even the Laperouse expedition suffered from a tsunami in this bay—a two-mast schooner of the squadron with a crew of 21 men was shattered by an 'unusual wave' against the cliffs of the island in 1787.

It is interesting that landslide tsunamis can occur not only in oceans and seas, but also in large rivers. The description of one such event, which took place in river Volga in 1597, is presented in [Didenkulova et al. (2007)]. We have succeeded in finding reference to another river tsunami, which took place in river Irtysh in 1885. Here, we quote the book of travel notes by K. M. Stanyukovich, the well-known Russian writer on the Sea, 'To far lands' (Collection of works in 10 volumes, Vol. 1 – Moscow: Pravda, 1977):

The right sandy bank of the Irtysh, being constantly washed out, once in while caves in, and, then, as the Siberians say, the 'landslides', that fall from the height into the water with a crash and noise, happen to cause accidents and misfortunes. Such a misfortune occurred just three weeks before we passed there. About two hundred versts[1] *from the estuary of Irtysh we saw a schooner lying helplessly on its side in the sands. It had been passing one verst from the right bank, precisely when the bank caved in. Such a mass of earth falling together with century-old trees caused the water to shrink back from the bank, thus giving rise to agitation so strong that it capsized the flat-bottomed schooner, which most likely had no appropriate ballast, and threw it toward the left bank. The barge, towed by the schooner, withstood the wave and remained unharmed. Of the crew and passengers of the schooner several peoples died in the river, several were crippled. A day after the catastrophe, cries for help were heard on the 'Reytern', that was passing by. The steamboat stopped and took the people, asking help, on board.*

Studies of landslide tsunamis have a long history; however, until recently publications devoted to investigation of this phenomenon were quite rare. One of the first attempts at detailed investigation of tsunami waves, caused by underwater landslides, was made by N. L. Leonidova [Leonidova (1972)]. This investigation was based on earlier works of B. Gutenberg [Gutenberg (1939)] and R. Mitchell [Mitchel (1954)], however, precisely this work laid the foundation for modern ideas concerning the problem of landslide tsunamis. One must also mention the experimental study, well known to specialists, carried out by R. Wigel [Wiegel (1955)], which was devoted to investigating wave generation in a channel, when hard bodies of different shape were made to slide along the channel bottom.

The recent enhancement of interest in studies of landslide tsunamis was initiated by the catastrophic events in Papua New Guinea and Indonesia. The wave that demolished the coast of Papua New Guinea on July 17, 1998 was 15 m high. It was due to a relatively moderate earthquake of $M_w = 7.1$, accompanied by a local underwater landslide [Tappin et al. (1998); Heinrich et al. (2000); Imamura et al. (2001)]. The earthquake that took place on December 12, 1992, with magnitude $M_w = 7.7$ on island Flores (Indonesia) also gave rise to an underwater landslide and subsequent tsunamis of heights up to 26 m.

The landslide process is usually the result of a prolonged accumulation of sedimentary material during tens and hundreds of years. With time the sedimentary masses on slopes lose stability. Numerous factors can provoke a landslide [Ren et al. (1996); Kulikov et al. (1998)]:

- Sudden surge of river silt during a freshet
- Erosion of sedimentary layer on a steep underwater slope
- Coastal construction projects
- Prolonged rain, resulting in saturation of coastal land
- Uncovering of coast during pronounced low tide

Recently, the role of gas hydrates in provoking underwater landslides is also discussed [Parlaktuna (2003)]. Earthquakes, naturally, serve as most important causes of landslides and collapses. Volcanic eruptions, also, happen not to play the last part in initiating landslide processes and collapses.

Sedimentary masses, deposited on underwater slopes during many decades, accumulate huge potential energy. As they lose stability, they are capable of moving over

---

[1] 1 verst = 1.067 km.

the ocean bottom with high velocities, transferring part of the accumulated potential energy to tsunami waves. Precipitations, annually accumulated in some canyons, amounts to 10^6–10^9 m^3, while the bottom slopes often exceed 0.1. Precipitations on slopes of the ocean bottom often exhibit thixotropic properties, i.e. they are capable of becoming fluid in the case of sharp enhancement of the threshold pressure due to blows, shaking and vibrations. The unstable friable sedimentary material, possessing a high content of subcolloidal fractions, may, when losing stability, form dense suspension (muddy, turbidite) flows. Moving down a bottom slope with a velocity exceeding 10 m/s, such a flow leads to waves of the tsunami type being generated at the water surface, and it also severs underwater cables. The strong earthquake that destroyed the city of Messina on December 28, 1908, gave rise to a landslide or muddy flow, that severed seven underwater cables connecting continental Italy and Sicily.

It must be noted that well-known large underwater canyons were located inside the source areas of some strong tsunamis: the Lisbon canyon (tsunami of 1755), the Messina canyon (tsunamis of 1783 and 1908), the Kamchatka canyon (tsunamis of 1791, 1923, 1937) and others. N. L. Leonidova was, evidently, the first to note that most aftershocks of strong tsunamigenic earthquakes, even when approximately equal in force to the main shock, do not cause noticeable tsunami waves. Thus, the well-known Kanto earthquake, that destroyed Tokyo in 1923, gave rise in the Sagami bay to a tsunami wave 12 m high, while its aftershock, that originated in about the same place and with practically the same energy, was accompanied by waves less than 0.3 m high. Measurements showed that the volume of the landslide, provoked by the first earthquake, amounted to about $7 \cdot 10^{10}$ m^3, the average width of the flow was 2 km, its length 350 km, it power (thickness) 100 m, the flow velocity in the canyon was estimated to be 25 m/s. The potential energy of the landslide that covered a path from a depth of 1,500 m (the average position of the landslide body at the beginning of its movement) down to 7,000 m (the bottom of the deep-water depression) can be estimated to have been 10^{18} J. The energy of the tsunami waves generated was of the order of 10^{16} J.

Note that after the earthquake of December 26, 2004 ($M_w = 9.3$) that gave rise to a catastrophic tsunami with run-ups as high as 35 m, another strong earthquake took place in March 2005 ($M_w = 8.7$) approximately in the same region, but caused quite a weak tsunami with heights up to 2 m.

Much of the information on ground or underwater landslides, avalanches and cliff collapses indicate that the models, in which the movement of a landslide is considered just forward displacement of a solid body, not subject to deformation, are too simplistic and do not describe the character of these processes adequately. The idea of a landslide representing a flow of a heavy viscous fluid is much closer to the true nature of landslide dynamics. In the region of river estuaries the sedimentary silt masses usually consist of diluted fractions, which after the breakdown of an unstable sedimentary mass form a dense dirt (mud) flow, behaving like a viscous fluid.

In problems concerning landslide tsunami generation the notion of a landslide in the form of a flow of a heavy viscous fluid has started to be applied only quite recently. Such an approach was first proposed in [Jiang, LeBlond (1992), (1994)]. Numerical methods, based on this approach, were successfully applied in analysing landslide tsunamis in Nizza of 1979 [Assier-Rzadkiewicz et al. (2000)], in Skagway Harbour of 1994 [Fine et al. (1998); Rabinovich et al. (1999)], in Papua New Guinea of 1998 [Heinrich et al. (2000); Titov, Gonzalez (2001); Imamura et al. (2001)]. In these studies, it was shown that the notion of a landslide in the form of a flow of a heavy viscous fluid provides reasonable agreement with data of in situ observations.

The version of the model described here is based on [Jiang, LeBlond (1994)] and [Fine et al. (1998)]. We shall consider the horizontal scales of surface waves to significantly exceed the basin depth, and the thickness of the landslide to be much smaller than its width and length. In this case it is possible to apply the longwave (hydrostatic) approximation both in the case of water and in the case of the fluid forming the landslide. The Coriolis force is usually neglected in such problems.

The scheme of the model is presented in Fig. 4.1. The origin of the Cartesian reference system $0xyz$ is placed on the unperturbed free surface, the $0z$ axis is directed vertically upward. The upper layer of water has a density ρ_1, a free-surface displacement $\eta(x,y,t)$, **u** is the horizontal velocity vector with components x and y; t is time. The lower layer (the landslide body) has a density characteristic of sedimentary deposits, ρ_2, ν is the kinematical viscosity and **U** is the horizontal velocity vector of the fluid in the lower layer with components U and V. The slope of the ocean bottom and the landslide surface are considered small, so that the fluid can be considered to undergo purely horizontal movement. The landslide body is limited by the bottom surface $z = -h(x,y,t)$, while its upper surface is given by its thickness $D(x,y,t) = h_s(x,y) - h(x,y,t)$.

The main assumptions concerning landslide properties, substantiated in [Jiang, LeBlond(1992), (1994)], are adopted in the form:

1. A landslide consists of an incompressible viscous fluid, and the sea water is considered an incompressible liquid of zero viscosity.
2. The difference between the density of the landslide and the density of water must be large, $(\rho_2 - \rho_1) \geqslant 0.2 \text{ g/cm}^3$.
3. The flow of a viscous fluid is laminary and quasistationary. For describing the movement of a viscous fluid over an underwater slope it is, generally

Fig. 4.1 Geometry of the model: reference system and notation. The shaded part shows the body of a viscous landslide. Adapted from [Rabinovich et al. (2003)]

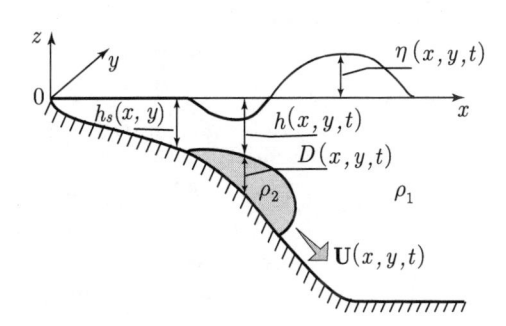

speaking, necessary to consider two modes—inertial and viscous (quasistationary) [Simpson (1987)]. After the landslide body has suddenly become free (from its initial state) the flow of the fluid forming the landslide undergoes transition from the inertial-mode state to the viscous-mode state, when the vertical profile of the flow has already been established. In the given model we assume the transition time to be negligible, and the flow to be constantly in the quasistationary-mode, adapting relatively slowly, in the process of movement, to the shape of the bottom relief.

4. In this model we neglect mixture effects on the landslide–water boundary. This means that no exchange of mass takes place between the flow of sedimentary material and the water.

Owing to the condition of adhesion the tangential velocity component at the ocean bottom must turn to zero. At the upper boundary of the landslide absence is assumed of tangential tensions, i.e. the normal component of the velocity gradient turns to zero. Under such conditions the horizontal velocity of the stationary flow of a fluid exhibits a parabolic vertical profile,

$$\mathbf{U}(x,y,z,t) = \mathbf{U}_m(x,y,t)(2\xi - \xi^2), \quad (4.1)$$

where $\xi = (z+h_s)/D$ is the dimensionless vertical coordinate.

The continuity and momentum balance equations for a viscous flow in a landslide, obtained from the equations of hydrodynamics by integration along the vertical coordinate with account of formula (4.1), have the following form:

$$\frac{\partial D}{\partial t} + \frac{2}{3}(\nabla \cdot D\mathbf{U}) = 0; \quad (4.2)$$

$$\frac{2}{3}\frac{\partial \mathbf{U}}{\partial t} - \frac{2}{15}\frac{\mathbf{U}}{D}\frac{\partial D}{\partial t} + \frac{8}{15}(\mathbf{U}\cdot\nabla)\mathbf{U} = -\frac{g}{\rho_2}\left((\rho_2-\rho_1)\nabla(D-h_s)+\rho_1\nabla\eta\right) - \frac{2\nu\mathbf{U}}{D^2}. \quad (4.3)$$

Here the condition must be fulfilled that the landslide flow across the boundary of the coastal line G always be zero, and that during its movement the landslide does not cross the external (free) boundary Γ.

The upper layer of the fluid (water) is described by non-linear equations of motion in the approximation of shallow water:

$$\frac{\partial(h+\eta)}{\partial t} + [\nabla \cdot (h+\eta)\mathbf{u}] = 0; \quad (4.4)$$

$$\frac{\partial \mathbf{u}}{\partial t} + (\mathbf{u}\cdot\nabla)\mathbf{u} = -g\nabla\eta. \quad (4.5)$$

Actually, the generation of surface waves by a moving landslide body is only due to the continuity equation (4.4). The waves further propagate under the condition that boundary conditions and the non-linear equation (4.5) be satisfied.

On the open external boundary of the region, Γ, the one-dimensional emission condition for outgoing waves is applied: $u_n = \eta(g/h)^{1/2}$, where u_n is the velocity

4.1 Tsunami Generation by Landslides

component, normal to the boundary Γ. Along the coastal line fulfilment is assumed of the noflow condition through a vertical wall: $u_n = 0$ on G.

As the initial conditions of the problem, the landslide and the water layer are assumed to be at rest at time moment $t = 0$, i.e. all the velocities and the displacement of the free water surface are equal to zero.

The set of equations (4.2–4.5) can be resolved by the explicit finite-difference method. Usually, the staggered leap-frog scheme in space and time is used [Imamura, Gica (1996)] in calculations.

To suppress the instability of the numerical scheme approximation is used of advective terms in the equations of motion according to the scheme with upstream differences [Roache (1976)]. To avoid generation of small-scale parasitic oscillations (network noise), it is expedient to set the time step (Δt) equal to 1.3 of the value given by the Courant condition.

We shall make use of a hypothetical event in the Malaspina strait (British Columbia, Canada) [Rabinovich et al. (2003)] as an example in considering the peculiarities of landslide tsunami formation. Numerical simulation of this tsunami is performed applying the mathematical model, described above. Studies of hypothetical tsunamis are important for estimation of the risk of tsunami hazard. For the demonstration of research methods of this kind we shall first give a detailed description of the primary geophysical (geomorphological) information and, then, present the results of simulation.

The Malaspina strait (Fig. 4.2), located between the continental coast of British Columbia and Texada island, is about 50 km long and 5 km wide. Its depth along the axis of the strait varies between 300 and 375 m. In the central part of the strait there is a thick ($\sim$100 m) sedimentary layer, mainly consisting of silt carried out from the estuary of Fraser River.

In 1946 an earthquake in the central part of Vancouver island, British Columbia, gave rise to a series of landslides and slumps in the coastal area of Malaspina strait.

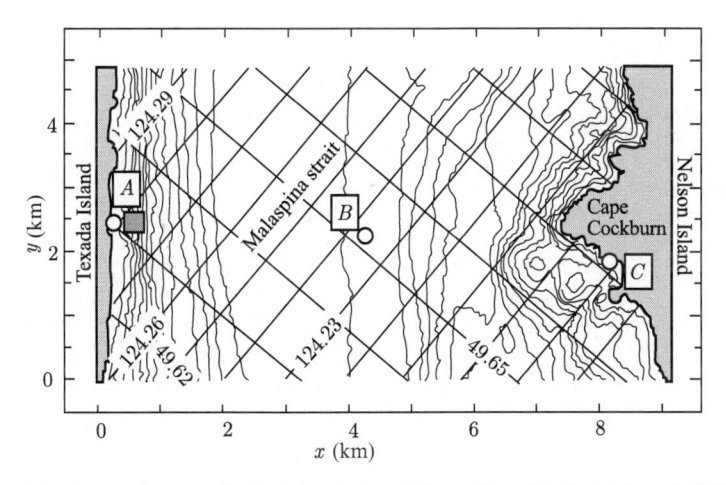

Fig. 4.2 Malaspina strait: map of calculation region. Adapted from [Rabinovich et al. (2003)]

In the northern part of the strait banks were observed to crumble and cave in, underwater cable lines were damaged.

Geophysical studies, performed by the Geological Survey of Canada, revealed the existence of two separate deposition zones of bottom sediments, located between isobaths 30 and 120 m. Identification of these areas was carried out with the aid of echo sounding gear for lateral observation and equipment for seismic profiling of high resolution. Special underwater video shooting, performed in 1996, revealed that lower down the slope there are a number of blocks of well-consolidated sediments several meters thick, which obviously resulted from their breaking away from the main mass and falling down the slope under the force of gravity. The main block of sediments has corresponding areas with a very steep edges left after partial collapse of the sedimentary mass. The lower boundary of this sedimentary layer exhibits a very steep inclination everywhere—practically like a precipice.

The northern zone of the sedimentary cover is up to 38 m thick, and the inclination of the inner boundary toward the sea is, on the average, 7.5°. The layer above it is inclined at approximately 16° relative to the boundary of the base layer and extends about 400 m along the slope, exhibiting a thickness of about 300 m.

The hypothetical scenario of tsunami generation assumes that the earthquake causes all the mass of sediments, accumulated in the northern zone of the sedimentary cover, to break off from and to slide down the steep slope of the basalt boundary of the bottom. Owing to a lack of geotechnical data on the properties of sediments, simulation of the movement of the landslide is based on a broad set of parameters of the material. The possibility of the southern and northern sedimentary layers collapsing at the same time is not considered. Such a joint scenario would, naturally, lead to the generation of waves of higher amplitudes.

Calculations were performed on a difference mesh with 365×197 nodes and steps $\Delta x = \Delta y = 25$ m. In the initial state, the landslide body was considered to have a rectangular shape with a parabolic profile over the thickness in both directions. Calculations were performed for the following landslide parameters:

Volume:	1,250,000 m^3
Width:	200 m
Average width	30 m
Coordinates of centre:	49°37.94′ N, 124°16.80′ W
Average depth:	80 m
Density (ρ_2):	2.0 g·cm^{-3}
Kinematic viscosity (ν):	0.01 m^2·s^{-1}

Figure 4.3 shows fragments of numerical calculations of the movements of the landslide and of tsunami waves in the strait. Unlike a solid-state body, retaining its size and shape, the viscous landslide moves along the slope, spreading out and assuming the form of a sickle. Displacement of the landslide takes place mostly along the normal to the west bank. Movement of the landslide gives rise to radially diverging surface waves. The leading wave (positive) moves towards the continent, while the negative wave (depression) moves in the opposite direction towards Texada island. The leading wave crosses the Malaspina strait and

4.1 Tsunami Generation by Landslides   161

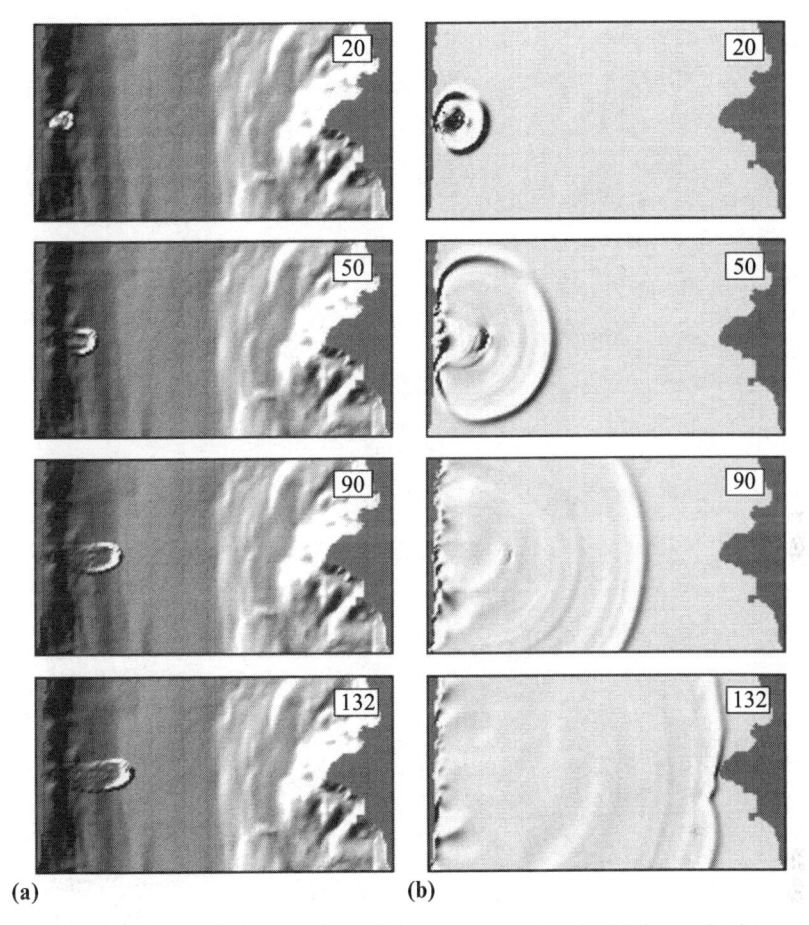

Fig. 4.3 Results of numerical simulation of the movement of a landslide on the bottom of Malaspina strait (**a**) and of the resultant surface waves (**b**) for times 20, 50, 90 and 132 s after the landslide collapses. Adapted from [Rabinovich et al. (2003)]

reaches Cape Cockburn on Nelson island in approximately 132 s after the slilde starts moving. As a result, reflected waves form, and the general picture of roughness in the Malaspina strait becomes complex, reminding standing oscillations. The waves leaving through the open boundaries of the strait leads to rapid dampening of the amplitudes of level oscillations.

Figure 4.4 presents examples of calculations of level variations at points A, B and C, the locations of which are indicated in Fig. 4.2. The maximum wave amplitude is observed at site A, the closest to the landslide zone, and the minimum amplitude turns out to be in the middle of the strait (point B). The tsunami starts with negative phase at point A and positive phase at points B and C.

The amplitude distributions of a tsunami wave at its crest and its depression along the west and east coasts of the strait are shown in Fig. 4.5. According to these calculations the maximum level reduction (down to -5 m) is observed in the immediate

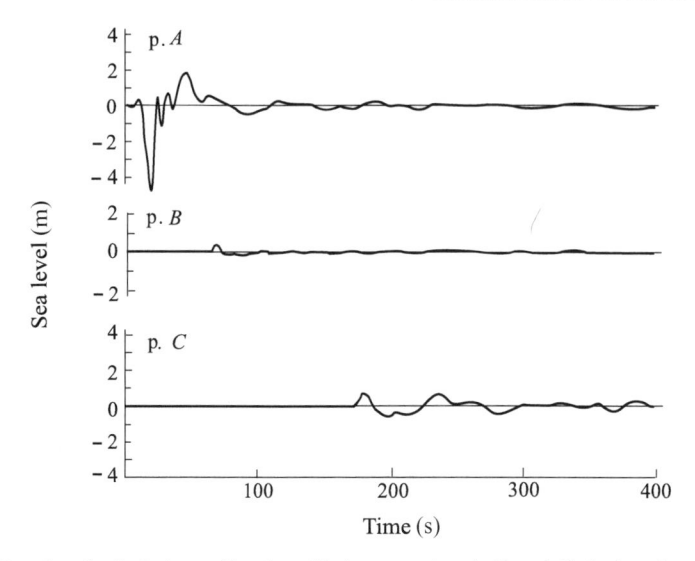

Fig. 4.4 Results of calculations of level oscillations at points A, B and C, the locations of which are indicated in Fig. 4.2. Adapted from [Rabinovich et al. (2003)]

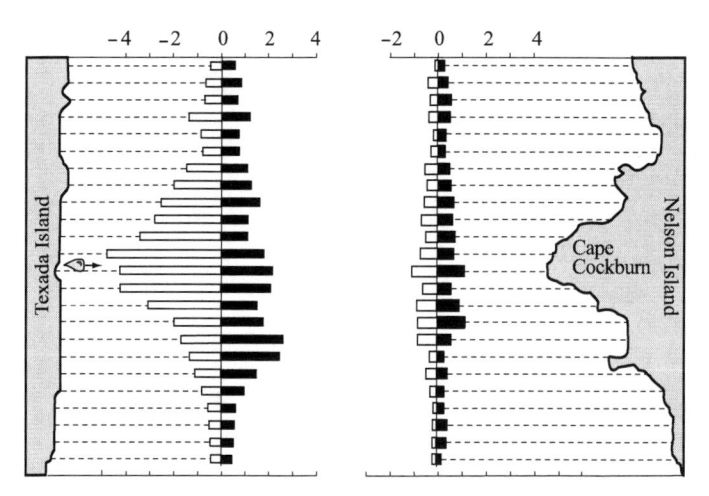

Fig. 4.5 Maximum level deviations at the crest and in the depression of a tsunami wave, calculated for the west and east coasts of Malaspina strait. Adapted from [Rabinovich et al. (2003)]

vicinity of the source. Toward the North and South the wave height rapidly dies out. The maximum at the wave crest on the west boundary is smaller, than in the depression, and amounts to +2.7 m. At the opposite west coast these maxima are significantly smaller (by approximately ±1 m).

Numerical calculations were performed for a wide range of density values of the landslide material ($1.6 \leqslant \rho_2 \leqslant 2.2 \, \text{g/cm}^3$), its viscosity coefficient ($10^{-3} \leqslant \nu \leqslant 1 \, \text{m}^2/\text{s}$) and the initial positions of the landslide on the slope. It turned out that the

4.1 Tsunami Generation by Landslides

results of calculations are least sensitive to variations in the landslide viscosity. An increase of the viscosity from 0.001 up to 0.1 m$^2\cdot$s^{-1} leads to a change in the wave amplitudes by merely 1%. Their sensitivity to changes in the density of the material turned out to be much higher. Enhancement of the density ρ_2 from 1.6 up to 2.0 g/cm^3 leads to an increase of the tsunami amplitude by 20%. The most important characteristic of a landslide, affecting the formation of tsunami waves, turned out to be its initial position on the slope. For example, displacement of the centre of a landslide by 100 m closer to the coast (the depth over the landslide amounts to about 30 m, here) results in the amplitudes of the tsunami waves increasing by 85%. Displacement of the landslide centre towards the sea (a change of depth from 80 down to 118 m) reduces wave heights by 70%. Additional test calculations have shown that the amplitudes of surface waves generated are approximately proportional to the volume of the landslide. Summing up these results, one can conclude that the energy of a landslide tsunami depends, first of all, on the potential energy of the landslide (its density, location on the slope and volume). The viscosity of the landslide body causes no noticeable dissipation of the landslide energy, which leaves a certain freedom in the choice of viscosity coefficient.

The character of interaction between a landslide body and surface waves depends on the relationship between the motion velocities of the landslide and of the surface waves (see Sect. 2.3.3). Actually, the process of wave generation by a landslide is similar to the formation of waves accompanying a ship, when it moves. The effect of resonance excitation of the accompanying wave is well known in the case, when a vessel moves towards shallow water with a velocity $c = \sqrt{gh}$, where h is the depth of the liquid. In this case the wave resistance increases sharply, and the wave amplitude starts to grow. In this manner, also, a landslide moving on the sea bottom gives rise to perturbations of the water surface, which remind the wave accompanying a ship. Here, a measure of 'closeness' to resonance conditions can be the Froud number $\mathrm{Fr} = U/c$, where U is the velocity of the landslide, and $c = \sqrt{gh(x,y)}$ is the velocity of gravitational waves on the variable relief of the bottom. The value $\mathrm{Fr} = 1$ corresponds to a resonance. For a landslide, representing a solid-state body, not subject to deformation, the notion of 'velocity of motion' is unambiguous. But in the case of the flow of a viscous fluid, the particles of which move with differing velocities, it is not simple to introduce the concept of 'landslide velocity'. Owing to the condition of 'adhesion' to the bottom, the velocity in the lower part of the landslide is much smaller, than at its surface. Moreover, while the landslide body moves, it 'spreads', essentially changing its form. We shall estimate the velocity of a landslide, U_f, as the velocity, with which its *front* moves (the corresponding Froud number $\mathrm{Fr} = U_f/c$).

The maximum velocity $U_{f\max}$, obtained in numerical calculations of a landslide in the Malaspina strait, amounted to 19.5 m/s at a distance of about 1 km from the coast of Texada island. The plot in Fig. 4.6a shows the variations in velocity for a gravitational wave, $c = \sqrt{gh(x)}$, and for the motion of a viscous slide front along the horizontal coordinate (across the strait). Figure 4.6b shows the dependence of the corresponding Froud number. The maximum value of the Froud number $\mathrm{Fr}_{\max} = 0.46$ is achieved at a distance of 0.85 km from the coast.

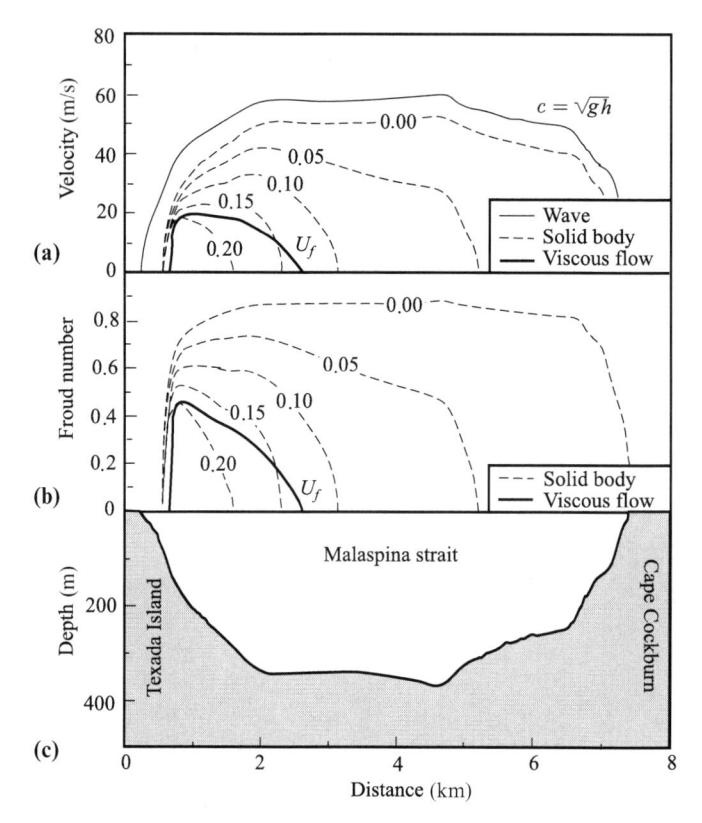

Fig. 4.6 Velocities of long gravitational waves, of a solid slide and of the front of a viscous slide in the Malaspina strait. Velocities of the solid slide are calculated for various friction coefficients (from 0 to 0.2) (**a**). Froud numbers, corresponding to these velocities (**b**). Depth profile in the strait (**c**). Adapted from [Rabinovich et al. (2003)]

For comparison, calculations were performed for the movement of a landslide in the form of a solid body sliding down the slope under the influence of the forces of gravity and of friction (between the landslide and the bottom). The friction coefficient k was set within the range from 0 to 0.2. The results of calculations are presented in Fig. 4.6.

The motion dynamics of a solid body on an inclined plane under the influence of the force of gravity with account of friction is such that there exists a 'critical' inclination of the bottom, ψ, at which the down-pulling force is balanced by the force of friction, and the landslide moves without acceleration, $k = \tan \psi$. It can be considered that the 'break-off' and subsequent sliding down of the landslide body takes place precisely, when the 'holding' forces (of friction) weaken (for instance, owing to erosion at the edge of the sedimentary layer) so much as to allow the down-pulling force to start to exceed the force of friction. The characteristic slope of the bottom near Texada island is $\psi \approx 6°$, which corresponds to $k = 0.1$. For this value, the maximum velocity of the solid landslide amounts to the value $U_{max}^k = 33.1$ m/s at a distance $x = 1.86$ km from the coast, and here the Froud number $\text{Fr}_{\max} = 0.61$ turns out to reach its maximum at a distance $x = 0.95$ km, i.e. much closer to the coast. From the figure it is seen that the Froud number rapidly reaches its maximum along the initial length of sliding down and gradually decreases as the landslide reaches the gently sloping bottom. Ultimately, the landslide stops, when its potential energy has already been spent on friction and wave generation. The 'path length' of a solid landslide depends directly on the friction coefficient; thus, for $k = 0.05$, 0.10, 0.15 and 0.20 the 'path length' amounts, respectively, to $x_s = 5.19, 3.13, 2.29$ and 1.58 km.

When $k < 0.15$, the velocity of motion and the Froude number for a solid landslide exceed the velocity of motion of the front of a viscous flow everywhere. Here, the 'path length' of the viscous landslide amounts to 1.6 km.

We draw attention to the velocity of motion of underwater landslides being, as a rule, smaller, than the velocity of long gravitational waves (i.e. $\text{Fr} < 1$). This fact follows from elementary physical arguments. If the motion of a landslide is considered without account of the wave resistance and of friction ($k = 0$), then its velocity is determined by the formula

$$U = \sqrt{2g \frac{\rho_2 - \rho_1}{\rho_2} \Delta h}, \qquad (4.6)$$

where Δh is the change in vertical position of the centre of mass of the landslide. Suppose that, going down the slope, the landslide reaches a certain depth h. Clearly, for an underwater landslide the inequality $\Delta h < h$ is always satisfied. Comparing the velocity of the landslide, determined by formula (4.6), and the velocity of long waves, $\sqrt{gh}$, it is not difficult to arrive at the conclusion that equality of these two quantities is possible, only if $\rho_2 > 2.0$ g/cm^3, i.e. for well-consolidated sediments and rock. If the force of friction is taken into account, the required density of the landslide body will be even greater. Thus, a 'resonance' ($\text{Fr} = 1$) is possible only for landslides, consisting of very dense materials, or when the landslide enters the water with a certain initial velocity. In the latter case, one can speak of both subaerial (partially submerged) landslides and of landslides sliding into water from a 'dry' coastal slope.

4.2 Tsunami Excitation Related to Volcanic Eruptions

Explosions of volcanic islands (collapses of calderas), explosive (explosion-like) eruptions of underwater volcanoes and pyroclastic flows landing in water, all these phenomena are capable of giving rise to waves, which in their destructive strength are in no way inferior to tsunamis of seismotectonic origin. At present, 66 tsunamis of volcanic origin are known merely in the Pacific region, and in 10 of the events wave heights amounted to 10 m and more (up to 55 m).

One of the most striking historical examples of volcanogenic tsunamis is represented by the waves caused by the activity of the Krakatau volcano in August of 1883 [Choi et al. (2003)]. On August 26, at 17:00, local time, a series of loud explosions took place, and the volcano ejected an ash cloud to a height of up to 25 km. A small tsunami 1–2 m high formed. In the morning of August 27 three colossal explosions took place. The first explosion (at 5 o'clock 28 min) destroyed mountain Perboewatan on Krakatau island, which was 130 m high. The caldera produced was immediately filled up with sea water, leading to the generation of a small tsunami. At 6.36 mountain Danan, which was 500 m high, exploded and collapsed, which gave rise to a tsunami wave up to 10 m high. The main (third) explosion took place at 9.58. It literally blew apart what remained of Krakatau island (Rakata island). The volcano threw out 9–10 km^3 of tephra (solid material) and 18–21 km^3 of pyroclastic deposits that were distributed over an area of about 300 km^2 with an average thickness of 40 m. Ash covered a territory of approximately $2.8 \cdot 10^6$ km^2. In the place of the island there emerged a caldera 6 km in diameter and 270 m deep. The third explosion came with the most strongest noise ever heard by mankind. Air-blasts circumvented the globe seven times. The energy released during the main eruption of the Krakatau volcano amounted to $8.4 \cdot 10^{17}$ J. The waves that resulted from the third most strong explosion were 42 m high; they arrived 5 km inland. The average height of waves on the coast of the Sunda straits (separating islands Java and Sumatra) was about 15 m. At least 36,000 people died. About 300 villages were destroyed. The tsunami caused by explosion of the Krakatau volcano was noticed everywhere. Waves were recorded by many mareographs not only in the Indian Ocean, but also in the Pacific and Atlantic. Far from the coasts of Indonesia wave amplitudes were relatively small.

On the basis of the diameter and depth of the caldera formed as a result of the explosion of the Krakatau volcano, it is not difficult to estimate the volume of the initial perturbation—the 'local depression' of the ocean level. It amounts to $\sim$7 km^3. It is interesting to note that this volume approximately corresponds to the volume of water ousted by ocean bottom deformations in the case of strong earthquakes (100×100 km$^2 \times 1$ m $= 10$ km^3). The potential energy corresponding to the initial perturbation that can be estimated by formula (2.2) amounts to $\sim 6 \times 10^{15}$ J, which is of the order of 1% of the eruption energy.

Another frequently discussed event took place in the Bronze Age (around 35 centuries ago) in the Aegean sea. There exists a hypothesis that explosion of the volcanic Thera island (the Santorini volcano) and the resulting tsunami caused the death of mythical Atlantis, while the explosive eruption itself contributed to destruction of the Cretan-Mycenaean culture. At any rate, geological traces of this tsunami have been found along the coastlines of Greece and Turkey [Minoura et al. (2003)].

Of the 933 volcanoes, known to be active on Earth, 195 are underwater volcanoes. Numerous works are devoted to the study of tsunami generation, related to volcanic eruptions [Basov et al. (1981); Egorov (1990), (2007); Pelinovsky (1996); Waythomas, Neal (1998); Belousov et al., (2000); Tinti et al., (2003); Ward, Day (2001), (2003); Kurkin, Pelinovsky (2004); Mader, Gittings (2006)]. The main physical mechanisms of volcanogenic tsunami generation comprise the following:

4.2 Tsunami Excitation Related to Volcanic Eruptions

1. The discharge into water of a large volume of matter (from slow lava flows to explosive eruptions)
2. The collapse of a caldera (explosion of a volcanic island)
3. Pyroclastic flows, landslides etc.
4. Volcanic earthquakes

In the case of underwater volcanoes the first two mechanisms are prevalent. The third (landslide) mechanism may be more peculiar to volcanoes on coasts, although the possibility cannot be excluded of underwater landslides and mudslides initiated by an underwater eruption.

In certain cases, volcanic eruptions can provocate enormous collapses. Thus, for example, in [Ward, Day (2001)] the possibility is indicated for part of the La Palma island (Canary islands) to collapse during the next eruption of the volcano, located there. Geological estimates reveal that the volume of such a collapse may amount to 500 km^3. A tsunami wave caused by such a colossal collapse would be capable of crossing the Atlantic Ocean and reaching the coasts of America with a height exceeding 10 m.

In this section we shall only deal with those original mechanisms of tsunami formation that are peculiar precisely to volcanic eruptions. Thus, here we shall not consider wave generation by volcanic earthquakes, as well as by volcanogenic landslides.

We shall first dwell upon certain peculiarities of tsunami formation in the case of a caldera collapsing and being subsequently filled up with water. If one speaks of an underwater volcano, then the description of the waves generated fully reduces to the problem of tsunami generation by deformation of the ocean bottom, which has been investigated in detail in Chapters 2 and 3. Truly, in the case of a collapsing caldera, the amplitude of the 'bottom deformation', $\eta_0 \sim 10^2$–10^3 m, and the horizontal dimension of the deformation area, $D \sim 10^3$–10^4 m, may turn out to be comparable to the ocean depth. Note that underwater volcanoes may be located both at small (shelf) and at large (abyssal) depths. If the eruption of a volcanic island takes place, then the water filling up the caldera, like the waters surrounding the island, are evidently characterized by shelf depths ($\sim 10^2$ m).

A suddenly generated caldera (in total absence of obstacles to water entering it) will be filled with water in a time $T \sim D/\sqrt{gH}$, where D is the diameter of the caldera and H is its characteristic depth. Taking advantage of the aforementioned ranges of these parameters, we obtain that the quantity T varies within the limits of 30–300 s. In the case of the most probable development of events, when obstacles to the arrival of water do exist, the time the caldera will take to fill up may increase significantly.

Thus, the caldera collapsing results in a source of waves (a flow of mass) with a characteristic action time of 10^2–10^3 s. The volume of water, taking part in the process, can be estimated as $V \sim \pi D^2 H/4 \sim 0.3$–30 km^3. The obtained characteristics of the source are quite in agreement with the values for a seismotectonic tsunami source. Caldera collapses are capable of generating powerful long-period tsunami waves. Simulation of tsunami wave propagation, due to activity of the Krakatau volcano in 1883, performed within the framework of longwave theory [Choi et al. (2003)], demonstrated good agreement between the model and observed arrival times of waves at various points of the World Ocean distant from the source. This result testifies that the wave front indeed propagated with a velocity close to the velocity of long waves, $\sqrt{gH}$, i.e. the waves were sufficiently long. Note that, in the case of abyssal depths, strongly dispersive waves are formed in the area of the caldera collapse, which rapidly die out with the distance from the source.

Further, we shall deal with the most 'original' of the tsunami generation mechanisms peculiar to volcanoes. What is meant is the release of a large volume of matter in the case of an underwater eruption. First, consider the case of a slow outflow of matter. An adequate model for describing the tsunami generation process will consist of a set of hydrodynamic equations with a source of mass (volume). Assume eruption of the underwater volcano to proceed slowly: a volume V_0 is released in time τ from the crater. On the basis of general physical arguments it is not difficult to estimate the amplitude and energy of surface gravitational waves, caused by such an underwater 'eruption'. Consider the ocean depth H to be fixed, and the area S of the crater to be small, and the condition $\sqrt{S} \ll H$ to be satisfied. Of course, the model of an ocean of constant depth is limited (the crater of the volcano is usually situated on top of a cone), but for presenting general physical regularities of the process such a simplified model is quite applicable.

The volume thrown out will oust an identical volume of water. This volume will spread over the area of a circle of radius, equal to the distance, which a long wave has time to cover during the eruption time $r = \tau\sqrt{gH}$. As a result, we have the amplitude of the initial water elevation

$$\xi_0 = \frac{V_0}{\pi \tau^2 gH}. \tag{4.7}$$

The potential energy of such an initial elevation, calculated by formula (2.2),

$$W_p = \frac{\rho V_0^2}{2\pi \tau^2 H}. \tag{4.8}$$

From formulae (4.7) and (4.8) the amplitude and especially the energy are seen to increase with the rate V_0/τ, at which volcanogenic material is released from the crater. An increase of the ocean depth reduces the efficiency of tsunami excitation.

For more accurate description of the waves caused by a flow of material from a hole of radius R in the ocean bottom, it is possible to apply the general solution of the problem, (2.43), (2.30), (2.31), obtained in Sect. 2.2.2 within the framework of linear potential wave theory. Formulation of the axially symmetric problem is schematically presented in Fig. 4.7. In the case dealt with the boundary condition on the bottom, (2.31), assumes the following form:

$$\frac{\partial F}{\partial z} = w(r,t) = w_0\big(1 - \theta(r-R)\big)\big(\theta(t) - \theta(t-\tau)\big), \qquad z = -H, \tag{4.9}$$

where $w_0 = V_0/(\tau \pi R^2)$ is the outflow velocity of material from the crater. Displacement of the free surface, caused by the flow, released from the ocean bottom, is determined by formula

4.2 Tsunami Excitation Related to Volcanic Eruptions

Fig. 4.7 Mathematical formulation of the problem of tsunami generation by an underwater eruption

Fig. 4.8 Perturbation of free surface caused by underwater eruption. The calculation is performed at the time moment, when the eruption finishes, $t = 10\sqrt{H/g}$, for various ratios of the crater radius and the ocean depth, R/H (indicated in the figure). The x-axis is normalized to the quantity $r_0 = \tau\sqrt{gH}$, the y-axis to $\xi_0 = V_0/(\pi\tau^2 gH)$

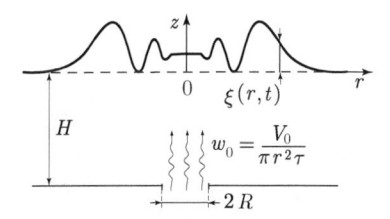

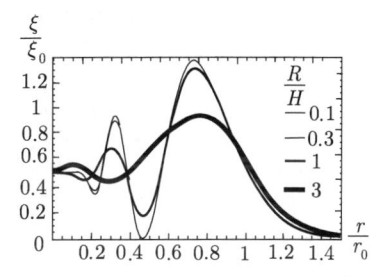

$$\xi(r,t) = \theta(t)\zeta(r,t) - \theta(t-\tau)\zeta(r,t-\tau),$$

$$\zeta(r,t) = \frac{V_0}{\pi R\tau}\int_0^\infty dk \frac{J_0(rk)J_1(Rk)\sin\left(t(gk\tanh(kH))^{1/2}\right)}{\cosh(kH)(gk\tanh(kH))^{1/2}}. \quad (4.10)$$

The form of the free-surface displacement at the moment, when the eruption finishes ($t = \tau$), calculated by formula (4.10) for $\tau = 10\sqrt{H/g}$ and various radii of the crater, $R/H = 0.1, 0.3, 1$ and 3 is shown in Fig. 4.8. The curves are presented in dimensionless coordinates. The x-axis is normalized to the distance covered by a long wave during eruption time $\tau\sqrt{gH}$, the y-axis is normalized to the free-surface displacement, determined by estimation formula (4.7). From the figure it is seen, that the form and amplitude of the free-surface perturbation depend little on the radius of the crater, when $R/H < 1$. Moreover, the quantity ξ_0, determined by formula (4.7), is indeed seen to represent a good estimate for the surface displacement amplitude.

Note that application of the theory of incompressible liquids imposes natural limits on the outflow velocity of material from the crater, $w_0 < c$, where c is the velocity of sound in water, and on the relationship between the eruption duration and the ocean depth, $\tau > 4H/c$.

For estimates we take advantage of the modest, as compared to the 1883 event (Krakatau), eruption of an underwater volcano, located at a depth $H = 1,000$ m. Let the release of material amount to $V_0 = 1$ km^3, and the eruption duration $\tau = 100$ s. For the indicated duration of the process the perturbation (elevation) radius of the water surface amounts to $r \approx 10$ km. Its height, calculated in accordance with formula (4.7), amounts to the significant value $\xi_0 \approx 3$ m. And the potential energy, calculated by (4.8) is $W_p = 1.7 \cdot 10^{13}$ J. A tsunami, generated by such an initial elevation will evidently represent a serious threat.

In the model, described above, we assumed the eruption to be a slow process. This provided grounds for applying linear theory and considering water to be incompressible. But the eruption of an underwater volcano may exhibit an explosive character. In such a case the products of eruption form a gaseous bubble in the water, which contains high-temperature gases and water vapour at high pressures. The expansion and floating-up of the bubble leads to the formation of a cupola or plume—an elevation on the water surface. An analogue of this process is the formation of a plume in the case of an underwater explosion. It must be stressed that the formation of a gaseous bubble at large depths is not always possible, owing to the colossal hydrostatical pressure.

In this case description of the wave generation process is a difficult task. But one can select an equivalent source and use it as the initial perturbation in calculating tsunami waves. Reasonable agreement with reality (explosions in water for energies within the range of $2 \cdot 10^6 - 3 \cdot 10^{10}$ J) is achieved for the following form of the initial displacement of the water surface [Kurkin, Pelinovsky (2004)]:

$$\xi_0(r) = H_S \left(2\left(\frac{r}{R_s}\right)^2 - 1 \right) (1 - \theta(r - R_s)), \qquad (4.11)$$

where R_s is the source radius, H_S is the amplitude of the water level displacement at the source. Both parameters, characterizing the source, can be estimated via the equivalent energy of the explosion (or volcanic eruption) [Le Mehaute, Wang (1996)].

In the case of an ocean of constant depth H, evolution of the initial free-surface perturbation, exhibiting radial symmetry, is described by the following expression (see general theory in Sect. 2.2.2):

$$\xi(r,t) = \int_0^\infty k dk A(k) J_0(kr) \cos(\omega(k)t), \qquad (4.12)$$

$$A(k) = \int_0^\infty r dr \, \xi_0(r) J_0(kr), \qquad (4.13)$$

where $\xi_0(r)$ is a function describing the form of the initial perturbation, J_n is the Bessel function of the first kind of nth order. The relationship between the cyclic frequency and the wave number is determined by the known dispersion relation for gravitational waves on water, $\omega^2 = gk \tanh(kH)$. For an initial elevation, exhibiting the form, determined by formula (4.11), we have

$$A(k) = -\frac{H_S R_S J_3(kR)}{k}. \qquad (4.14)$$

For large times we represent the integral in expression (4.12) with the aid of the stationary-phase method,

4.3 Meteotsunamis

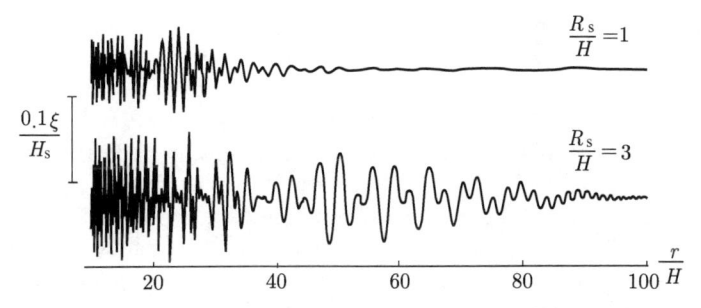

Fig. 4.9 Profile of waves excited by underwater eruption. The calculation is performed for $t = 100\sqrt{H/g}$ for two different ratios of the initial perturbation radius and the ocean depth, R_S/H (the values are indicated in the figure)

$$\xi(r,t) \cong \sqrt{\frac{2\pi}{t|S''(k_0)|}} k_0 A(k_0) J_0(k_0 r) \cos\left(\omega(k_0)t - \frac{\pi}{4}\right), \qquad (4.15)$$

where k_0 is the extremum of function $S(k) = \sqrt{gk\tanh(kH)} - kx/t$, which exists under the condition $x < t\sqrt{gH}$.

As an example, Fig. 4.9 shows the profiles of waves generated as a result of evolution of the initial perturbation (4.11). The calculation is performed by formula (4.15) for the moment of time $t = 100\sqrt{H/g}$ for two different radii of the initial perturbation. The waves are seen to be strongly dispersive, therefore, the propagation velocity of the wave packet depends strongly on the radius of the initial perturbation. The existence of such a dependence leads to an interesting effect (dispersion amplification), which was first noted in [Mirchina, Pelinovsky (1987)]. Consider two or more successively amplified eruptions taking place. The radius R_S of the perturbation created on the water surface increases with the strength (energy) of the eruption. In accordance with the growth of the radius the propagation velocity of the wave packet also increases. Consequently, during the process of wave propagation the superposition is possible of wave packets from different eruptions, which may lead to significant amplification of the tsunami amplitude. We recall that in August 1883, three eruptions occurred of the Krakatau volcano, each of which was stronger than the preceding one.

4.3 Meteotsunamis

Long waves, similar in characteristics to tsunami waves of seismotectonic origin, can be formed as a result of the influence of various atmospheric processes upon the water layer. These waves are conventionally termed anemobaric waves or meteotsunamis. The term 'meteotsunami' was apparently coined in [Nomitsu (1935)].

The main causes for meteotsunamis to arise are moving inhomogeneities of the atmospheric pressure or the tension of wind friction. We at once note that, unlike other tsunami generation mechanisms, in this case an important role is played by resonance effects, revealed, when the propagation velocity of atmospheric perturbations and their period turn out to be close to the velocity of long waves and to the period of eigen oscillations of the acquatorium, respectively.

Like tsunamis of seismotectonic origin, meteotsunamis represent quite a rare phenomenon. Similarly to tsunamis not being excited by each individual earthquake, not every cyclone, atmospheric front, train of internal gravitational waves or other atmospheric perturbation leads to the formation of meteotsunamis. A great number of examples are known, when quite strong atmospheric perturbations were not accompanied by any generation of long waves. Nevertheless, only in the Pacific region 36 events have been registered, which are classified as tsunamis of meteorological origin.

The extent, to which the parameters of tsunamis due to meteorological and seismotectonic causes are identical, is such that in a number of cases it is difficult to determine the actual cause of wave generation. Thus, for example, the group of long waves about 60 cm high with a period of 24–60 min, registered at the coast of South Africa on May 11, 1981, was initially identified as a seismotectonic tsunami and described in the September issue of 'Tsunami Newsletter'. Later, these waves were classified as a meteotsunami caused by a deep cyclone and atmospheric waves related to it.

From general arguments it is clear that an intensification of atmospheric processes, for example, in the case of tropical cyclones should lead to perturbations of the water layer and to the generation of long waves. The passage of cyclones is nearly always accompanied by significant oscillations of the atmospheric pressure, enhancement of the wind, development of storm agitation. Extreme values of pressure and wind velocity in tropical cyclones reach 870 hPa (cyclone 'TIP', October 1979) and 82 m/s (cyclone 'LINDA', September 1997), respectively. Part of the energy of such intense atmospheric processes, doubtless, must be transformed into the energy of long waves. But an analysis of synchronous measurements of ocean level oscillations and of atmospheric pressure fluctuations reveal that the direct relationship between these processes, with the exception of individual cases, is not essential [Munk (1962); Kovalev et al. (1991); Rabinovich, Monserrat (1996)].

At the same time, there exist numerous examples of the observation of long waves, the formation of which is unambiguously related to atmospheric processes. Thus, for example, in [Bondarenko, Bychkov (1983)] a description is given of the generation of long waves with a period of about 23 min, caused by internal gravitational waves with the same period that propagated over the Caspian Sea in the region of Svinoi island. Several cases are known of catastrophic waves arising on the Great lakes [Donn, Ewing (1956)]. On May 5, 1952, June 26 and July 6, 1954 sharp jumps in the atmospheric pressure that propagated with velocities of 20–40 m/s led to the formation of strong long waves, which caused significant destruction on the coast and even the death of people.

The anomalous character of seiche oscillations in Nagasaki Bay (Kyushu island) is renown [Rabinovich (1993)]. Oscillations of amplitudes ~ 0.5 m and periods of about 30 min in this bay represent quite a typical phenomenon. It is known by the local term 'abiki'. In a number of cases abiki waves are capable of achieving significant amplitudes. Thus, for example, on March 31, 1979 the maximum height of waves that caused significant damage and the death of three persons, amounted to 4.78 m. In [Hibiya, Kajiura (1982)] this event was shown to be caused by the passage of a jump of atmospheric pressure from 2 to 6 hPa over the western part of the East-Chinese Sea. The propagation velocity of the pressure jump was about 30 m/s.

Characteristic depths of the sea in between the region where the perturbation originates and Kyushu island lie between 50 and 150 m. The propagation velocities of long waves, corresponding to them vary within the limits 22–28 m/s, which is close to the propagation velocity of the atmospheric perturbation. The period of the waves that arrived in Nagasaki Bay also turned out to be close to the period of eigen oscillations. As a result of the double resonance the height of waves increased by more than a factor of 100. Thus, a jump in pressure of only several hectopascal (hPa) caused the formation of abiki waves in Nagasaki Bay several meters high.

Another well-known example consists in seiche oscillations with periods from several minutes up to several tens of minutes, which are observed regularly in summertime off the south-east coast of Spain in the region of the Balearic islands. From this point of view, the Ciutadella Bay, located in the north-west part of Menorca island, is the most renown. The bay is of the order of 1 km long, about 90 m wide with a practically flat bottom at a depth of 5 m. In certain cases seiches with typical periods of about 10 min reach heights of 4 m, here, leading to serious damage to ships and coastal structures. This calamity has received the local name 'rissaga' [Monserrat et al. (1991)]. An analysis of synchronous measurements of the atmospheric pressure and of long waves, performed in [Rabinovich, Monserrat (1996)], has permitted to reveal a series of cases, when strong level oscillations were caused by perturbations of the atmospheric pressure. One of such events is the formation of seiche oscillations in Ciutadella Bay with an amplitude of 0.87 m. Formation of the waves resulted from the passage of a train of intense internal gravitational waves in the atmosphere with an amplitude of about 2 h Pa and propagation velocity of about 30 m/s. The period of atmospheric waves was of the order of 1 h, while their length over 100 km, which essentially exceeds the period of proper oscillations of the bay and its dimensions. These differences exclude a possible resonance response of the bay. Most likely, formation of the long waves was a result of resonance effects in open sea, after which they approached the coast and caused strong oscillations in the bay. A fact, favouring this assumption, consists in that the velocity of the long waves on the external shelf with depths $\sim$100 m were in good accordance with the propagation velocity of atmospheric perturbations.

The meteotsunami phenomenon has much in common with so-called storm surges. In the monograph of [Murty (1984)] storm surges are defined as sea-level oscillations in the coastal zone or within internal basins with periods from several minutes up to several days, and caused by atmospheric influence. Note that this definition excludes wind waves and choppy sea, since they are characterized by periods smaller than a minute. Actually, the terms meteotsunami and storm surge denote phenomena of the same scope, which are caused by the same reason—by influence of the atmosphere. The only formal difference between a storm surge and a meteotsunami consists in the difference between their maximum periods. The maximum period for a tsunami does not exceed several hours, while storm surges may last several days.

Strong storm surges of heights up to 5 m are observed off the coast of China in the northern part of the Yellow Sea. This phenomenon results in colossal calamities for the Republic of Bangladesh—only during recent decades it has brought about the death of several hundreds of thousands of people. Storm surges are also known in Europe. The catastrophic storm surge that occurred in the North Sea in the period between January 31 and February 2, 1953 destroyed protective coastal structures, flooded an area of 25,000 km^2 and killed 2,000 people in Great Britain and Holland [Gill (1982)]. The famous inundations of St. Petersburg are nothing, but storm surges. Besides St. Petersburg, strong storm surges in Russia also take place off the coasts of the Azov and Okhotsk seas, and the Sea of Japan.

The physical mechanism of meteotsunami formation can be related to the influence upon the water surface of atmospheric pressure and tangential tensions, created by the wind. In principle, there exists, also, the possibility of non-linear energy transfer from the relatively short wind (storm) waves to the longwave components, but we shall not deal with this mechanism, here.

From the point of view of mathematical description, the influence of the atmosphere upon a water layer is taken into account by the boundary condition on the free water surface. Instead of the traditional condition of constant pressure on the free water surface for tsunami problems, we shall now assume this quantity to be variable in space and time, $p_{atm} = p(x,y,t)$. Besides the pressure, acting upon the water surface along the normal direction, there also exists a tangential tension of friction, caused by the wind. The tangential tension per unit surface area, $\mathbf{T}$, is related to the speed of the wind, $\mathbf{U}$, by the following approximate relationship:

$$\mathbf{T}_S = C\, \rho_{atm} \mathbf{U} |\mathbf{U}|,$$

where ρ_{atm} is the density of air, C is a dimensionless empirical coefficient, the value of which usually lies within limits from 0.0012 up to 0.003 [Lichtman (1970)]. A similar formula relates the velocity of the water flow near the bottom, $\mathbf{v}$, and the tension of friction, acting on the water column from the bottom,

$$\mathbf{T}_B = -C_B \rho \mathbf{v} |\mathbf{v}|,$$

where ρ is the density of water, C_B is a dimensionless empirical coefficient, the value of which is usually set equal to 0.0025 [Murty (1984)]. We recall that in the case of tsunami generation by bottom displacements tangential tensions on the bottom are not taken into account (owing to the short duration of a displacement). But in the case considered the action of tangential tension of the wind may turn out to be prolonged (up to several days) and to transfer significant momentum to the water column.

The presence of tangential tensions on the free water surface and on the bottom is accompanied by the formation of a pronounced vertical flow structure, which in real natural conditions is usually turbulent. Owing to the turbulence, the solution of

4.3 Meteotsunamis

the problem should, evidently, not be based on the Navier–Stokes equations, but on Reynolds equations. The existence of a vertical flow structure complicates transition from the general non-linear equations of hydrodynamics to the long-wave equations. But, if the non-linear term $(\mathbf{v}, \nabla)\mathbf{v}$ is neglected, then the Reynolds equations can be integrated over the vertical coordinate from the bottom, $z = -H$, up to the free water surface, $z = \xi$. As a result, a set of equations will be obtained, which will contain flow velocities averaged over the depth, while the term, describing the vertical turbulent momentum transfer, will be expressed as the difference between tensions on the bottom and on the free surface,

$$\int_{-H}^{\xi} \frac{\partial}{\partial z}\left(K_z \frac{\partial \mathbf{v}}{\partial z}\right) dz = \frac{1}{\rho}(T_S - T_B).$$

Without going into the details of obtaining the equations, expounded, for example, in the monograph of [Murty (1984)], we present a version of the set of equations applied in practice for calculating meteotsunami generation and propagation [Vilibic et al. (2004)],

$$\frac{\partial u}{\partial t} + u\frac{\partial u}{\partial x} + v\frac{\partial u}{\partial y} - fv$$
$$= -g\frac{\partial \xi}{\partial x} - \frac{1}{\rho}\frac{\partial p_{atm}}{\partial x} + \frac{(T_S - T_B)_x}{\rho(H+\xi)} + K_L\left(\frac{\partial^2 u}{\partial x^2} + \frac{\partial^2 u}{\partial y^2}\right),$$

$$\frac{\partial v}{\partial t} + u\frac{\partial v}{\partial x} + v\frac{\partial v}{\partial y} + fu$$
$$= -g\frac{\partial \xi}{\partial y} - \frac{1}{\rho}\frac{\partial p_{atm}}{\partial y} + \frac{(T_S - T_B)_y}{\rho(H+\xi)} + K_L\left(\frac{\partial^2 v}{\partial x^2} + \frac{\partial^2 v}{\partial y^2}\right),$$

$$\frac{\partial \xi}{\partial t} + \frac{\partial}{\partial x}((H+\xi)u) + \frac{\partial}{\partial y}((H+\xi)v) = 0,$$

where u, v are velocity components averaged over the depth, ξ is the free-surface displacement from equilibrium position, $f = 2\omega \sin\varphi$ is the Coriolis parameter and K_L is the constant horizontal turbulence viscosity coefficient. In principle, the quantity K_L may be variable, and then it should be present under the derivative sign. In [Vilibic et al. (2004)] the assumption was made that $K_L = 15$ m²/s.

Now, consider the main physical regularities of the meteotsunami generation process, taking advantage of the example of waves, caused by moving perturbations of atmospheric pressure. For clarity the problem will be considered within the framework of the simple one-dimensional model. Let $|\xi|$ be the absolute value of the free-surface displacement from equilibrium, and consider the ocean depth $H = $ const and the horizontal scale of atmospheric perturbation a to be related as follows: $|\xi| \ll H \ll a$. With account of such assumptions the meteotsunami formation process can be described by linear equations of long-wave theory,

$$\frac{\partial u}{\partial t} + g\frac{\partial \xi}{\partial x} = -\frac{1}{\rho}\frac{\partial p_{\text{atm}}}{\partial x}, \quad (4.16)$$

$$\frac{\partial \xi}{\partial t} + H\frac{\partial u}{\partial x} = 0. \quad (4.17)$$

If the atmospheric pressure is constant in time, but depends on the space coordinate ($\partial/\partial t = 0$), then from equation (4.16) immediately follows the so-called 'inverse barometer law'

$$\xi(x) = -\frac{p_{\text{atm}}(x)}{\rho g}. \quad (4.18)$$

In accordance with this law, the local enhancement of atmospheric pressure 'presses down' the free sea surface, forcing the water to occupy those regions, where the atmospheric pressure is lower. And, contrariwise, in the region of local reduction of pressure, for example, in cyclones, an enhancement of the water level should be observed. Extreme variations of atmospheric pressure are observed in tropical cyclones. The pressure at the centre of such a gigantic whirlwind can drop by a value of ~ 100 hPa, which amounts to about 10% of normal atmospheric pressure. A local elevation of the level by ~ 1 m corresponds to such a depression. But in the case of most tropical cyclones and of other atmospheric processes the amplitude of pressure perturbations and, consequently, the amplitude of the free water surface deviation will be by 1–3 orders of magnitude smaller. Variations of atmospheric pressure with amplitudes exceeding 10% can, most likely, arise only in the case of powerful explosions of natural (volcanoes, meteorites) or of artificial origin. In such cases the pressure perturbation will, naturally, not be motionless, but will propagate in the atmosphere, most probably, like a shock wave.

We, now, introduce the dimensionless variables (the asterisk * will be further omitted)

$$x^* = \frac{x}{H}, \quad t^* = t\sqrt{\frac{g}{H}}, \quad V^* = \frac{V}{\sqrt{gH}},$$

$$p^*_{\text{atm}} = \frac{p_{\text{atm}}}{\rho g H}, \quad \xi^* = \frac{\xi}{H}, \quad u^* = \frac{u}{\sqrt{gH}} \quad (4.19)$$

Note that the dimensionless velocity is the well-known Froud number $\text{Fr} = V/\sqrt{gH}$. With account of transformations (4.19) the set of equations (4.16) and (4.17) is easily reduced to the inhomogeneous wave equation

$$\frac{\partial^2 \xi}{\partial t^2} - \frac{\partial^2 \xi}{\partial x^2} = \frac{\partial^2 p_{\text{atm}}}{\partial x^2}. \quad (4.20)$$

If motion exists only at times $t > 0$, then for zero initial conditions the solution of equation (4.20) is determined by the formula [Tikhonov, Samarsky (1999)]

$$\xi(x,t) = \frac{1}{2}\int_0^t dT \int_{x-(t-T)}^{x+(t-T)} dX \frac{\partial^2 p_{\text{atm}}}{\partial X^2}. \quad (4.21)$$

4.3 Meteotsunamis

Let the propagating perturbation of atmospheric pressure be described by the formula

$$p_{\text{atm}}(x,t) = p(x - Vt)\theta(t), \qquad (4.22)$$

where p is an arbitrary function determining the space distribution of the pressure, θ is the steplike Heaviside function and V is the propagation velocity of the perturbation. The dynamics of the atmospheric process, described by formula (4.22), is such that at time moment $t = 0$ the atmospheric pressure perturbation is 'switched on' and starts movement unlimited in time with constant velocity V in the positive direction of axis $0x$. In the case considered the integrals in expression (4.21) are calculated analytically, and the solution of the problem is given by the formula

$$\xi(x,t) = \frac{p(x - Vt)}{V^2 - 1} - \frac{p(x-t)}{2(V-1)} + \frac{p(x+t)}{2(V+1)}. \qquad (4.23)$$

From formula (4.23) it follows that the wave perturbation on the water surface has three components. One of them propagates with the velocity V, following the area of altered pressure. The other two components correspond to free waves travelling along axis $0x$ in the positive and negative directions, respectively, with the velocity of long waves. The amplitude of waves on the water surface depends strongly on the propagation velocity of the atmospheric perturbation. Here, the amplitude of waves travelling in the same direction as the atmospheric perturbation may undergo a sharp increase, when $V \approx 1$. When the equality $V = 1$ is satisfied exactly, the growth of amplitude is without limit, within the model considered. This effect is known as the 'Proudman resonance'. The amplitude of waves travelling in the negative direction of axis $0x$ exhibit no such peculiarities. Always remaining a relatively small quantity, it monotonously decreases as the velocity V increases.

It is possible to determine the behaviour of a wave perturbation on a water surface in resonance conditions by calculating the limit of expression (4.23), when $V \to \infty$. The resonance effects involve the first two terms of expression (4.23). In the case of resonance, each of these terms tends towards infinity, but their sum has a finite limit. We shall, now, expand function $p(x - Vt)$ in a Taylor series at point $z_0 = x - t$ with an accuracy up to the linear term,

$$p(z_0) \approx p(z_0) + p'(z_0)(z - z_0).$$

Upon performing elementary transformations we obtain an expression, describing the free-surface displacement in the case of resonance,

$$\xi_{\text{res}}(x,t) = \lim_{V \to 1}\left(\xi(x,t)\right) = -\frac{p'(x-t)t}{2} - \frac{p(x-t)}{4} + \frac{p(x+t)}{4}. \qquad (4.24)$$

From formula (4.24) it follows that, when resonance conditions are fulfilled, the wave perturbation comprises three components. The first component represents a wave of amplitude, increasing linearly with time, and the growth rate of the amplitude is proportional to the derivative of the distribution of pressure in space. The other two components describe waves of insignificant and fixed amplitudes.

For definiteness, we shall further consider the distribution of pressure in space in formula (4.22) to have a Gaussian form:

$$p(z) = p_0 \exp\left\{-\frac{z^2}{a^2}\right\}, \qquad (4.25)$$

where p_0 is the pressure amplitude. Figure 4.10 presents the example of the movement of an atmospheric perturbation (in the region of a local enhancement of pressure) and of the evolution of waves, generated by this perturbation. The calculation is performed in accordance with formula (4.23) for three different velocities of the perturbation. From the figure it is seen, that below the critical velocity ($V = 0.75$), immediately under the atmospheric perturbation, a similar in shape,

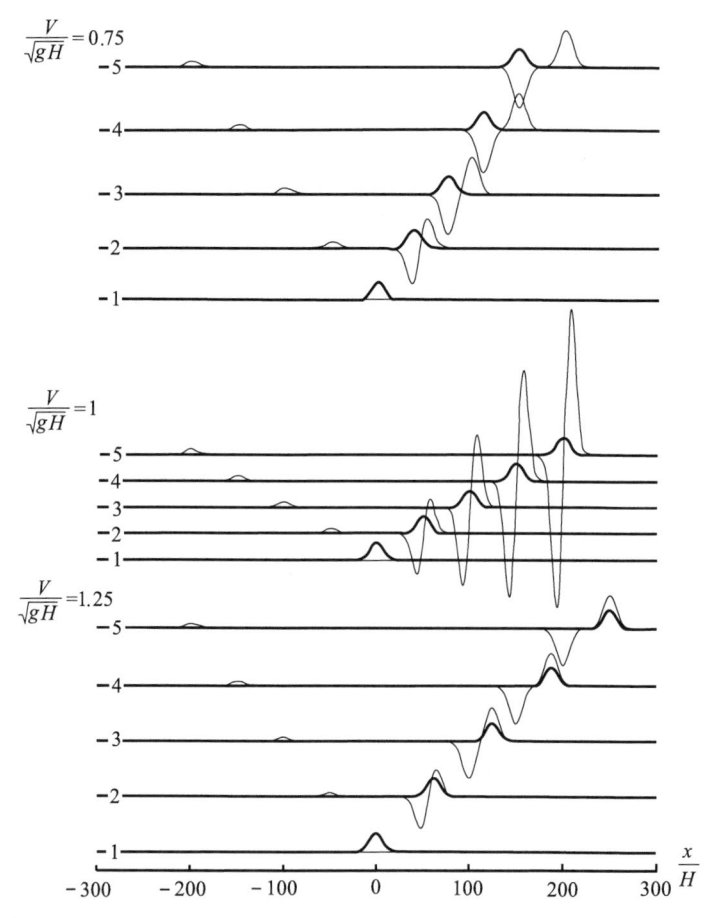

Fig. 4.10 Profiles on free water surface of waves (thin line), formed by perturbation of atmospheric pressure (thick line), travelling with a velocity V. The calculation is performed at $a = 10$ for fixed time moments $t\sqrt{g/H} = 0, 50, 100, 150, 200$ (curves 1–5)

4.3 Meteotsunamis

but opposite in sign, perturbation forms of the water surface—an induced wave. Moreover, there arise two free waves, travelling in opposite directions, and the one propagating in the same direction as the atmospheric perturbation has a larger amplitude, while its polarity is opposite to the polarity of the induced wave. When $V \ll 1$, the amplitude and polarity of the induced wave are in accordance with the values determined by the inverse barometer law. Practically, all atmospheric processes in open ocean serve as natural prototypes for slowly propagating atmospheric perturbations. Thus, for instance, the propagation velocity of a tropical cyclone usually amounts to $V \sim 5$–10 m/s, which is significantly inferior to the propagation velocity of long waves at large depths $\sqrt{gH} \sim 200$ m/s (for $H = 4{,}000$ m).

In the case of resonance ($V = 1$) only two waves are observed on the water surface. The induced wave follows the atmospheric perturbation, linearly increasing its amplitude with time. The second wave is free. It travels in the opposite direction, and its amplitude is small. We underline that within the framework of the model problem considered the amplitude of the induced wave grows without limit. Fulfilment of the resonance conditions is possible, for example, in shallow water ($H \sim 10$–100 m), where the velocity of long waves ($\sqrt{gH} \sim 10$–30 m/s) may turn out to be close to the typical propagation velocity of atmospheric perturbations.

If the velocity V exceeds the critical velocity (in Fig. 4.10 the case of $V = 1.25$ is presented), then the induced wave turns out to be similar in shape and sign to the perturbation of atmospheric pressure. The polarity of the free wave, travelling in the same direction as the atmospheric perturbation, differs in polarity from the induced wave. The free wave, travelling in the opposite direction, here, like in all other cases, repeats the polarity of the atmospheric perturbation. From formula (4.23) it is seen that at high velocities V the amplitude of the free surface response tends asymptotically toward zero. Note that the similar dependence (2.85) for waves generated by a running displacement of the ocean bottom exhibits a somewhat different character: at high velocities V the surface displacement tends towards a constant, instead of zero. In reality, the pressure perturbations, corresponding to the velocity range $V \gg 1$, may be related, for example, to acoustic waves in the atmosphere or to the propagation of atmospheric internal waves above shallow-water areas.

Figure 4.11 illustrates the character of variation of the maximum amplitude of waves on the water surface versus the distance covered by the atmospheric perturbation, $L = Vt$. The maximum range is calculated by the formula

$$A_{\max}(x) = \max_t \left(\xi(x,t) \right) - \min_t \left(\xi(x,t) \right). \tag{4.26}$$

From the figure it is seen that for a noticeable increase in the amplitude it is necessary that the resonance condition be fulfilled along a path the length of which holds several horizontal extensions of the atmospheric perturbation. If the velocity $V \neq 1$, then the growth of the amplitude is limited. At any rate, at the initial stage of wave formation, when $V \approx 1$, the growth rate of the amplitude does not differ strongly from the resonance case. Therefore, if the velocity of the atmospheric perturbation, V, varies within limits $\pm 10\%$ of the resonance velocity, then a tenfold

Fig. 4.11 Amplitude (range) of waves on water surface versus distance $L = Vt$, covered by perturbation of atmospheric pressure of amplitude p_0 and with horizontal dimension a. The calculation is performed for different propagation velocities of the atmospheric perturbation. The numbers indicate values of dimensionless velocity $V/\sqrt{gH}$ for respective curves

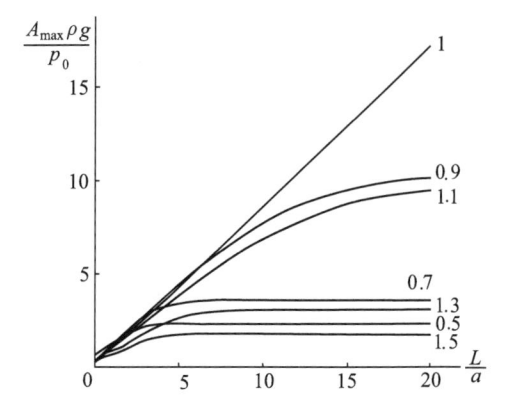

increase is possible of the wave perturbation amplitude as compared with the value determined by the inverse barometer law.

One of the most important properties of meteotsunamis is the proportionality of the wave amplification coefficient to the ratio of the length of the 'resonance' area of water and the horizontal size of the atmospheric perturbation. Taking advantage of this property, it is possible to determine in advance the sections of the coast, potentially endangered by meteotsunamis. To this end, it is necessary to analyse the littoral bathymetry and to reveal extended shelf zones, within which the resonance conditions can be fulfilled. For this work it is, naturally, necessary to know the characteristic propagation velocities of atmospheric perturbations.

Since typical propagation velocities of atmospheric perturbations amount are from units to tens of meters per second, fulfilment of the Proudman resonance conditions is most probable in shallow-water areas of the ocean. But, when meteotsunamis of significant amplitude are excited within shallow-water areas, linear theory is no longer applicable. Therefore, it is expedient to consider the problem of wave generation by atmospheric perturbations within the framework of non-linear theory of long waves. We shall now assume that the displacement amplitude of the free water surface may be comparable to the basin depth, i.e. the main parameters of the problem are related as follows: $|\xi| \sim H \ll a$. We shall write the equations of non-linear theory of long waves in dimensionless variables, bearing mind the formulae (4.19),

$$\frac{\partial u}{\partial t} + u\frac{\partial u}{\partial x} + \frac{\partial \xi}{\partial x} = -\frac{\partial p_{\text{atm}}}{\partial x}, \qquad (4.27)$$

$$\frac{\partial \xi}{\partial t} + \frac{\partial}{\partial x}((1+\xi)u) = 0. \qquad (4.28)$$

It is not possible to resolve the complete non-linear problem (4.27) and (4.28) analytically. But, when movement of the atmospheric perturbation is not limited in time ($-\infty < t < +\infty$), it is possible to obtain an analytical relationship between the free-surface displacement in the induced wave and the perturbation of atmospheric pressure.

4.3 Meteotsunamis

Like in the linear problem, we shall consider a perturbation of the atmospheric pressure (deviation from a certain standard value), which propagates with a constant velocity V in the positive direction of axis $0x$,

$$p_{atm}(x,t) = p(x-Vt). \tag{4.29}$$

We shall assume the response of the water column to represent an induced wave, travelling with the velocity V in the same direction,

$$u(x,t) = u(x-Vt), \tag{4.30}$$
$$\xi(x,t) = \xi(x-Vt). \tag{4.31}$$

Successive differentiation with respect to time and integration over space of functions with arguments of the form $(x-Vt)$ is similar to multiplication by the quantity $-V$,

$$\int \frac{\partial}{\partial t} f(x-Vt)\,dx = -V f(x-Vt).$$

Making use of this fact, we pass from differential equations (4.27) and (4.28) to the algebraic relations [Pelinovsky et al. (2001)]

$$-Vu + \frac{u^2}{2} + \xi = -p_{atm}, \tag{4.32}$$
$$-V\xi + (1+\xi)u = 0. \tag{4.33}$$

Generally speaking, expressions (4.32) and (4.33) are correct with an accuracy up to certain integration constants. We have chosen the values of these constants so as to have zero flow velocities and zero free-surface displacements $u=0$, $\xi=0$, respectively, to correspond to zero perturbation of the atmospheric pressure.

Excluding the quantity u from the set of equations (4.32) and (4.33), we obtain the relationship between the moving perturbation of atmospheric pressure and the free water surface response, corresponding to it,

$$p_{atm} = V^2 \left(\frac{\xi}{1+\xi} - \frac{\xi^2}{2(1+\xi)^2} \right) - \xi. \tag{4.34}$$

When $\xi \ll 1$, the problem considered reduces to the linear problem. In this case formula (4.34) can be written in the form

$$\xi = \frac{p_{atm}}{V^2 - 1}, \tag{4.35}$$

which fully corresponds to the first term in the analytical solution of the linear problem (4.23).

The free-surface displacement ξ can also be expressed explicitly in the nonlinear case via the perturbation of pressure p_{atm}. Equation (4.34) has three solutions, and some of them have no physical sense in the case of certain values of quantities p_{atm} and V. Moreover, the form of the solutions is determined by quite cumbersome formulae. In this connection, it is much more simple to select a certain free-surface displacement and, by applying the unambiguous functional relationship (4.34), to determine the pressure perturbation, which could have been caused by this displacement.

For definiteness we shall assume the free-surface displacement to be determined by the function

$$\xi(x,t) = A_{\max} \exp\left\{-\frac{(x-Vt)^2}{a^2}\right\}.$$

The perturbation of pressure, corresponding to it, which can be calculated by formula (4.34) in the vicinity of the resonance or for large values of the quantity $A_{\max}$, generally speaking has a form essentially differing from a Gaussian. Therefore, it has sense to introduce a certain quantity $p_{\max}$, characterizing the intensity of pressure variations. Let $p_{\max}$ denote the amplitude of pressure variations, understood in the sense of formula (4.26). The ratio of quantities $A_{\max}$ and $p_{\max}$ (in dimensionless form $A_{\max}\rho g/p_{\max}$) represents the 'amplification coefficient' of the wave amplitude.

In Fig. 4.12 the quantity $A_{\max}\rho g/p_{\max}$ is plotted against the propagation velocity V of the atmospheric perturbation. The dotted line in the figure shows the dependence, corresponding to linear theory and calculated with the use of formula (4.35). The amplification coefficient is seen to depend, in accordance with non-linear theory, not only on the velocity V, but also on the sign of the wave produced. For positive waves the Proudman resonance point turns out to be shifted to the right as compared with the linear case, while in the case of negative perturbations to the left. Moreover, the growth of the wave amplitude for any fixed values of velocity V turns out to be limited. This fact is a most important manifestation of non-linearity in the problem dealt with. Let us briefly dwell upon its physical interpretation. Consider the resonance condition $V^2/gH = 1$ to be fulfilled, starting from a certain moment of time. Then, at the initial stage, in accordance with linear theory, an increase in the amplitude of the free-surface perturbation will take place. But, as soon as the quantity ξ reaches sufficiently high values, the actual basin depth, present in

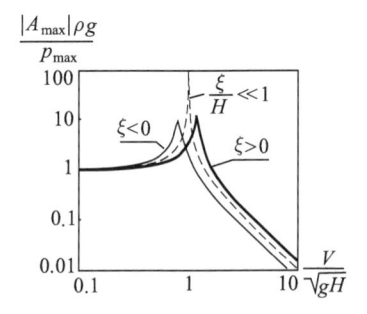

Fig. 4.12 Ratio of free-surface displacement amplitude $A_{\max}$ and amplitude of atmospheric pressure perturbation, $p_{\max}$, versus propagation velocity of atmospheric perturbation. The calculation is performed for positive ($A_{\max}/H = 0.25$), negative ($A_{\max}/H = -0.25$), and infinitesimal (dotted line) displacements of the free surface

the resonance condition, will change from H to $H \pm \xi$. As a result, the resonance condition will be violated, and growth of the amplitude will stop. Further enhancement of the amplitude is possible in certain cases, when the resonance conditions are continuously corrected by variation of the velocity V or by realization of a certain particular profile of depths along the route.

4.4 Cosmogenic Tsunamis

Recently, particular interest has been shown in the possibility of catastrophic tsunami waves arising due to bodies falling into the ocean from outer space. Such waves are conventionally termed cosmogenic tsunamis in modern scientific literature. Geological structures, reminiscent in shape of craters and found on all continents, have been understood to be traces of collisions of meteor bodies with the Earth only during the past 30–40 years. Such ring structures called astroblemes—star scars, contain rock that was produced under huge pressures (up to a million atmospheric pressures) and exhibit signs of shock wave transformations of the mineral components and are often related to diamond deposits. At present, on Earth already 150 such objects have been found with characteristic dimensions from 1.2 km (Arizona crater, USA) up to 100 km (Popigai astroblem, East Siberia, Russia). The World Ocean occupies approximately two thirds of the surface of our planet, therefore a great part of meteorites falls precisely into the ocean, the bottom of which guards the traces of many such collisions, that in the past caused catastrophes of a planetary scale [Kharif, Pelinovsky (2005)].

Since in this section we are entering into a field far from oceanology, we shall define some concepts. *Meteorites* are the remains of *meteor bodies* that survived transition through the atmosphere and that have fallen to the Earth's surface from outer space. According to their composition, meteorites are divided into three main classes: stony (93.3%), stony-iron (1.3%) and iron (5.4%). When a meteor body enters the atmosphere, air resistance causes the body to heat up strongly and to shine brightly (the *bolide* phenomenon). According to modern ideas, meteorites are fragments of parent bodies—asteroids. *Asteroids* are considered small planets of diameters approximately between 1 and 1,000 km.

The task of describing cosmogenic tsunamis can be divided into three stages. First, it is necessary to determine the characteristics (dimensions, density and velocity) of meteorites, that can fall into the ocean, and to estimate the probability of such an event. Second, the essentially non-linear process of interaction of a meteorite with the water column must be described, and the relationship between parameters of the initial perturbation of the water column and characteristics of the celestial body must be revealed. At the third stage an analysis must be performed of the peculiarities of cosmogenic tsunami propagation in open ocean and of waves running up shores. All three stages are connected with numerous uncertainties that arise primarily because no cosmogenic tsunami has been recorded yet.

Estimates made by specialists reveal that several thousand large objects (asteroids and comets) of diameters over 1 km have a potential possibility of colliding with our planet [Solem (1999)]. A cosmic body of size superior to 2 km colliding with the Earth will result in a global catastrophe [Paine (1999)]. Luckily, the probability of such a collision is extremely small, and in all written history of the existence of mankind no such catastrophe occurred. Objects of relatively small dimensions regularly bombard the Earth, but most of them are destroyed and already burn up in the upper layers of the atmosphere. The critical for a stony meteor body size, with which it is capable of reaching the Earth's surface, amounts to about 100 m in diameter. In the case of iron objects this critical size is significantly smaller ($\sim$1 m), but they are encountered very rarely, so they will not be dealt with.

From Fig. 4.13 one can judge the collision probability of a celestial body with the Earth depending on the radius of the object. This dependence has been obtained in [Ward, Asphaug (2000)]. Actually, it represents a straight line (in logarithmic scale) that passes through the two points indicated in the figure by stars. The first point is based on existing material—observations from geostationary satellites of meteor bodies with dimensions $\sim$1 m exploding in the atmosphere. These data permit to assert that on the average about 25 events of this kind take place in a year [Nemtchinov et al. (1997)]. The second point is based on estimations of the collision frequency of the Earth with large objects ($\sim$1 km), made in refs.

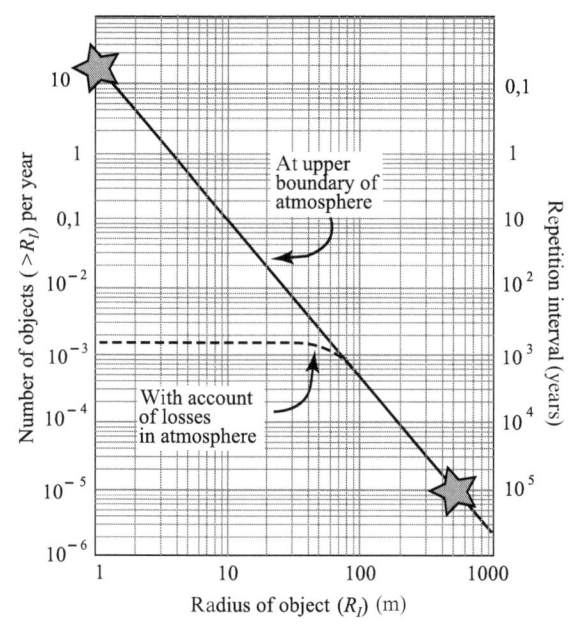

Fig. 4.13 Number of asteroids colliding with the Earth per year versus their radius. Dotted line—number of objects reaching the Earth's surface (with account of losses in the atmosphere). Stars show the results of [Nemtchinov et al. (1997); Shoemaker et al. (1990)]. Adapted from [Ward, Asphaug (2000)]

4.4 Cosmogenic Tsunamis

[Shoemaker et al. (1990); Toon et al. (1994)]. Large objects collide with our planet approximately once every 100,000 years. The authors of [Ward, Asphaug (2000)] point out that the dependence presented in Fig. 4.13 is not quite accurate, but that it is the best estimate of all that could be made for objects with radiuses of 1–1,000 m, using presently available information. The actual collision frequency of the Earth with celestial bodies within the range of dimensions indicated may differ by a factor of 3 as compared to the dependence proposed.

The dotted line in Fig. 4.13 shows the number of objects that are not destroyed in the atmosphere and that are capable of reaching the Earth's surface. From the point of view of tsunami generation we will be interested precisely in these objects, since they are capable of effectively influencing the water column. Of course, large meteor bodies, exploding in the atmosphere at small heights (such as the Tungus meteorite, 1908) over the surface of the ocean are probably also capable of causing gravitational waves, but, most likely, their energy will be insufficient for exciting dangerous tsunami waves.

The typical density of stony asteroids amounts to about 3,000 kg/m^3, their velocity to 20 km/s. Assuming the object to have a spherical shape, it is easy to estimate its kinetic energy. Thus, for example, a meteor body of diameter 100 m will have a kinetic energy $\approx 3 \cdot 10^{17}$ J. This value corresponds to the energy of a very strong seismotectonic tsunami (see Fig. 3.1). Now, if the diameter of the object amounts to 1 km, then its energy will be colossal, $\approx 3 \cdot 10^{20}$ J. This value is already many times greater than the energy of the source of the strongest earthquake of the twentieth century that occurred in Chile in 1960. It is not difficult to estimate that such an energy is sufficient to evaporate 10^{11} m^3 of water (the heat required to evaporate water is $2.3 \cdot 10^6$ J/kg). It is interesting to note that precisely such a volume of water is ousted by ocean bottom deformations in the case of very strong earthquakes (source area of $1,000 \times 100$ km, average vertical deformation of bottom of 1 m). At any rate, a relatively small part of the energy is, most likely, spent on the evaporation of water.

Following [Ward, Asphaug (2002)], we shall assume a meteorite falling into the ocean to create, at the initial stage, a radially symmetric cavity, described by the following function:

$$\xi_0(r) = D_C \left(\frac{r^2}{R_C^2} - 1 \right) \left(1 - \theta(r - R_D) \right), \tag{4.36}$$

where D_C is the depth of the cavity, R_C and R_D are its internal and external radii, respectively and θ is the Heaviside step function. The case, when $R_D = R_C$ corresponds to water being released into the atmosphere (or to its evaporation). The initial perturbation, here, represents a depression (Fig. 4.14a). When $R_D = R_C \sqrt{2}$, the water ejected from the cavity, forms an external circular structure (a splash or circular swell), the volume of which is exactly equal to the volume of the water, ejected from the cavity (Fig. 4.14b).

From the shape of the cavity it is possible to estimate the tsunami energy as the potential energy of the initial elevation,

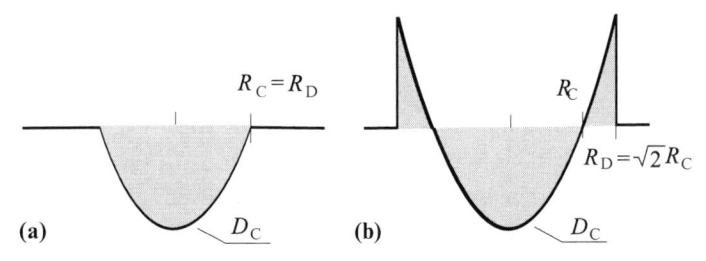

Fig. 4.14 Model shape of perturbation (cavity), resulting from a meteorite falling into the ocean. D_C—depth of cavity, R_C and R_D—internal and external radii of the cavity (Adapted from [Ward, Asphaug (2000)])

$$E_T = \frac{\pi \rho_w g}{2} (D_C R_D)^2 \left(1 - \frac{R_D^2}{R_C^2} + \frac{R_D^4}{3R_C^4}\right), \quad (4.37)$$

where ρ_w is the density of water, g is the acceleration of gravity. When $R_D = R_C\sqrt{2}$, the general formula (4.37) assumes the more simple form

$$E_T = \frac{\pi \rho_w g}{3} (D_C R_C)^2. \quad (4.38)$$

Only a fraction ε of the meteorite's kinetic energy E_I is transformed into the tsunami energy, so we can write

$$E_T = \varepsilon E_I = \varepsilon \frac{\rho_I (4\pi/3) R_I^3 V_I^2}{2}, \quad (4.39)$$

where ρ_I, R_I and V_I are the meteorite density, radius and velocity, respectively. The part of the meteorite energy transferred to the tsunami is not, generally speaking, a constant, but depends on the properties of the water column and of the falling body.

Comparison of expressions (4.38) and (4.39) permits to express the depth of the cavity as follows:

$$D_C = \left(\frac{2\varepsilon \rho_I R_I^3 V_I^2}{\rho_w g R_C^2}\right)^{1/2}. \quad (4.40)$$

We further assume the relationship between the depth of the cavity and its radius to be of the form

$$D_C = q R_C^\alpha, \quad (4.41)$$

where q and α are coefficients related to the properties of the meteorite and of the water column. Substitution of relation (4.41) into formula (4.40) permits to express the radius of the cavity as follows:

$$R_C = R_I \left(2\varepsilon \frac{V_I^2}{gR_I}\right)^\delta \left(\frac{\rho_I}{\rho_w}\right)^{1/3} \left(\left(\frac{\rho_w}{\rho_I}\right)^{1/3-\delta} \left(\frac{1}{qR_I^{\alpha-1}}\right)^{2\delta}\right), \quad (4.42)$$

4.4 Cosmogenic Tsunamis

where $\delta = 1/(2\alpha + 2)$. The form of formula (4.42) corresponds to the known relation for the radius (diameter) of a crater [Schmidt, Holsapple (1982)]

$$R_C^{SH} = R_I \left(\frac{1}{3.22} \frac{V_I^2}{gR_I} \right)^\beta \left(\frac{\rho_I}{\rho_T} \right)^{1/3} \left(\frac{C_T}{1.24} \right), \qquad (4.43)$$

where β and C_T are parameters depending on the properties of the target (water, in this case). For water their values are $\beta \approx 0.22$ (i.e. $\alpha = 1/(2\beta) - 1 \approx 1.27$), $C_T \approx 1.88$. By comparison of formulae (4.42) and (4.43) one can note that about 16% of the kinetic energy of the falling body is transformed into tsunami energy ($\varepsilon = 1/(2 \cdot 3.22) \approx 0.16$). Of course, this is an approximate estimate, and it is correct only if the quantity ε is actually not subject to strong variations.

The quantity q present in formula (4.41) varies weakly with the size of the falling celestial body, R_I, and of the density ratio ρ_I/ρ_w. By comparison of formulae (4.42) and (4.43) it is not difficult to obtain the following approximate dependence:

$$q \approx 0.39 \left(\frac{\rho_w}{\rho_I} \right)^{0.26} \frac{1}{R_I^{0.27}}. \qquad (4.44)$$

In the case, when the density of the celestial body is three times that of the density of water, $(\rho_I/\rho_w = 3)$, the value of q varies between 0.1 ($R_I = 50$ m) and 0.054 ($R_I = 500$ m).

To simplify the calculations it is possible, instead of the cumbersome expressions (4.41) and (4.43) to use approximate formulae that are valid for $V_I = 20$ km/s and $\rho_I/\rho_w = 3$ [Ward, Asphaug (2002)],

$$R_C \approx 98 \cdot R_I^{3/4}, \qquad (4.45)$$
$$D_C \approx 0.64 \cdot R_C. \qquad (4.46)$$

In Fig. 4.15 the dependences (4.41) and (4.43) are shown by solid lines, the approximate relationships (4.45) and (4.46) by dotted lines. The cavity diameter is usually 2.5–3 times greater than its depth. Thus, for example, a cosmic body of radius 200 m falling into the ocean creates a cavity of diameter about 10 km and depth of the order of 3.5 km. Note that in the case of celestial bodies of radius $R_I > 300$ m the calculated cavity depth D_C will, as a rule, exceed the ocean depth H. In this case, a crater will not only form in the water, but also in the ocean bottom. To avoid overcomplicating the problem we shall further assume an effective cavity depth D_C^{eff}, equal to the ocean depth, to be applicable, when $D_C > H$. The cavity radius is calculated as previously, in this case.

Figure 4.16 presents a comparison of cavity shapes calculated using the proposed parametrization (4.41) and (4.43) and obtained by detailed numerical simulation of the process performed in [Crawford, Mader (1998)]. The complex non-linear model and the parameterizations proposed are seen to be in quite reasonable agreement. Noticeable divergence is only observed in the external circular structure, but for preforming tsunami calculations at a level of estimations it is not too important.

188　　　　　　　　　　　　　　　　　　　4 Tsunami Sources of Nonseismic Origin

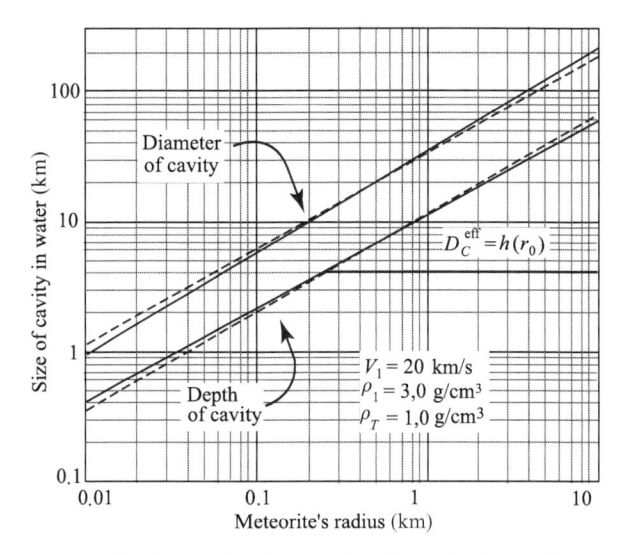

Fig. 4.15 Size of cavity (depth and diameter), produced in water by a falling meteorite, versus the meteorite's radius (Adapted from [Ward, Asphaug (2002)])

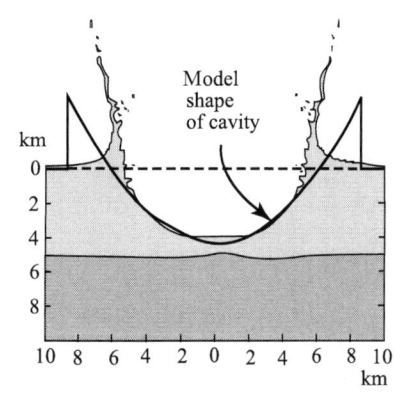

Fig. 4.16 Shape of cavity formed in water by a falling meteorite: comparison of results of numerical simulation [Crawford, Mader (1998)] and of idealized model (formula (4.36) with account of relations (4.41) and (4.43)). Calculations are performed for a time moment of 25 s, an asteroid diameter of 500 m, a fall velocity of 20 km/s, a density of 3.32 g/cm^3 and ocean depth of 5 km (Adapted from [Ward, Asphaug (2000)])

Owing to dispersion and dissipation, the short-wave components making up the external circular structure will not play any noticeable role at large distances from the source.

At the next stage we must describe the evolution of waves from the initial perturbation (4.36), generated by the celestial body. We assume that at the moment, when the cavity and circular swell behind it have formed, the velocity of motion of

4.4 Cosmogenic Tsunamis

water particles can be neglected. To describe the waves we take advantage of linear potential theory. In this problem it is essential to take into account phase dispersion, so application of the long-wave approximation is not permitted. Note that owing to the large wave amplitudes (comparable to the depth) the application of linear theory is also not quite correct. But for approximate estimation of the properties of a cosmogenic tsunami such an approach is quite justified.

In the case of an ocean of constant depth, evolution of the initial free-surface perturbation, exhibiting radial symmetry, is described by the following expression (see general theory in Sect. 2.2.2):

$$\xi(r,t) = \int_0^\infty k\,dk A(k) J_0(kr) \cos(\omega(k)t), \qquad (4.47)$$

$$A(k) = \int_0^\infty r\,dr \xi_0(r) J_0(kr), \qquad (4.48)$$

where $\xi_0(r)$ is the function describing the form of the initial perturbation, J_0 is the Bessel function of the first kind of 0th order. The relation between the cyclic frequency and the wave number is determined by the known dispersion relation for gravitational waves on water, $\omega^2 = gk\tanh(kH)$. For an initial perturbation, determined by formula (4.36), the Fourier–Bessel transformation (4.48) yields the following form for the dependence of the amplitude of space harmonics upon the wave number:

$$A(k) = D_C \frac{R_D\left((R_D^2 - R_C^2)kJ_1(kR_D) - 2R_D J_2(kR_D)\right)}{R_C^2 k^2}. \qquad (4.49)$$

Figure 4.17 presents the example of a calculation of waves caused by the fall into an ocean 4 km deep of a celestial body of radius 100 m, density 3,000 kg/m^3, moving with a velocity of 20 km/s. It is seen that during the first minutes after the fall the waves in the immediate vicinity of the incidence point may reach colossal heights of the order of 1 km and more. Figure 4.18 presents in dimensionless coordinates the wave number dependences of phase and group velocities of surface gravitational waves on water. The same plot shows the amplitude distribution of space harmonics over the wave numbers, calculated in accordance with the form of the initial perturbation (4.36) for an internal cavity radius equal to the ocean depth. The amplitude distribution is determined by function $|kA(k)|$, where $A(k)$ is given by formula (4.49). The position of the space spectrum on the wave number axis is related in an evident manner to the cavity radius. Therefore, on the basis of calculations for $R_C = H$, from which the maximum turns out to be located at the value $kH \approx 2.97$, it is possible to write the formula determining the position of the maximum as function of the cavity radius, $k_{max} \approx 2.97/R_C$. Recalculation for the wave lengths reveals this to correspond to $\lambda_{max} \approx 2.12 R_C$, which somewhat exceeds the cavity diameter. Taking advantage of the dispersion relation, it is not difficult to determine the period corresponding to the maximum of the spectrum,

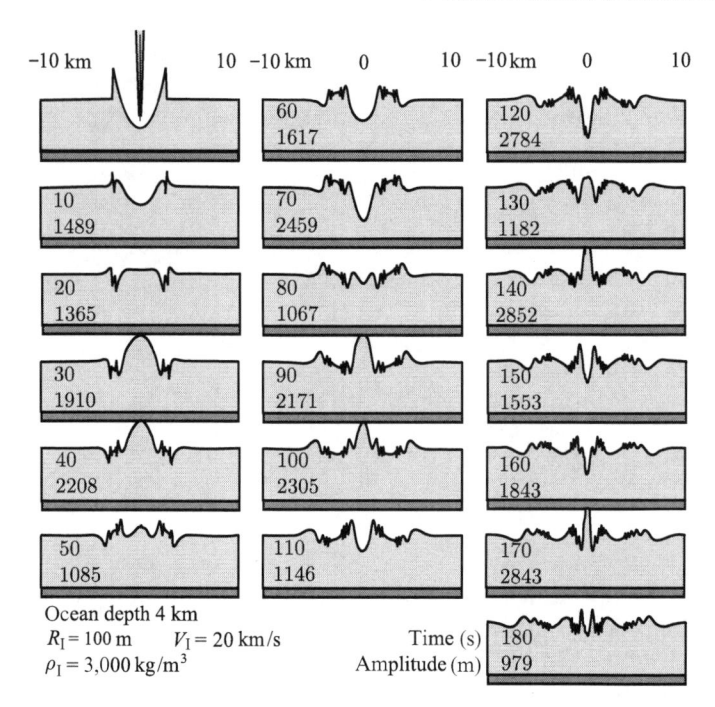

Fig. 4.17 Dynamics of wave perturbation in the vicinity of the incidence point of a meteorite during the first 3 min (Adapted from [Ward, Asphaug (2000)])

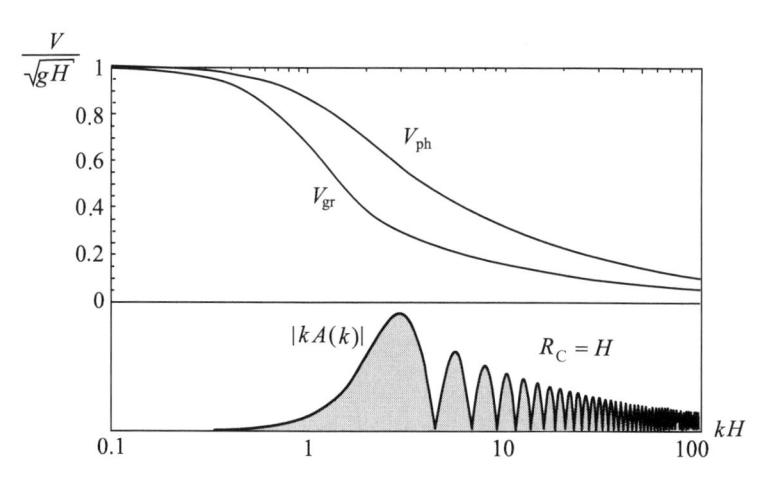

Fig. 4.18 Phase and group velocities of gravitational waves on water versus the wave number, space spectrum (absolute value of function $kA(k)$)

$T_{max} = 2\pi/\sqrt{gk_{max}\tanh(k_{max}H)}$. For typical values of $R_C = H = 5$ km the period amounts to $T_{max} \approx 83$ s. We recall that the range of tsunami wave periods is 10^2–10^4 s. The orders of magnitude of wave periods caused by falling meteorites can be

4.4 Cosmogenic Tsunamis

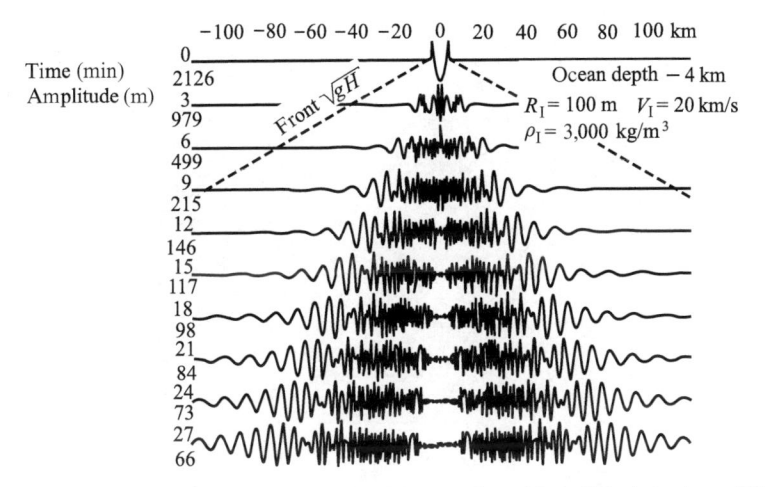

Fig. 4.19 Dynamics of wave perturbation at long times (Adapted from [Ward, Asphaug (2000)])

seen to correspond to the most short-period region of the tsunami spectrum. This fact is essentially reflected in the character of cosmogenic tsunami propagation. Unlike waves of seismotectonic origin, cosmogenic tsunamis have no fronts propagating with the velocity of long waves, $\sqrt{gH}$. From Fig. 4.18 their spectrum is seen just not to contain components of the necessary wavelength ($kH < 0.1$). As to the components carrying energy they will propagate significantly slower (about two times slower in the case of $R_C = H$), than usual seismotectonic tsunamis.

Straightforward calculation of waves for long times, the results of which are shown in Fig. 4.19, confirm the arguments presented above. At the point, corresponding to the position of the front of a long wave, no visible signal is present. The wave packet, in which the long-wave components lead at long times, propagates with a velocity more than two times inferior to the velocity of long waves. Here, the amplitude of waves rapidly decreases with time and distance from the area of origin. Thus, while at the 3rd minute after the impact of the celestial body the amplitude amounts to over 2,000 m, in 27 min it no longer exceeds 70 m. And then, waves of noticeable amplitude only have time to cover 100 km. Note that in 27 min long waves cover distances of over 300 km.

Analysis of numerous calculations have permitted the authors of [Ward, Asphaug (2000)] to propose a formula describing the damping of cosmogenic tsunamis with distance in an ocean of constant depth. These calculations show that variation in the wave amplitude is only related to geometrical factors and phase dispersion. Dissipative factors and the Earth's sphericity are not taken into account. Moreover, the assumption is made that, independently of the characteristics of the celestial body falling into the ocean, the initial wave amplitude cannot exceed the ocean depth.

$$\xi_{\max}(r) = \min(D_C, H)\left(\frac{1}{1 + r/R_C}\right)^\gamma, \qquad (4.50)$$

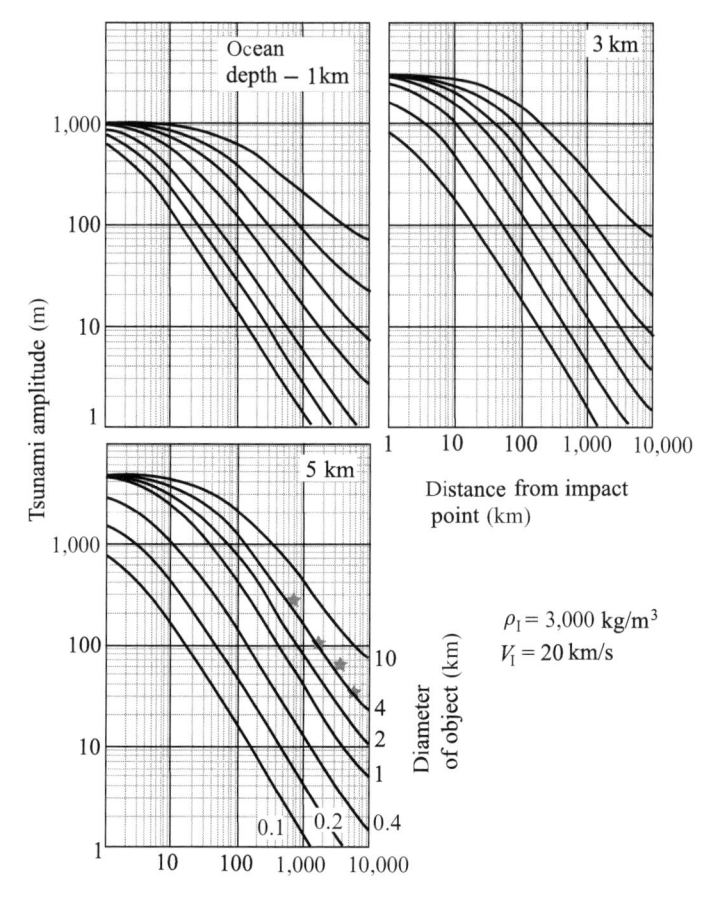

Fig. 4.20 Decrease of cosmogenic tsunami amplitude with distance from impact point of object. Calculations are preformed for various sizes of object and ocean depths (Adapted from [Ward, Asphaug (2002)])

where $\gamma = 1/2 + 0.575 \exp\{-0.035 R_C/H\}$. The dependence (4.50) is shown in Fig. 4.20. It is calculated for various sizes of falling celestial bodies and ocean depths. The data, shown in the figure, permit to estimate the degree of danger (the height of waves) represented by tsunamis of cosmogenic origin at different distances from the point of impact. When one approaches the coast, the wave amplitude (like in all tsunami cases) will increase by several times, owing to a decrease in depth.

In conclusion, we once again draw attention to the probability of a large meteorite, capable of causing significant tsunami waves, falling on Earth being extremely small. But, if such an event actually does take place, then among other catastrophic consequences tsunami waves will certainly not play the last part.

References

Assier-Rzadkiewicz S., Heinrich P., Sabatier P. C. Savoye B., Bourillet J. F. (2000): Numerical modeling of landslide-generated tsunami: The 1979 Nice event. Pure Appl. Geophys. **157** 1707–1727

Basov B. I., Dorfman A. A., Levin B. W., Kharlamov A. A. (1981): On perturbations of the ocean surface, excited by the eruption of an underwater volcano. Vulcanologia i Seismologia (in Russian). No 1 93–98

Belousov A., Voight B., Belousova M., Muravyev Y. (2000): Tsunamis generated by subaquatic volcanic explosions: unique Data from 1996 eruption in Karymskoye Lake, Kamchatka, Russia. Pure Appl. Geophys. **157** 1135–1143

Bondarenko A. L., Bychkov V. S. (1983): Marine baric waves. Meteorologia i gidrologia (in Russian). No 6 86–91

Choi B. H., Pelinovsky E., Kim K. O., Lee J. S. (2003): Simulation of the trans-oceanic tsunami propagation due to the 1883 Krakatau volcanic eruption. Nat. Hazard. Earth Syst. Sci. **3** 321–332

Crawford D. A., Mader C. (1998): Modeling asteroid impact and tsunami. Sci. Tsunami Hazard **16** 21–30

Didenkulova I. I., Kurkin A. A., Pelinovsky E. N. (2007): Run-up of solitary waves on slopes with different profiles. Izvestiya RAN, Atmos. Ocean. Phys. **43**(3)

Donn W. L., Ewing M. (1956): Stokes edge waves in Lake Michigan. Science **124** 1238–1242

Egorov Yu. A. (1990): Hydrodynamic model of tsunami wave generation by eruption of submarine volcano. Natural catastrophes and disasters in the Far-East region. Publishing House of the Far-East branch of the USSR Academy of Sciences (in Russian), Vladivostok, **1** 82–93

Egorov Y. (2007): Tsunami wave generation by the eruption of underwater volcano. Nat. Hazard. Earth Syst. Sci. **7** 65–69

Fine I. V., Rabinovich A. B., Kulikov E. A., Thomson R. E., Bornhold B. D. (1998): Numerical modeling of landslide-generated tsunamis with application to the Skagway Harbor tsunami of November 3, 1994. In: Proc. Intern. Conf. on Tsunamis, Paris, May 26–28, 211–223

Gill A. E. (1982): Atmosphere and Ocean Dynamics. Academic Press, New York

Gusiakov V. K. (2001): 'Red', 'green' and 'blue' Pacific tsunamigenic earthquakes and their relation with conditions of oceanic sedimentation. In: Tsunamis at the End of a Critical Decade. Edited by G. Hebenstreit, 17–32. Kluwer, Dordrecht Boston MA London

Gutenberg B. (1939): Tsunamis and earthquakes. Bull. Seismol. Soc. Am. **29**(4) 517–526

Harbitz C. B. (1992): Model simulations of tsunamis generated by the Storegga slides. Mar. Geol. **105** 1–21

Heinrich, P., Piatensi, A., Okal, E., Hébert, H. (2000): Near-field modeling of the July 17, 1998 tsunami in Papua New Guinea, Geophys. Res. Lett. **27** 3037–3040

Hibiya T., Kajiura K. (1982): Origin of Abiki phenomena (a kind of seiches) in Nagasaki Bay. J. Oceanogr. Soc. Jpn. **38**(3) 172–182

Jansen E., Befring S., Bugge T., et al. (1987): Large submarine slides on the Norwegian continental margin: sediments, transport, and timing. Mar. Geol. **78** 77–107

Jiang L., LeBlond P. H. (1992): The coupling of a submarine slide and the surface waves which it generates. J. Geophys. Res. **97**(C8) 12731–12744

Jiang L., LeBlond P. H. (1994): Three-dimensional modeling of tsunami generation due to a submarine mudslide. J. Phys. Oceanogr. **24**(3) 559–572

Imamura, F., Gica, E.C. (1996): Numerical model for tsunami generation due to subaqueous landslide along a coast. Sci. Tsunami Hazard. **14**(1) 13–28

Imamura F., Hashi K., Imteaz Md. M. A. (2001): Modelling for tsunamis generated by landsliding and debris flow. In: Tsunami Research at the End of Critical Decade (ed. G. T. Hebenstreit), pp. 209–228. Kluwer, Dordrecht

Kharif Ch., Pelinovsky E. (2005): Asteroid impact tsunamis. Comptes Rendus Physique **6** 361–366

Kovalev P. D., Rabinovich A. B., Shevchenko G. V. (1991): Investigation of long waves in the tsunami frequency band on the southwestern shelf of Kamchatka. Nat. Hazard. **4** 141–159

Le Mehaute B., Wang S. (1996): Water waves generated by underwater explosion. World Sci., Singapoure

Kulikov E. A., Rabinovich A. B., Fine I. V., Bornhold B. D., Thomson R. E. (1998): Tsunami generation by slides on the Pacific coast of North America and the role of tides (in Russian). Oceanology **38**(3) 361–367

Kurkin A. A., Pelinovsky E. N. (2004): Freak waves: facts, theory and modelling (in Russian). Publishing house of Nizhegorod. State Techical University, N. Novgorod

Lander J. F. (1996): Tsunamis Affecting Alaska, 1737-1996. US Dept. Comm. Boulder

Leonidova N. L. (1972): On the possibility of exciting tsunami waves by muddy flows (in Russian). Works of SakhKNII of Far-East Scientific Center of USSR Academy of Sciences. Tsunami waves. No 29, pp. 262–270, Yuzhno-Sakhalinsk

Lichtman D. L. (1970): Physics of the atmospheric boundary layer (in Russian). Hydrometizdat, Leningrad

Mader C. L., Gittings M. L. (2006): Numerical model for the Krakatoa hydrovolcanic explosion and tsunami. Sci. Tsunami Hazard. **24**(3) 174–182

Miller D. J. (1960): The Alaska Earthquake on July 10, 1958: Giant wave in Lituya Bay. Bull. Seismol. Soc. Am. **50**(2) 253–266

Minoura K., Imamura F., Kuran U., et al. (2003): Tsunami hazard associated with explosion-collapse processes of a dome complex on Minoan Thera. In: Submarine Landslides and Tsunamis, pp. 229–236 Kluwer, Dordrecht

Mirchina N. P., Pelinovsky E. N. (1987): Dispersive amplification of tsunami waves (in Russian). Oceanology **27**(1) 35–40

Mitchel R. (1954): Submarine landslips of the coasts of Puerto-Rico and Barbados, West-Indies. Nature **173** 4394

Monserrat S., Ibbetson A., Thorpe A. J. (1991): Atmospheric gravity waves and the 'rissaga' phenomenon. Quart. J. Roy. Meteor. Soc. **117** 553–570

Munk W. H. (1962): Long ocean waves. In: The Sea. Ideas and Observations on Progress in the Study of the Sea, pp. 647–663 Wiley, New York

Murty T. S. (1977): Seismic sea waves – tsunamis. Bull. Fish. Res. Board Canada **198**, Ottawa

Murty T. S. (1984): Storm surges. Meteorological ocean tides. Department of Fisheries and Oceans, Bulletin 212, Ottawa

Nemtchinov, I. V., Svetsov V. V., Kosarev I. B., et al. (1997): Assessment of kinetic energy of meteoroids detected by satellite-based light sensors. Icarus **130** 259–274

Nomitsu T. (1935): A theory of tsunamis and seiches produced by wind and barometric gradient. Met. Coll. Sci. Imp. Univ. Kyoto A **18**(4) 201–214

Paine M. P. (1999): Asteroid Impacts: The Extra Hazard Due to Tsunami. Sci. Tsunami Hazard. **17**(3) 155–166

Parlaktuna M. (2003): Natural Gas Hydrates as a Cause of Underwater Landslides: a Review. In: Submarine Landslides and Tsunamis, pp. 163–169. Kluwer, Dordrecht

Pelinovsky E. N. (1996): Hydrodynamics of tsunami waves (in Russian). Institute of Applied Physics, RAS, Nizhnii Novgorod

Pelinovsky E., Talipova T., Kurkin A., Kharif C. (2001): Nonlinear mechanism of tsunami wave generation by atmospheric disturbances. Nat. Hazard. Earth Sys. Sci. **1** 243–250

Rabinovich A. B. (1993): Long gravitational waves in the ocean: capture, resonance, irradiation (in Russian). Hydrometeoizdat, St. Petersburg

Rabinovich A. B., Monserrat S. (1996): Meteorological tsunamis near the Balearic and Kuril Islands: descriptive and statistical analysis. Nat. Hazard. **13** 55–90

Rabinovich A. B., Thomson R. E., Kulikov E. A., et al. (1999): The landslide-generated tsunami of November 3, 1994 in Skagway Harbor, Alaska: a case study. Geophys. Res. Lett. **26**(19) 3009–3012

Rabinovich A. B., Thomson R. E., Bornhold B. D., et al. (2003): Numerical modelling of tsunamis generated by hypothetical landslides in the Strait of Georgia, British Columbia. Pure Appl. Geophys. **160**(7) 1273–1313

Ren, P., Bornhold, B. D., Prior, D. B. (1996): Seafloor morphology and sedimentary processes, Knight Inlet, British Columbia. Sediment. Geol. **103** 201–228

Roache P. J. (1976): Computational Fluid Dynamics. Hermousa, Albuquerque, N.M.

Schmidt R. M., Holsapple K. A. (1982): Estimates of crater size for large-body impacts: Gravitational scaling results. GSA Special Paper. **190** 93–101. GSA, Boulder

Shoemaker, E. M., Wolfe R. F., Shoemaker C. S. (1990): Asteroid and comet flux in the neighborhood of Earth. In: Global Catastrophes in Earth History (edited by V. L. Sharpton, P. D. Ward,). GSA Special Paper 247. GSA. Boulder co, pp. 155–170

Simpson J. E. (1987): Gravity currents: in the environment and laboratory. Halsted Press, England

Solem J. C. (1999): Comet and Asteroid Hazards: threat and mitigation. Sci. Tsunami Hazard. **17**(3) 141–153

Tappin D. et al. (1998): Sediment slump likely caused 1998 Papua New Guinea Tsunami. EOS **80** 329, 334, 340

Tikhonov A. N., Samarsky A. A. (1999): Equations of mathematical physics (in Russian). Publishing house of Moscow University, Moscow

Tinti S., Pagnoni G., Piatanesi A. (2003): Simulation of tsunamis induced by volcanic activity in the Gulf of Naples (Italy). Nat. Hazard. Earth Syst. Sci. **3** 311–320

Titov V. V., Gonzalez F. I. (2001): Numerical study of the source of the July 17, 1998 PNG tsunami. In: Tsunami Research at the End of a Critical Decade (ed. Hebenstreit G. T.), pp. 197–207. Kluwer, Dordrecht

Toon O. B., Zahnle K., Turco R. P., Covey C. (1994): Environmental perturbations caused by asteroid impacts. In: Hazards due to Comets and Asteroids (ed. Gehrels T.), pp. 791–826. University of Arizona Press, Tucson Az.

Vilibic I., Domijan N., Orlic M., Leder N., Pasaric M. (2004): Resonant coupling of a traveling air pressure disturbance with the east Adriatic coastal waters. J. Geophys. Res. 109 **C10001**, doi:10.1029/2004JC002279

Ward S. N., Asphaug E. (2000): Asteroid impact tsunami: a probabilistic hazard assessment. Icarus **145** 64–78

Ward S. N., Asphaug E. (2002): Impact tsunami—Eltanin. Deep-Sea Res. Part II **49** 1073–1079

Ward S. N., Day S. (2001): Cumbre Vieja Volcano—potential collapse and tsunami at La Palma, Canary Islands. Geophys. Res. Lett. **28** 397–400

Ward S. N., Day S. (2003): Ritter Island Volcano—lateral collapse and the tsunami of 1888. Geophys. J. Int. **154** 891–902

Waythomas C. F., Neal C. A. (1998): Tsunami generation by pyroclastic flow during the 3500-year B P caldera-forming eruption of Aniakchak Volcano, Alaska. Bull. Volcanol. **60** 110–124

Wiegel R. L. (1955): Laboratory studies of gravity waves generated by the movement of a submerged body. Trans. Am. Geophys. Union **36**(5)

Wiegel R. L., Noda E. K., Kuba E. M., et al. (1970): Water waves generated by landslides in reservoirs. J. Waterway. Harbour Coastal Eng., ASCE, **96** 307–333

Chapter 5
Propagation of a Tsunami in the Ocean and Its Interaction with the Coast

Abstract Traditional ideas of tsunami propagation in the open ocean are dealt with. The significance is estimated of manifestations of phase and amplitude dispersions. Classical problems are considered, concerning variation of the amplitude of a long wave in a basin with gently varying depth (Green's law) and the reflection of a wave from a step and from a rectangular obstacle. Formulae of the ray method are presented in Cartesian and spherical coordinate systems. Phenomena of long-wave refraction and capture by underwater ridges and the shelf are described. Estimation is performed of linear (viscous) and non-linear (turbulent) dissipation of the energy of long waves. The effect of a wave amplitude being reduced by scattering on bottom irregularities is considered. Approaches to the numerical simulation of tsunami wave propagation are described. Conventionally applied equations of nonlinear long-wave theory, taking into account the Coriolis force and bottom friction, are presented both in Cartesian and spherical coordinate systems. The technique for formulating initial and boundary conditions in the tsunami propagation problem is described. Brief information is given on certain tsunami models (codes) that are actively applied at present. Features of transoceanic wave propagation are considered, taking advantage of the 26 December 2004 tsunami as an example. The main results, due to investigation of the issues of a tsunami run-up on the shore, are presented.

Keywords Tsunami propagation · tsunami run-up · phase dispersion · amplitude dispersion · long-wave theory · Green's law · refraction · reflection · scattering · dissipation · numerical tsunami models · non-linear long-wave theory · bottom friction · Coriolis force · Manning coefficient · Cartesian reference system · spherical coordinates · initial elevation · boundary conditions · bathymetry · topography · waveguide · ridge · breaking waves · run-up height · laboratory experiments · numerical simulations

We have already noted that the following three stages are traditionally distinguished in the life of a tsunami: generation of the wave, its propagation in open ocean and its interaction with the coast (its uprush or run-up). The most simple task is to describe tsunami propagation in the open ocean. In this case, the wave's amplitude $A \sim 10^{-1}$–10^0 m is significantly smaller than the ocean depth $H \sim 10^3$ m, while the depth, in turn, is much smaller than the wavelength $\lambda = T(gH)^{1/2} \sim 10^4$–$10^6$ m. These two facts permit to apply with success the simplest linear theory of long waves. At any rate, the manifestations of amplitude and phase dispersion will be insignificant. To describe the run-up of waves is already a more complicated problem pertaining to the class of non-linear problems in a region with moving boundaries. Indeed, as one approaches the coast, the ocean depth decreases, while the tsunami amplitude increases, so the non-linearity parameter A/H is no longer a small quantity. Moreover, currents associated with the wave become turbulent, the influence of friction on the sea floor increases, processes resulting in suspension of bottom sediments are activated. A problem of no less complexity is presented by the fact that as the wave propagates over the land, the boundaries of the region, in which the hydrodynamic problem is resolved, alter quite essentially. First of all, this naturally concerns the advancement of the shoreline, but the 'water–air' and the 'water–bottom' (owing to erosion) boundaries also move. In this chapter, the main physical regularities, determining tsunami propagation in the open ocean and the run-up of waves on a shore, will be dealt with. Significant attention will be devoted to mathematical models, applied in numerical tsunami simulation.

5.1 Traditional Ideas Concerning the Problem of Tsunami Propagation

Right up till the last quarter of the twentieth century all measurements of tsunami waves were performed exclusively by coastal stations. Only during the past decades has the development of engineering reached a level that provides for the possibility of reliable tsunami registration in the open ocean and even at the very source during generation. Measurements of wave parameters, done with the aid of pressure sensors at the ocean bottom [Jacques, Soloviev (1971); Gonzalez et al. (1987); Kulikov, Gonzalez (1995); Milburn et al. (1996)], and satellite radio-altimeters [Okal et al. (1999); Kulikov et al. (2005)] permit to claim with certainty that the amplitude of a tsunami in open ocean, as a rule, lies between several centimeters and several tens of centimeters. In the most strong cases the amplitude of the free water surface displacement in the vicinity of the source may, apparently, reach several meters.

In any case, at great distances from the coast the tsunami amplitude A turns out to be essentially smaller than the ocean depth H. The value of H, in turn, is essentially inferior to the wave length λ. These two facts permit, in a first approximation, to consider tsunami as long (not subject to dispersion) linear waves, the velocity of which is determined by the simple formula $c = \sqrt{gH}$, where g is the acceleration of gravity. The period of tsunami waves, T, lies within the range 10^2–10^4 s. With account of the relationship $\lambda = T\sqrt{gH}$ it is possible to rewrite the condition $\lambda \gg H$ as $T\sqrt{g/H} \gg 1$. It is easy to verify that for the range of periods indicated this condition is always quite satisfied at small (shelf) depths. But for short-period tsunamis, propagating in the open ocean, fulfilment of this condition is not so evident.

5.1 Traditional Ideas Concerning the Problem

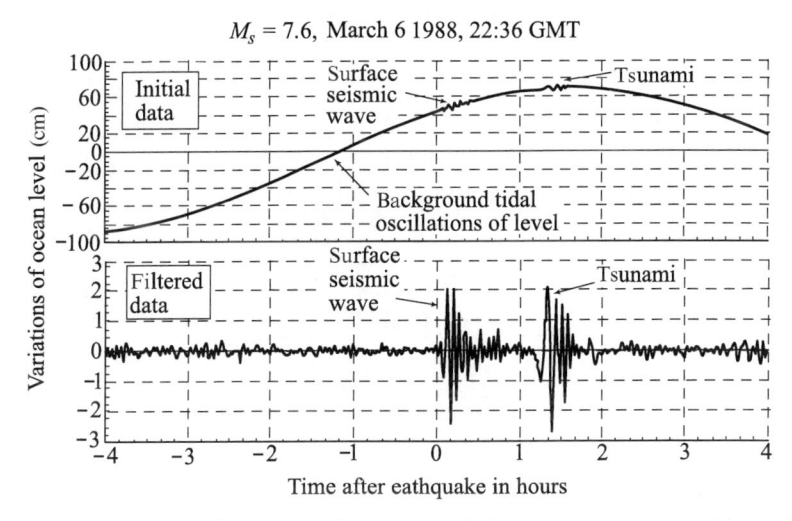

Fig. 5.1 Example of tsunami registration in open ocean by bottom pressure sensor (Adapted from [Milburn et al. (1996)])

The linear theory of long waves, representing the most simple version of a theory of gravitational surface waves on water, serves as quite an applicable approximation for describing the process of tsunami propagation in the open ocean along short routes. But, when waves cover long distances, dispersion and non-linear effects that exhibit the property of accumulating are capable of essentially altering not only the amplitude, but the very structure of the wave perturbation, also.

Manifestations of phase dispersion are well observed in measurements of tsunami waves in the open ocean by bottom pressure sensors. In Fig. 5.1 the example is presented of a record of ocean-level variations occurring during the passage of a weak tsunami, caused by an underwater earthquake in Alaska bay on March 6, 1988 [Milburn et al. (1996)]. The distance between the earthquake epicentre and the registration point was 978 km. The upper part of the figure shows the original record, in which tidal oscillations of the level prevail. The lower part of the figure presents the filtered signal, from which the tidal component has been removed. From the record two groups of waves are well distinguished. The first group, taking effect practically immediately after the earthquake, represents the response of the pressure sensor to the surface seismic wave. The second group of waves, delayed by over an hour, represents oscillations of the ocean level due to passage of the tsunami. From the tsunami record it is well seen that at the beginning there are long-period components of the signal, and only subsequently there appear short-period oscillations. This fact is established well by spectral-time analysis, the results of which are shown in Fig. 5.2. The spectral composition of the signal not only changes with time, but the behaviour of these changes satisfy the dispersion law for gravitational waves on water [Kulikov, Gonzalez (1995)]. The dispersion law is shown in the figure by the solid line, calculated by formula $t(\omega) = L/C_{gr}(\omega)$, where L is the distance from the sensor to the earthquake epicentre, $C_{gr}(\omega)$ is the group velocity, which is a function of the cyclic frequency ω.

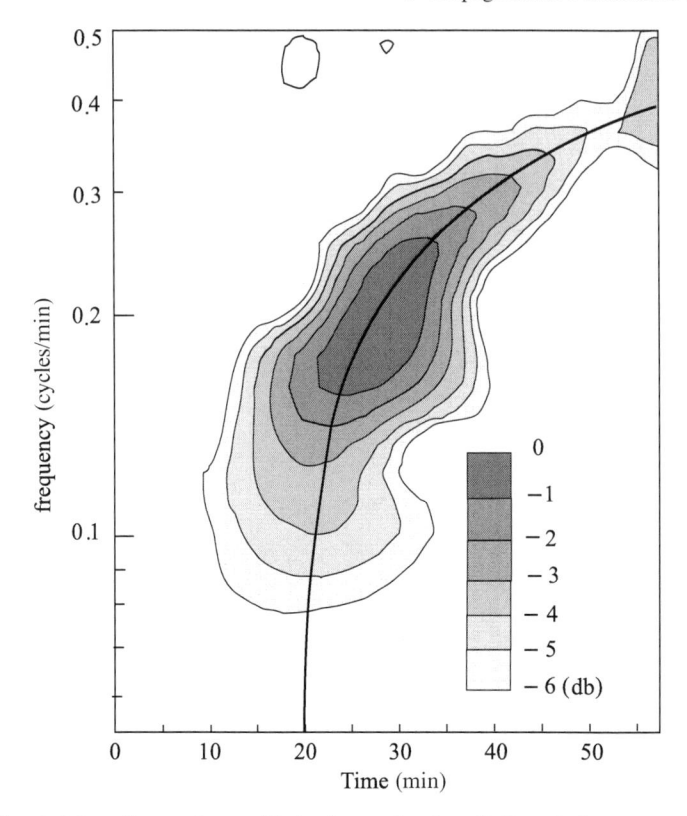

Fig. 5.2 Spectral-time diagram for amplitude of ocean-level oscillations during passage of tsunami of March 6, 1988, tsunami (the record is shown in Fig. 5.1). The isolines are drawn steps of 1 dB. The solid line shows theoretical calculation of time spectral components take effect, performed in accordance with the dispersion law for gravitational waves (Adapted from [Kulikov, Gonzalez (1995)])

Manifestations of tsunami wave dispersion have also been observed during analysis of the space structure of ocean level oscillations using the data of satellite altimetry (tsunami of December 26, 2004). These data are presented in Sect. 6.3.

We shall now estimate the distance, at which manifestations of dispersion effects should turn out to be quite significant. We shall take advantage of the dispersion relation for gravitational surface waves in a liquid, $\omega^2 = gk \tanh(kH)$, according to which the group velocity is determined by the formula

$$C_{gr} = \frac{\partial \omega}{\partial k} = \frac{g\left(\frac{kH}{\cosh^2(kH)} + \tanh(kH)\right)}{2\sqrt{gk \tanh(kH)}}.$$

The distance of dispersive destruction of a wave, L_{cd}, can be determined as the product of the velocity of long waves by the time, required for a wave packet to lag behind the front at a distance equal to the wavelength [Kulikov et al. (1996)],

5.1 Traditional Ideas Concerning the Problem

Fig. 5.3 Distance of dispersive destruction of a tsunami wave as function of period T and ocean depth (numbers at curves). The dotted line shows a distance equal to the Earth's equator, as a measure of a limit distance, that can be covered by a tsunami wave

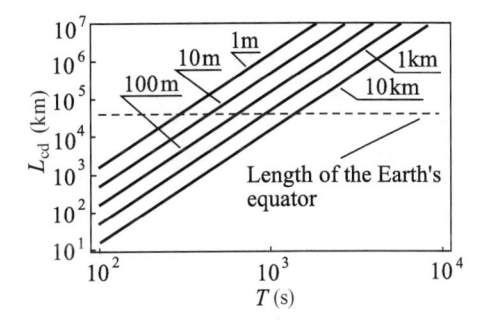

$$L_{cd} = \frac{\lambda(\omega)\sqrt{gH}}{\sqrt{gH} - C_{gr}(\omega)}. \tag{5.1}$$

The following approximate relation follows from formula (5.1), when $\lambda \gg H$:

$$L_{cd} \sim \lambda \left(\frac{\lambda}{H}\right)^2. \tag{5.2}$$

The dependence (5.1) is presented in Fig. 5.3. The periods of tsunami waves, that vary within limits of 10^2–10^4 s, are plotted along the x-axis. Calculations are performed for different depths of the water column (numbers near the curves). The dotted line in the figure shows the Earth's equator, indicating a measure of the limit distance, which can be covered by a tsunami wave. For typical depths of the open ocean the whole range of tsunami wave periods can be divided into two intervals. 'Short-period' waves ($T < 10^3$ s), for which the manifestation of dispersion may turn out to be significant, correspond to the first interval. In the second interval ($T > 10^3$ s), along routes not longer than the Earth's equator, no significant manifestation of dispersion will be observed. In those cases, when the wave periods exceed 100 s only slightly, the manifestation of dispersion will already be noticeable at relatively short distances of the order of 100–1,000 km.

Similar estimation can be performed in the case of transformation of a wave packet due to amplitude dispersion, arising as a consequence of non-linearity. Consider a wave with a crest of height A. The propagation velocity of the crest will differ from the velocity of linear long waves, its value can be estimated as $\sqrt{g(H+A)}$. By analogy with the distance of dispersive destruction, we introduce the distance of 'non-linear destruction' of a wave,

$$L_{cn} = \frac{\lambda\sqrt{gH}}{\sqrt{g(H+A)} - \sqrt{gH}}. \tag{5.3}$$

If $A/H \ll 1$, then the following approximate relation will be valid:

$$L_{cn} \sim \lambda \frac{H}{A}. \tag{5.4}$$

From formula (5.4) it is seen that, even when the tsunami wave height in open ocean is quite significant, $A = 1$ m, in the case of a typical depth $H = 4$ km and wavelength $\lambda = 100$ km, the value of L_{cn} will amount to about 400,000 km, which exceeds the length of the Earth's equator by an order of magnitude. Therefore, non-linear effects, in the case of tsunami propagation in the open ocean, can indeed be neglected.

The ratio of the quantities L_{cd} and L_{cn}, determined in accordance with the approximate formulae (5.2) and (5.4), gives the Ursell number $Ur = A\lambda^2/H^3$ [Pelinovsky (1996)], known in the theory of non-linear-dispersion waves on water. In open ocean, as a rule, $Ur \ll 1$, which means that phase dispersion prevails over non-linear effects. Near the coast (in shallow water), if microtsunamis are not considered, the parameter $Ur \gg 1$, i.e. non-linear effects become predominant. Estimation of the distances of dispersive and non-linear destruction, yielding formulae similar to (5.2) and (5.4), can be found in the book [Pelinovsky (1982)].

From the above analysis it follows that in simulating tsunamis, even along extended routes, the application of linear theory is quite justified. Moreover, long-wave theory is also quite appropriate for long-period waves. In this connection, it will be expedient to dwell upon certain partial results, following from the linear theory of long waves.

The ocean depth is the only variable quantity entering into the formula for the velocity of long waves, $c = \sqrt{gH}$. Therefore, many effects of tsunami propagation and run-up are related to the relief of the ocean bottom.

Consider the one-dimensional problem of the propagation of a long wave in a basin, the depth of which varies along the horizontal coordinate. We consider depth variations to be sufficiently smooth, so the reflection of waves from inclined sections of the bottom can be neglected. For definiteness, we shall consider a sine wave of length λ. Within the linear model, the kinetic and potential energies of the wave are equal to each other, therefore, the total energy attributed to a single space period (and to unit front length) can be calculated as twice the potential energy,

$$W = \rho g \int_0^\lambda \xi^2 dx, \qquad (5.5)$$

where ξ is the free water surface displacement from the equilibrium position and ρ is the density of water.

Since in a linear system the perturbation frequence remains unchanged, while the wavelength may change during propagation, it is worthwhile to perform in formula (5.5) integration over time, instead of space,

$$W = \rho g \sqrt{gH} \int_0^T \xi^2 dt = \text{const} \cdot \xi_0^2 \sqrt{H}. \qquad (5.6)$$

From energy conservation (we neglect dissipation, here) it follows that the quantity $\xi_0^2 \sqrt{H}$ must be conserved along the route of the wave propagation. In other words, if the ocean depth decreases, as the wave propagates, then the wave

5.1 Traditional Ideas Concerning the Problem

amplitude will increase by the law $\xi_0 \sim H^{-1/4}$. The relationship obtained is termed the Green's law or the 'one quarter' rule. This law, for instance, explains why the tsunami amplitude increases as it approaches the coast. Owing to a decrease in depth and, consequently, in propagation velocity, the wave packet shrinks in space, but boosts its amplitude.

Another 'classical' effect of the interaction of long waves with the relief consists in their transformation in the region of abrupt changes in the ocean depth. In those cases, when the ocean depth changes over distances much shorter than the wavelength, the distribution of depths is expediently represented in the form of a step (Fig. 5.4a). Such a situation is dealt with in many sectors of classical wave theory (optics, acoustics), and it is known as wave refraction and reflection at the boundary of two media. We shall only consider normal incidence of waves (the one-dimensional problem). We shall determine the amplitude coefficients for reflection, R, and transmission, T. To this end, following the classical book [Lamb (1932)], we take advantage of the continuity conditions for the free-surface displacement, ξ, and the water release (Hu) at the depth jump point. The resulting reflection and transmission coefficients R and T, respectively, are

$$R = \frac{\sqrt{H_1/H_2} - 1}{\sqrt{H_1/H_2} + 1}, \tag{5.7}$$

$$T = \frac{2\sqrt{H_1/H_2}}{\sqrt{H_1/H_2} + 1}. \tag{5.8}$$

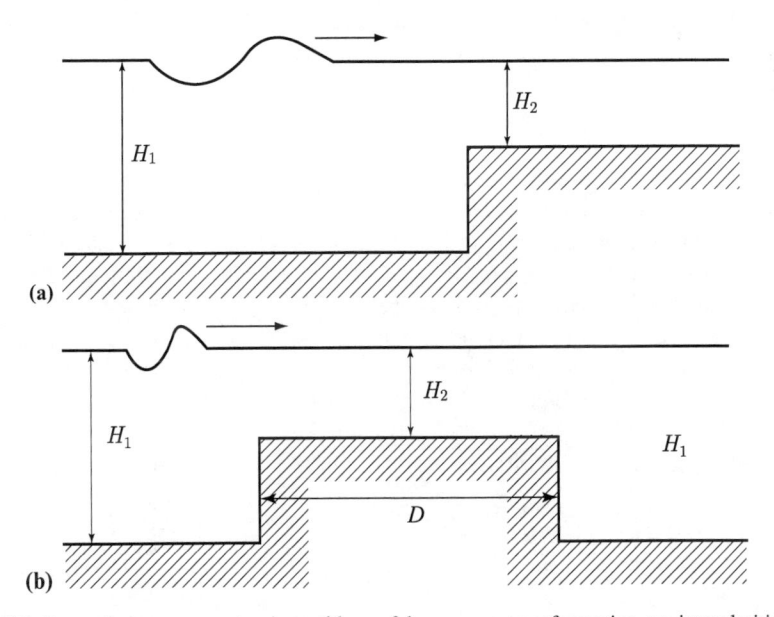

Fig. 5.4 Ocean bottom geometry in problem of long-wave transformation on irregularities of the bottom relief: step (**a**); rectangular obstacle (**b**)

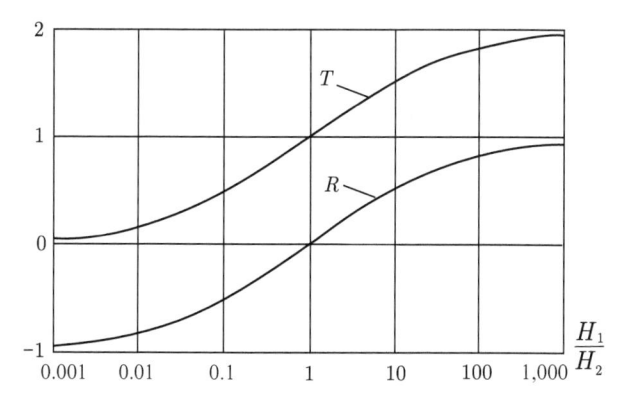

Fig. 5.5 Amplitude coefficients for transmission, T, and reflection, R, versus the depth ratio in the case of wave transformation on a step

The dependences (5.7) and (5.8), calculated within a wide range of depth ratios H_1/H_2, are shown in Fig. 5.5. If $H_1 > H_2$, then the waves that are transmitted through and reflected from, respectively, a step, will have the same polarity as the incident wave, and the amplitude of the transmitted wave will increase. In the case of transformation on a step the wave amplitude cannot be more than twice the initial value. When $H_1 < H_2$, the reflected wave changes polarity, and the amplitude of the wave, reaching deep water, is reduced.

Now, consider the 'classical' problem, akin to the previous one, of transformation of a long wave above a rectangular obstacle (Fig. 5.4a), of length D and height $|H_2 - H_1|$. The role of the obstacle can be assumed both by a local elevation of the bottom and by a depression. We note, right away, that this problem cannot be reduced to two consecutive independent acts of wave transformation on the front and back edges of the obstacle, i.e. on two steps. Anyhow, in the case of a solitary wave, with a length much shorter than the length of the obstacle, such an approach is quite adequate [Nakoulima et al. (2005)].

In the general case, a correct description of wave transformation above a rectangular obstacle requires the examination of a constrained system comprising five waves. Consider a sine wave incident upon the obstacle and travelling in the positive direction of axis $0x$. Then, in the regions $x < 0$ (before the obstacle) and $0 < x < D$ (above the obstacle) there exist two wave perturbations, propagating in both the positive and negative directions, while in the region $x > D$ there is only one perturbation, running in the positive direction. From the continuity condition for the free-surface displacement ξ and the water release (Hu) at points $x = 0$, D the following expression is obtained for the amplitude transmission coefficient [Mofjeld et al. (2000)]:

$$T = \frac{T_{\min}}{\sqrt{T_{\min}^2 \cos^2 \beta + \sin^2 \beta}}, \qquad (5.9)$$

5.1 Traditional Ideas Concerning the Problem

where $T_{\min} = \dfrac{2\sqrt{H_1/H_2}}{1+H_1/H_2}$, $\beta = k_2 D$ is the phase difference between the boundaries of the obstacle, k_2 is the wave number over the obstacle. In the case of transformation of a long wave on a step the transmission and reflection coefficients were only determined by the depth ratio and did not depend on any parameters of the wave. In the case of wave transformation above the rectangular obstacle the transmission coefficient turns out to depend on the wave frequency. The phase difference β is related to the wave number and, consequently, to the wave frequency, $\beta = \omega D/\sqrt{gH_2}$.

The dependence (5.9) is presented in Fig. 5.6. Its important peculiarity consists in the existence of a minimum transmission coefficient $T_{\min}$, the value of which is only determined by the depth ratio H_1/H_2, but does not depend on the width of the obstacle or the wavelength. The transmission coefficient is quite weakly related to the quantity H_1/H_2. The less the width of the obstacle, i.e. the smaller the ratio D/λ, the weaker this relationship happens to be.

If the width of the obstacle is small as compared to the tsunami wavelength ($D/\lambda < 0.2$), then an increase in the dimensions (width and height) of the obstacle unambiguously results in a decrease of the transmission coefficient. As soon as

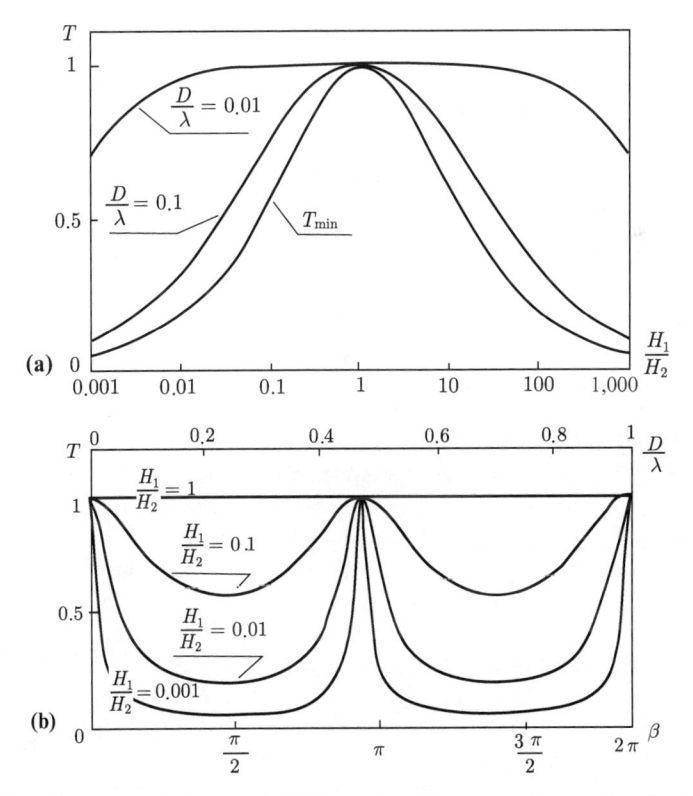

Fig. 5.6 Amplitude transmission coefficient for a long wave on a rectangular obstacle versus the depth ratio (**a**) and versus the phase difference between the edges of the obstacle (**b**)

the width of the obstacle is commensurable with the wavelength, interference effects start to become apparent. When the phase difference β (Fig. 5.6b) changes, the values of coefficient T change periodically from $T_{\min}$ up to 1. These changes are explained as follows. The interference between the waves, reflected from the front and rear boundaries of the obstacle, leads to mutual cancelling of the waves and, consequently, to amplification of the transmitted wave intensity. We recall that this effect is widely applied for producing antireflection optics. As to applications to the tsunami problem, we are, first of all, interested in wave transformation on obstacles of small sizes, $D \ll \lambda$. The point is that the transformation of waves by large-scale bottom irregularities, $D \gtrsim \lambda$, is automatically taken into account in numerical simulations. Therefore, it is practically important to estimate the contribution of small-scale (several kilometers and less) or so-called sub-net inhomogeneities, the size of which turns out to be smaller than the distance between the nodes of the mesh. We shall return to this estimate at the end of the section.

Small-scale inhomogeneities of the open ocean bottom exhibit heights significantly smaller than the thickness of the water column. Therefore, it is reasonable to introduce the relative height of an obstacle,

$$\alpha \equiv \frac{H_1 - H_2}{H_1},$$

that is a small quantity. When $\alpha \ll 1$ and $\beta \ll 1$, from formula (5.9) we obtain the simple approximate relation

$$T \approx 1 - \frac{(\alpha\beta)^2}{8}. \tag{5.10}$$

The expression obtained permits to assert that in the case of transformation of a wave passing above an obstacle, the decrease in amplitude is proportional to the square area of the obstacle.

We have hitherto considered influence of the bottom relief on tsunami waves within the framework of one-dimensional problems. Actually, tsunami propagation takes place in two-dimensional space: on a plane or on the surface of a sphere. Certain two-dimensional peculiarities of the bottom relief, such as underwater oceanic mountain ridges, the shelf, are capable, for example, of effectively capturing waves, thus creating priority directions for tsunami propagation and providing for prolonged 'sounding' of tsunamis at a coast. Phenomena of such kind are readily tracked making use of beam theory, which is also called an approximation of geometrical optics. Ray theory provides an effective instrument for operative calculation of tsunami arrival times. Its application permits to determine the contours of a tsunami source from data of the network of mareograph stations. Ray theory is extremely illustrative and permits to judge about the directions of tsunami energy propagation. Computational methods have been developed for calculating wave amplitudes on the basis of equations written 'along the ray' [Pelinovsky (1996)].

The velocity of a wave c being a function of two coordinates, x and y, the ray equations for nondispersive waves are written as follows [Lighthill (1978)]:

5.1 Traditional Ideas Concerning the Problem

$$\frac{dX}{dt} = c\frac{k_x}{\sqrt{k_x^2 + k_y^2}}, \tag{5.11}$$

$$\frac{dY}{dt} = c\frac{k_y}{\sqrt{k_x^2 + k_y^2}}, \tag{5.12}$$

$$\frac{dk_x}{dt} = -\frac{\partial c}{\partial x}\sqrt{k_x^2 + k_y^2}, \tag{5.13}$$

$$\frac{dk_y}{dt} = -\frac{\partial c}{\partial y}\sqrt{k_x^2 + k_y^2}, \tag{5.14}$$

where x and y are the coordinates of ray points, k_x and k_y are components of the wave vector and t represents time. The ray equations written in this form, permit not only to easily calculate the course of the wave ray, but also the evolution of the wave front. For computing the ray evolution the set of equations (5.11)–(5.14) must be supplemented by initial conditions consisting in determination of the initial coordinates and direction. It is not difficult to note that in the case of a fixed basin depth ($c = $ const) the rays will be straight lines.

Figure 5.7 presents two examples of the computation of ray behaviour, performed with the aid of formulae (5.11)–(5.14). The model bottom relief in the first case (Fig. 5.7a) imitates a mountain ridge in the middle of the ocean. Part of the rays emitted by a point-like source turn out to be captured, these rays further propagate

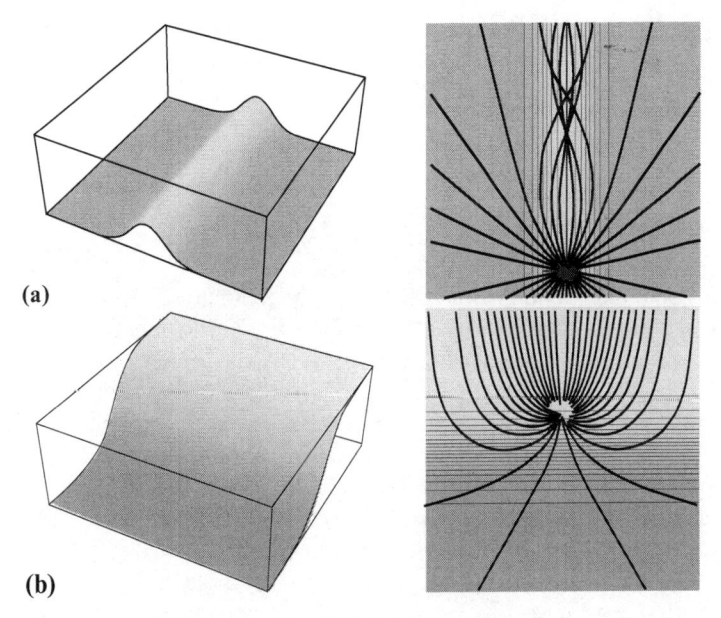

Fig. 5.7 Examples of influence of bottom relief on the course of wave rays emitted by a pointlike source: capture of rays by an underwater ridge (**a**); refraction and capture of waves in the shelf zone (**b**)

along the underwater eminence. Not all rays happen to be captured, but only those the direction of which does not differ strongly from the axis of the ridge, namely, in this case is the condition of total internal reflection realized.

Figure 5.7b shows the course of rays from a pointlike source, situated on the 'shelf'. It is well seen that part of the rays, curved at the beginning towards large depths, turn back to shallow water. Thus, refraction results in a significant part of the wave energy turning out to be captured by the shelf and to propagate along the coast. In Fig. 5.7b, one also observes a classical refraction effect: as the rays arrive in shallow water, they are turned around in the direction normal to the coastline.

A striking example of the role of waves, captured by the shelf is related to the tsunami that took place in Indonesia on December 12, 1992. This event is known as the tsunami of Flores island. At a distance of 5 km north of the Flores island coast there is a small island (the Babi island) of approximately circular shape and diameter about 2.5 km. The tsunami source was north of Babi island, however, the maximum run-up (7.1 m) was observed on the southern coast of the island. This effect is explained by the fact that the tsunami wave, having approached the island from the north, happened to be captured by the shelf, then turned round the island on both sides and provided maximum run-up on the coast of the back side of the island relative to the tsunami source. This effect has been studied with the use of laboratory simulation in [Yeh et al. (1994)].

In the analysis of real events, the use of equations (5.11)–(5.14) written in Cartesian coordinates is limited to small-scale areas of water. Beam computation for transoceanic routes requires taking into account the Earth's sphericity. Calculation of the path of a ray on a spherical surface is performed applying the following set of equations [Satake (1988)]:

$$\frac{d\theta}{dt} = \frac{\cos \varsigma}{nR}, \tag{5.15}$$

$$\frac{d\varphi}{dt} = \frac{\sin \varsigma}{nR \sin \theta}, \tag{5.16}$$

$$\frac{d\varsigma}{dt} = -\frac{\sin \varsigma}{n^2 R} \frac{\partial n}{\partial \theta} + \frac{\cos \varsigma}{n^2 R \sin \theta} \frac{\partial n}{\partial \varphi} - \frac{\sin \varsigma \cot \theta}{nR}, \tag{5.17}$$

where θ is the colatitude (supplement up to the latitude), φ is the ray longitude, t is time and the quantity $n = (gH)^{-1/2}$ is the inverse velocity of long waves, R is the Earth's radius, ς determines the ray direction counted anticlockwise from the direction toward the South. For computation of the evolution of wave rays knowledge is required of the distribution of ocean depths over latitude and longitude. At present, information on the global topography and bathymetry with a space resolution of 2 ang. min. (ETOPO2) is available on the site of the National Center of Geophysical Data (http://www.ngdc.noaa.gov/). An example of the application of equations (5.15)–(5.17) can be found, for example, in [Choi et al. (2003)].

Any wave motion in a nonideal (viscous) liquid is subject to dissipation. Tsunami waves also lose part of their energy as they travel, owing to its irreversible transformation into heat. We shall estimate the influence of dissipative factors on tsunami propagation.

5.1 Traditional Ideas Concerning the Problem

The damping of long gravitational waves in a viscous liquid is known to be due to energy dissipation within a thin bottom layer. The wave amplitude, here, decreases exponentially with time,

$$A \sim \exp\{-\gamma t\}.$$

The following formula has been obtained in the book [Landau, Lifshits (1987)] for the time decrement of amplitude damping for a wave propagating in a basin of constant depth H:

$$\gamma = \left(\frac{\nu\omega}{8H^2}\right)^{1/2},$$

where ν is the molecular kinematic viscosity of the liquid and ω is the cyclic frequency of the wave. If a long wave propagates only in the direction of axis Ox, then its amplitude will also decrease exponentially with the distance,

$$A \sim \exp\left\{-\frac{x}{L_1}\right\},$$

where

$$L_1 = \frac{\sqrt{gH}}{\gamma} = \left(\frac{8gH^3}{\nu\omega}\right)^{1/2}. \tag{5.18}$$

The physical meaning of quantity L_1 is the distance, over which the amplitude of a long wave in a viscous liquid becomes e times smaller. We shall call this quantity the viscous (linear) dissipation length.

Note that formula (5.18) is correct for any constant viscosity coefficient, which, generally speaking, can be both molecular and turbulent. One must, however, bear in mind that in the bottom boundary layer the turbulent viscosity, as a rule, depends strongly on the vertical component, i.e. is not a constant value. Therefore, it would not be quite correct to substitute into formula (5.18) any values of the turbulent viscosity coefficient. On the other hand, in a real ocean exchange of momentum does not proceed via molecular mechanisms, but by turbulence. Indeed, in spite of the relatively low flow velocity, characteristic of tsunami waves in open ocean, $u \approx A\sqrt{g/H} \sim 10^{-2}$ m/s, The Reynolds numbers turn out to be sufficiently large for the development of turbulence.

Determination of the quantity L_1 from formula (5.18) actually gives an idea of the minimum possible level of tsunami wave energy losses. Actually, owing to turbulence these losses may turn out to be more significant. We shall estimate the damping of tsunami waves on the basis of the known parameterization of frictional tension exerted by the ocean bottom on the water flowing along it with a velocity $\mathbf{v}$,

$$\mathbf{T}_B = -C_B \rho \mathbf{v} |\mathbf{v}|, \tag{5.19}$$

where ρ is the density of water, C_B is a dimensionless empirical coefficient, the value of which is usually set to 0.0025 [Murty (1984)]. The minus sign in formula (5.19) indicates that the water flow is hindered by a force directed against the flow velocity vector. The absolute value of the force of friction is proportional to the square flow velocity, therefore, the problem of wave damping under the action of bottom friction is certainly not linear, and, consequently, one can expect the damping not to be exponential in character.

For definiteness, we shall consider the one-dimensional problem of a sine wave propagating in the positive direction of axis $0x$ in a basin of fixed depth H,

$$u(x,t) = u_0 \sin(x - \sqrt{gH}t).$$

Assume the action of the friction force to be insignificant, so that the amplitude and shape of a wave covering a distance comparable with the wavelength λ undergo no significant changes. We shall estimate the wave energy per period in space (and per unit length of front), as twice the kinetic energy,

$$E = \rho H \int_0^\lambda u^2 dx = \frac{\rho H \lambda u_0^2}{2}. \tag{5.20}$$

Strictly speaking, such an estimate is only valid for linear waves, but, as we have already noted, we consider the non-linearity to be weak.

The losses of wave energy per unit time in the region from 0 up to λ, that are due to the action of bottom friction, are determined by the following formula (the point above the variable signifies differentiation with respect to time):

$$\dot{E} = \int_0^\lambda (T_B, u)\, dx = -C_B \rho \int_0^\lambda |u|^3 dx = -\frac{4 C_B \rho \lambda u_0^3}{3\pi}. \tag{5.21}$$

Now, pass in formulae (5.20) and (5.21) to specific energy per unit mass of liquid

$$b \equiv \frac{E}{\rho H \lambda} = \frac{u_0^2}{2}, \tag{5.22}$$

$$\dot{b} \equiv \frac{\dot{E}}{\rho H \lambda} = -\frac{4 C_B u_0^3}{3\pi H}. \tag{5.23}$$

Excluding the quantity u_0 in expression (5.23), with the aid of the constraint (5.22) one obtains the ordinary differential equation

$$\dot{b} = -\frac{8\sqrt{2} C_B b^{3/2}}{3\pi H}. \tag{5.24}$$

We recall, that we are tracing the energy of a sole space period of the wave. The ordinary differential equation (5.24) describes the variation of this quantity in time.

The solution of equation (5.24), written with respect to the wave velocity amplitude u_0, has the following form:

5.1 Traditional Ideas Concerning the Problem

$$u_0(t) = \frac{u_0(0)}{1 + \frac{4C_B u_0(0)}{3\pi H}t}, \quad (5.25)$$

where $u_0(0)$ is the velocity amplitude at time moment $t = 0$. Taking advantage of the relationship between the free-surface displacement and the flow velocity $u \approx \xi\sqrt{g/H}$ and taking into account that $x = t\sqrt{gH}$, we obtain an expression describing variation of the wave amplitude along the horizontal coordinate,

$$\xi_0(x) = \frac{\xi_0(0)}{1 + x/L_2}, \quad (5.26)$$

where $L_2 = \dfrac{3\pi H^2}{4C_B \xi_0(0)}$ is the distance, along which the wave amplitude becomes two times smaller.[1] We shall call this quantity the non-linear dissipation length.

We shall point out a number of special features, distinguishing viscous (linear) and non-linear damping of long waves from each other. First, the actual character of damping is different: in the first case it is exponential, while in the second it is hyperbolic. Second, the characteristic distance, along which noticeable wave damping occurs (L_1 and L_2), is related to different parameters of the problem. The quantity L_1 depends on the wave frequency and on the basin depth, while the quantity L_2 depends on the wave amplitude and depth. In both cases the distance L_i increases with the depth H, but in the case of non-linear damping this dependence is stronger.

Figure 5.8 presents the dependences of dissipation lengths L_1 and L_2 upon the ocean depth. Calculations are performed for characteristic ranges of tsunami wave frequencies (10^{-4}–10^{-2} Hz) and amplitudes (0.1–10 m) for the coefficient $C_B = 0,0025$, viscosity $v = 10^{-6}$ m^2/s. From the figure it is seen that for conditions of the open ocean, $H > 10^3$ m, viscous and non-linear friction cannot influence tsunami wave propagation in any noticeable way. For dissipative effects to

Fig. 5.8 Tsunami wave dissipation length versus ocean depth. Curves 1, 2—viscous (linear) dissipation, 3, 4—non-linear dissipation. The calculation is performed for $C_B = 0.0025$, $v = 10^{-6}$ m^2/s. Curve 1—10^{-4} Hz, 2—10^{-2} Hz, 3—0.1 m, 4—10 m. For comparison, the dotted line shows a distance equal to the length of the Earth's equator

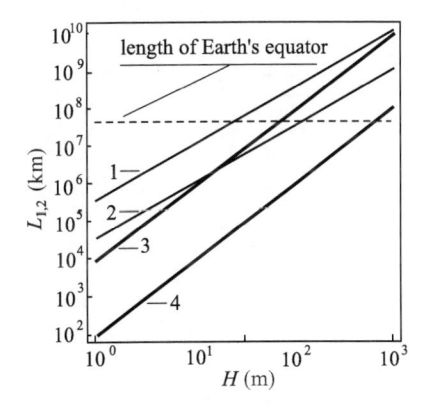

---

[1] With a precision up to a numerical coefficient, this quantity is in accordance with the result obtained in the book [Pelinovsky (1996)].

be manifested in a noticeable manner the wave must cover a distance exceeding the length of the Earth's equator, which is not possible in practice. From a purely theoretical point of view it is interesting that at large depths the quantities L_1 and L_2 become closer. Anyhow, this fact rather reflects the correct choice of coefficient C_B. The point is that viscous dissipation is not taken into account in tsunami models, applied in practice. At the same time non-linear dissipation is present in model equations. As it is seen, it provides approximately the same (albeit tiny) contribution to wave damping, which could have been provided by viscous dissipation.

Essential manifestations of dissipation are only possible at small depths $H < 10$ m. Here, the role of viscous linear wave dissipation turns out to be insignificant. Most likely, in shallow water the dissipation lengths will be related as $L_1/L_2 > 10$. Therefore, only taking into account non-linear dissipation, like it is presently done in practical models, can be considered justified. The role of classical linear dissipation can indeed be neglected. We stress, that one can speak of a noticeable influence of dissipation on a tsunami wave only in the case of very small depths. If we were to consider, for example, typical shelf depths $H \sim 100$ m, then the dissipation length would, most likely, exceed 1,000 km.

The wave amplitude decreasing as it propagates can be related not only to dissipation, but also to waves being scattered on small-scale irregularities of the ocean bottom. For estimating the significance of the scattering effect, we shall make use of the aforementioned results, concerning the transformation of long waves above a rectangular obstacle. Consider a 'comb' on the ocean bottom, consisting of rectangular obstacles with a repetition period in space of $2D$. Then, along a route of length x, the number of obstacles encountered by a wave will be $N = x/2D$. Each time interaction with an obstacle takes place, a decrease in the wave amplitude will occur determined by the transmission coefficient T. The law by which the wave amplitude A decreases with distance is written as

$$A(x) = A_0 T^N. \tag{5.27}$$

Formula (5.27) is expediently represented in a more customary exponential form,

$$A(x) = A_0 \exp\left\{-\frac{x}{L_3}\right\},$$

where $L_3 = -2D/\ln T$ is the characteristic distance, along which the wave amplitude is reduced by e times, owing to scattering on irregularities of the ocean bottom. In the case of small relative heights of the roughnesses α and a small phase difference β we obtain the simple formula

$$L_3 = \frac{16D}{(\alpha\beta)^2}.$$

Expressing parameter β through the wave length and the obstacle width, we ultimately obtain

$$L_3 = \frac{4}{\pi^2} \frac{\lambda^2}{\alpha^2 D}. \qquad (5.28)$$

Applying expression (5.28), we shall perform a simple estimation showing the negligible role of wave scattering by small-scale irregulartities of the ocean bottom. Let the tsunami wave length be 100 km and the ocean depth 4 km. Then, if the obstacle has a width $D = 1$ km and is 100 m high, then the quantity L_3 will amount to $6.5 \cdot 10^9$ m, which is equivalent to over 160 lengths of the Earth's equator.

5.2 Numerical Models of Tsunami Propagation

The headlong development of computational technologies, taking place in recent decades, has opened up new possibilities for numerical studies of problems of the mechanics of continuous media (MCM). The necessary computational facilities are now available to a wide range of researchers.

Description of tsunami evolution from the moment of generation to arrival of the wave on the shore represents one of the tasks of MCM. The application of analytical models for describing real tsunamis is limited, even if only for the complex topography of the ocean bottom. The only obvious alternative consists in numerical modelling. The efficiency of such means for studying tsunamis has long been unanimously acknowledged by the scientific community. Hopes of resolving the problem of tsunami prediction are also to a great extent related to the development of numerical models.

The 'age' of numerical simulation of real tsunamis started at the end of the 1960s. The first works in this direction were performed by Japanese researchers [Aida (1969), (1974); Abe (1978), (1979)]. One of the first numerical models developed in Russia was described in [Gusyakov, Chubarov (1982), (1987)] and [Chubarov et al. (1984)].

As a rule, numerical tsunami models are based on the theory of long waves (in shallow water), which deals with the equations of hydrodynamics, averaged over the vertical coordinate. Within the theory of long waves, the total three-dimensional (3D) problem reduces to the two-dimensional (2D) one, numerical resolution of which requires a relatively small volume of calculations.

The simulation of tsunami propagation at transoceanic scales within the complete 3D model is not only impossible, at present, but also obviously irrational. The resolution of such a 3D problem is only of purely scientific, but not practical, interest. The point is that in most cases tsunami wave propagation is quite satisfactorily described by the linear theory of long waves. Taking into account the insignificant manifestations of phase dispersion and non-linearity, which are peculiar to tsunami waves, can also be done within the framework of long-wave non-linear-dispersion models [Pelinovsky (1996); Satake, Imamura (1995); Rivera (2006); Horrillo et al. (2006)].

Note that the theory of long waves does not always describe processes at the tsunami source correctly. As it was shown in Chap. 3, three-dimensional formulation of the problem for describing processes at the source may happen to be quite essential.

In a Cartesian reference system the equations of non-linear theory of long waves, with account of the bottom friction and the Coriolis force, has the following form:

$$\frac{\partial U}{\partial t} + U\frac{\partial U}{\partial x} + V\frac{\partial U}{\partial y} = -g\frac{\partial \xi}{\partial x} - \frac{C_B U \sqrt{U^2 + V^2}}{D} + fV, \quad (5.29)$$

$$\frac{\partial V}{\partial t} + U\frac{\partial V}{\partial x} + V\frac{\partial V}{\partial y} = -g\frac{\partial \xi}{\partial y} - \frac{C_B V \sqrt{U^2 + V^2}}{D} - fU, \quad (5.30)$$

$$\frac{\partial \xi}{\partial t} + \frac{\partial}{\partial x}(DU) + \frac{\partial}{\partial y}(DV) = 0, \quad (5.31)$$

$$U = \frac{1}{D}\int_{-H}^{\xi} u\,dz, \quad V = \frac{1}{D}\int_{-H}^{\xi} v\,dz,$$

where U, V are the flow velocity components, averaged over the depth, along the axes $0x$ and $0y$, respectively, ξ is the free-surface displacement from the equilibrium position, $D(x,y,t) = H(x,y) + \xi(x,y,t)$ is the thickness of the water column, g is the acceleration of gravity, $f = 2\omega \sin\varphi$ is the Coriolis parameter, ω is the angular velocity of the Earth's rotation, φ is the latitude and C_B is a dimensionless empirical coefficient, which is usually set to 0.0025. There also exist more precise models, which take into account the dependence of quantity C_B on the thickness of the water column. Thus, for example, the following dependence is applied in [Titov et al. (2003)]:

$$C_B = \frac{gn^2}{D^{1/3}}, \quad (5.32)$$

where n is the Manning coefficient, the value of which depends on the roughness of the bottom surface. A typical value of the Manning coefficient for a coast free from dense vegetation, amounts to $n = 0.025$ s/m$^{1/3}$.

Note that formula (5.32) yields the value $C_B = 0.0025$ for a water column of $D \approx 15$ m. The dependence of $C_B(D)$ is weak ($C_B(1\,\text{m}) \approx 0.006$, $C_B(100\,\text{m}) \approx 0.0013$), therefore, the results of calculations of wave dynamics carried out assuming the coefficient C_B to be constant and with account of the dependence (5.32), should not differ strongly from one another. We recall that the bottom friction does practically not influence tsunami propagation at large depths.

In certain models another form is used for writing the non-linear equations of the theory of long waves ('in total fluxes'),

$$\frac{\partial M}{\partial t} + \frac{\partial}{\partial x}\left(\frac{M^2}{D}\right) + \frac{\partial}{\partial y}\left(\frac{MN}{D}\right) = -gD\frac{\partial \xi}{\partial x} - \frac{C_B M\sqrt{M^2 + N^2}}{D^2} + fN, \quad (5.33)$$

$$\frac{\partial N}{\partial t} + \frac{\partial}{\partial x}\left(\frac{MN}{D}\right) + \frac{\partial}{\partial y}\left(\frac{N^2}{D}\right) = -gD\frac{\partial \xi}{\partial y} - \frac{C_B N\sqrt{M^2 + N^2}}{D^2} - fM, \quad (5.34)$$

5.2 Numerical Models of Tsunami Propagation

$$\frac{\partial \xi}{\partial t} + \frac{\partial M}{\partial x} + \frac{\partial N}{\partial y} = 0, \tag{5.35}$$

where $M = UD$, $N = VD$ are components of the water release along the $0x$ and $0y$ axes, respectively. Transition from system (5.29)–(5.31) to system (5.33)–(5.35) is performed as follows. Equation (5.29) is multiplied by the quantity D, while equation (5.31), in which the partial derivative $\partial \xi/\partial t$ is replaced by the equivalent quantity $\partial D/\partial t$, is multiplied by the velocity component U. Upon adding up the obtained expressions and performing elementary transformations, we obtain equation (5.33). Equation (5.34) is derived in a similar manner. Transition from formula (5.31) to (5.35) is trivial and requires no comments.

Note that system (5.29)–(5.31) is not a rigorous consequence of the equations of hydrodynamics. First of all, this is due to the expression for the force of bottom friction having been obtained from an empirical dependence. Moreover, the stopping of the flow is due to tangential tension, acting only on the lower boundary. This circumstance hinders the rigorous derivation of non-linear equations for long waves. However, linear equations (without advective terms) can be obtained in a rigorous manner by integration of the linearized Reynolds equations along the vertical coordinate from the bottom up to the free water surface.

In calculating tsunami propagation along extended routes account must be taken of the curvature of the Earth's surface. The form of the surface of our planet can be considered spherical with a precision sufficient for our problem, therefore, it is expedient to write the equations of the theory of long waves in spherical coordinates,

$$\frac{\partial U}{\partial t} + \frac{1}{R \cos \varphi} \left(U \frac{\partial U}{\partial \psi} + V \cos \varphi \frac{\partial U}{\partial \varphi} \right) - \frac{UV \tan \varphi}{R}$$
$$= -\frac{g}{R \cos \varphi} \frac{\partial \xi}{\partial \psi} - \frac{C_B U \sqrt{U^2 + V^2}}{D} + fV, \tag{5.36}$$

$$\frac{\partial V}{\partial t} + \frac{1}{R \cos \varphi} \left(U \frac{\partial V}{\partial \psi} + V \cos \varphi \frac{\partial V}{\partial \varphi} \right) + \frac{U^2 \tan \varphi}{R}$$
$$= -\frac{g}{R} \frac{\partial \xi}{\partial \varphi} - \frac{C_B V \sqrt{U^2 + V^2}}{D} - fU, \tag{5.37}$$

$$\frac{\partial \xi}{\partial t} + \frac{1}{R \cos \varphi} \left(\frac{\partial (UD)}{\partial \psi} + \frac{\partial (VD \cos \varphi)}{\partial \varphi} \right) = 0, \tag{5.38}$$

where ψ is the longitude, φ is the latitude, U and V are the flow velocity components, averaged over the depth, along the parallel (west–east) and along the meridian (north–south), respectively, $R \approx 6{,}371$ km is the mean radius of the Earth.

The set of equations of long-wave theory, written in a Cartesian or spherical reference system, is usually resolved with initial conditions (initial elevation), representing a free-surface displacement, equivalent to vertical residual deformations of the ocean bottom, resulting from an earthquake. The initial field of flow velocities is assumed to be zero.

Here, we shall no longer deal with a certain physical incorrectness of the traditional approach, related to the effects of water compressibility being neglected.

This issue has been dealt with in detail in Sect. 2.1.1 and in Chap. 3. Following the existing practical techniques for tsunami calculations, we shall discuss the formulation of initial conditions assuming water to be an incompressible liquid.

The traditional approach, within which the initial elevation is set equal to the vertical residual bottom deformation, more or less adequately reproduces the main effect responsible for tsunami generation—ousting of the water. Moreover, this approach is readily realized in practice.

The imperfectness of the traditional approach is due to at least two reasons. First, even if the bottom is horizontal, and the ocean bottom deformations are instantaneous ($\tau \ll L(gH)^{-1/2}$, where L is the horizontal extension of the source), the displacement of the water surface at the moment the motion terminates, $\xi(x,y,\tau)$, and the vertical residual bottom deformation $\eta_z(x,y,\tau)$ will not be equal to each other. The point is that the water column smoothes out the inhomogeneities of function $\eta_z(x,y,\tau)$ of scale $\Delta L < H$, so that such inhomogeneities are weakly manifested on the water surface. An attempt at taking this effect into account has been done in [Tanioka, Seno (2001)] and [Rabinovich et al. (2008)]. Second, in the case of an inclined (rough) bottom the horizontal deformation components can also contribute significantly to function $\xi(x,y,\tau)$.

A logical development of the traditional approach is to determine the initial elevation $\xi(x,y,\tau)$ from the solution of the 3D problem taking into account all three components of the bottom deformation vector $\eta \equiv (\eta_x, \eta_y, \eta_z)$ and the distribution of depths $H(x,y)$ in the vicinity of the source. This problem can be resolved, for example, if formulated as follows:

$$\Delta F = 0, \tag{5.39}$$

$$F_{tt} = -gF_z, \quad z = 0, \tag{5.40}$$

$$\frac{\partial F}{\partial \mathbf{n}} = \left(\frac{\partial \eta}{\partial t}, \mathbf{n}\right), \quad z = -H(x,y), \tag{5.41}$$

where F is the potential of the flow velocity. The initial elevation sought is calculated by the formula $\xi(x,y,\tau) = -g^{-1}F_t(x,y,0,\tau)$. If the bottom deformation process turns out to be long ($\tau \sim L(gH)^{-1/2}$), then the initial conditions of the propagation problem must include, besides the initial elevation, the initial distribution of flow velocities, as well,

$$\{U(x,y,\tau), V(x,y,\tau)\} = \frac{1}{H(x,y)} \int_{-H(x,y)}^{0} dz \{F_x(x,y,z,\tau), F_y(x,y,z,\tau)\}.$$

In case the deformation can be considered instantaneous, the evolution problem (5.39)–(5.41) can be reduced to a more simple static problem:

$$\Delta \hat{F} = 0, \tag{5.42}$$

$$\hat{F} = 0, \quad z = 0, \tag{5.43}$$

5.2 Numerical Models of Tsunami Propagation

$$\frac{\partial \hat{F}}{\partial \mathbf{n}} = (\eta, \mathbf{n}), \quad z = -H(x,y), \tag{5.44}$$

where

$$\hat{F} = \int_0^\tau F \, dt.$$

The initial elevation is determined from the solution of problem (5.42)–(5.44) by the following formula:

$$\xi(x,y) = \hat{F}_z(x,y,0).$$

The vector of bottom deformations η, entering into formulae (5.41) and (5.44), is calculated making use of the analytical solution of the stationary problem of elasticity theory [Okada (1985)]. The calculation technique is described in detail in Sect. 2.1.3.

As a rule, the boundary conditions used for simulating tsunami propagation within the theory of long waves pertain to one of the following three types [Marchuk et al. (1983)]:

1. interaction with the coast
2. non-reflecting
3. perturbation, arriving from external area.

In the most simple case, the interaction of waves with the coast is described as total reflection from the coast. To this end one considers that on a certain fixed isobath (usually, 10–20 m) the flow velocity component normal to the coastline (or the chosen isobath) turns to zero,

$$V_\mathbf{n} = 0.$$

A direct consequence of this condition is the equality to zero of the component normal to the shoreline (or the chosen isobath) of the derivative of the free surface displacement,

$$\frac{\partial \xi}{\partial \mathbf{n}} = 0.$$

The condition of total reflection is usually applied in those cases, when the main goal is to investigate wave propagation in the open ocean. In analysing tsunami dynamics in the shelf zone a more detailed description is necessary of the interaction of waves with the coast. Here, it has sense to consider partial reflection of waves and to make use of the formula proposed by A. V. Nekrasov [Nekrasov (1973)],

$$V_\mathbf{n} = \frac{1-r}{1+r} \frac{\xi \sqrt{gH}}{H+\xi},$$

where the parameter r, characterizing the degree of reflection, varies within limits from 0 up to 1.

A more complex version of the description of tsunami interaction with the coast implies numerical simulation of waves running up the coast. We shall dwell upon methods for resolving this problem in Sect. 5.3.

In those cases, when detailed simulation of the tsunami dynamics within a restricted region is required, the necessity often arises to make use of boundaries that freely transmit incident waves. In other words, the amplitude of a wave, reflected from such a boundary, should be reduced to the minimum. The physical principle for realization of such a 'non-reflecting' boundary condition is quite simple. At each moment of time a boundary point is assigned that value, which should be brought to it by the wave incident upon the boundary. However, technical realization of the condition of free transmission turns out to be elementary only in the one-dimensional case. If one considers wave propagation along the $0x$ axis, then the condition of free transmission will be of the form

$$\frac{\partial u}{\partial t} = \pm c \frac{\partial u}{\partial x}, \qquad (5.45)$$

where $c = \sqrt{gH}$ is the velocity of long waves. The quantity u in formula (5.45) is understood to be any of the sought functions (the free-surface displacement or the flow velocity component).

A condition of the same form as (5.45) is also applicable in resolving two-dimensional problems, but it will no longer provide for ideal free transmission through the boundary $x = $ const of waves, travelling at a certain angle to the $0x$ axis. Regretfully, no success has been achieved in totally avoiding the reflected wave, when resolving the problem on a plane. It is possible to reduce the amplitude of waves, reflected by the boundary, by enhancing the order of the boundary condition approximation [Marchuk et al. (1983)],

$$c\frac{\partial^2 u}{\partial x \partial t} = \frac{\partial^2 u}{\partial t^2} - \frac{c^2}{2}\frac{\partial^2 u}{\partial y^2},$$

$$c\frac{\partial^2 u}{\partial t^2} - \frac{c^3}{4}\frac{\partial^3 u}{\partial y^2 \partial x} = \frac{\partial^3 u}{\partial t^3} - \frac{3c^2}{4}\frac{\partial^2 u}{\partial y^2 \partial t}.$$

There also exists another approach to realizing nonreflecting boundaries. It consists in the introduction of an absorbing layer in the vicinity of the boundary [Israeli, Orzag (1981); Kosloff (1986)].

The third type of boundary conditions is the most simple to realize. If a certain perturbation approaches the boundary from outside of the calculation region, then at all points of the boundary one must set the velocity components and the surface displacement to correspond to this perturbation. Depending on the concrete problem these quantities can either be determined from the solution of another numerical problem or be given by certain functions.

The key information, upon the reliability of which the precision of numerical tsunami calculations depends, comprises data on the ocean bottom bathymetry and on the topography of the coastal area. At present, free access is provided to a 2-min global database for the Earth's relief (ETOPO2, http://www.ngdc.noaa.gov/)

5.2 Numerical Models of Tsunami Propagation

and a 1-min digital atlas (GEBCO, British Oceanographic Data Centre, http://www.ngdc.noaa.gov/mgg/gebco/). For some regions data are available with a significantly improved space resolution, for example, like in the NGDC Tsunami Inundation Gridding Project (http://www.ngdc.noaa.gov/mgg/inundation/tsunami/). In run-up simulation it may turn out to be useful to take advantage of 3 arc second data, obtained by the Shuttle Radar Topography Mission (SRTM). SRTM successfully collected data over 80% of the Earth's land surface, for all the area between 60° N and 56° S latitude. The data are available at the site http://seamless.usgs.gov/.

At present, many numerical tsunami models have been developed that are based on the theory of long waves. Not claiming to present a full list, we shall only present several of the actively applied models.

The first model is 'MOST' (Method Of Splitting Tsunami), developed by V. V. Titov, a graduate of the Novosibirsk university, who is presently with the Pacific Marine Environmental Laboratory (PMEL) at Seattle (USA). Detailed information on the model is to found in [Titov, Synolakis (1995), (1998); Titov, Gonzalez (1997)] and [Titov et al. (2003)].

A widely renowned model is 'TUNAMI-N2' [Goto et al. (1997)]. Among specialists it is often called the Shuto (Shuto N.) model or the Imamura (Imamura F.) model, in spite of the fact that the initial version was developed by T. Takahashi [Takahashi et al. (1995)]. The model has been recommended by UNESCO for tsunami calculations, and nowadays it is applied in many countries. The first numerical simulation of the 2004 Indonesian tsunami, performed in Russia, made use of a software complex, that represented an improved version of TUNAMI-N2 [Zaitsev et al. (2005)].

The Zigmund Kowalik (Kowalik Z.) model, modified by Elena Troshina-Suleimani [Suleimani et al. (2003); Kowalik et al. (2005)], was developed at the Geophysical Institute of the Alaska University (Geophysical Institute University of Alaska Fairbanks).

The Antonio Baptista model [Myers, Baptista (1995)], based on realization of the method of finite elements, represents a modified version of the ADCIRC model of storm surges.

We shall further touch upon certain results of numerical simulation of the Indonesian catastrophic tsunami, that took place on December 26, 2004. The example of this tsunami will be used in describing characteristic features of tsunami wave propagation for demonstrating the possibilities of modern numerical models. We shall mainly adhere to results, obtained in [Titov et al. (2005)].

The tsunami of December 26, 2004, happened to be the first global event, for which there were high-quality measurements of the level supplemented with data from satellite altimeters. The first instrumental measurement of this tsunami appeared 3 h after the earthquake—the wave was registered by a station on the Coconut islands (Fig. 5.9) at about 1,700 km from the epicentre. According to these data, the first wave was only 30 cm high. The first wave was followed by prolonged level oscillations with a maximum amplitude not exceeding 53 cm. At the same time, at a number of coastal sites of India and Sri Lanka, located at approximately the same distance, waves ten times higher than on the Coconut islands were registered. Such a large difference in amplitude, confirmed by the results of numerical simulation, demonstrates a pronounced orientation of the wave energy emission. The data from other mareographs in the Indian Ocean showed wave amplitudes between 0.5 and 3 m, and no noticeable damping was observed as the distance from the source increased.

Note that the wave heights measured by mareographs are not always in good accordance with the tsunami run-up heights on the coast. Several records, obtained from regions with significant run-ups, registered wave heights 2–5 times smaller, than actually observed values. Thus, for example, the mareograph at Phuket showed 1.5 m, while the actual run-up height was from 3 to 6 m. This divergence strongly complicates determination of the true tsunami height on the coast. Moreover, many mareographs in the Indian Ocean were destroyed (Thailand) or happened to be strongly damaged (Colombo, Sri Lanka). Therefore, the true maximum wave heights may have remained unknown.

Data on tsunamis in the remote zone revealed that, unlike manifestations near the source, the maximum wave amplitude was not associated with the leading wave. In the North Atlantic and at the North of the Pacific Ocean maximum wave heights were observed with delays from several hours up to several days after the onset of the tsunami front (Fig. 5.9). It is interesting to note that at Callao (Peru), situated 19,000 km from the source, the waves were higher than on the Coconut islands, lying significantly closer (1,700 km). Moreover, the tsunami amplitude at Halifax (Nova Scotia, Canada) was also greater, while in this case the waves had to cross not only the Indian but also the Atlantic Ocean (longitudinally) and in doing so to cover over 24,000 km.

Model studies of tsunami propagation in the open ocean permit to obtain a picture of energy propagation, which cannot be reconstructed having only the data of

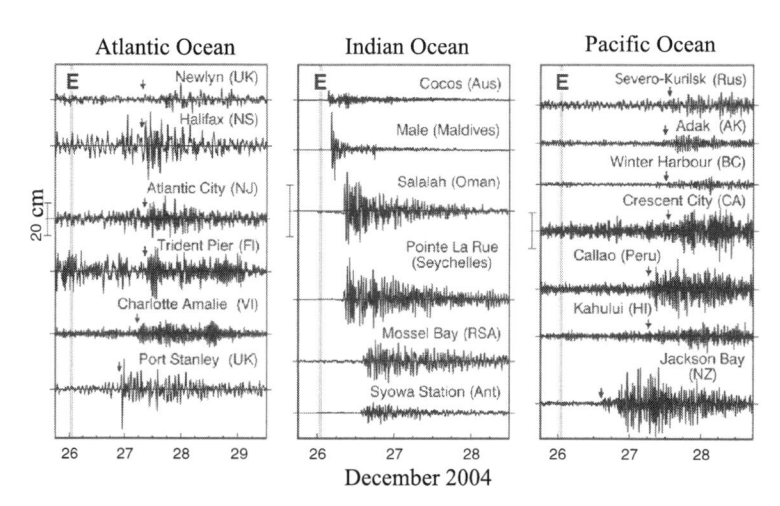

Fig. 5.9 Time series of tsunami wave heights (cm) as recorded at selected tide-gauge stations in the three major ocean basins. Arrows indicate first arrival of the tsunami (Reprinted from [Titov et al. (2005)] by permission of the publisher)

5.2 Numerical Models of Tsunami Propagation

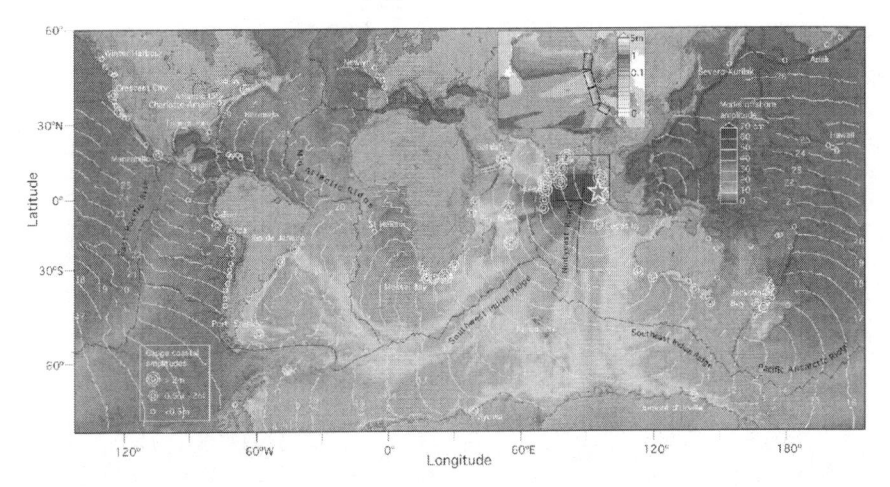

Fig. 5.10 Global chart showing energy propagation of the 2004 Sumatra tsunami calculated from MOST. Filled colors show maximum computed tsunami heights during 44 hours of wave propagation simulation. Contours show computed arrival time of tsunami waves. Circles denote the locations and amplitudes of tsunami waves in three range categories for selected tide-gauge stations. Inset shows fault geometry of the model source and close-up of the computed wave heights in the Bay of Bengal. Distribution of the slip among four subfaults (from south to north: 21 m, 13 m, 17 m, 2 m) provides best fit for satellite altimetry data and correlates well with seismic and geodetic data inversions (see also Plate 6 in the Colour Plate Section on page 314) (Reprinted from [Titov et al. (2005)] by permission of the publisher)

coastal measurements at one's disposal. Since the tsunami dynamics in the open ocean is linear, the height of a wave is proportional to the square root of its energy. Thus, the space distribution of calculated maximum wave heights, presented in Fig. 5.10, provides a clear picture of tsunami energy propagation. Numerous versions of calculations, performed for different values of bottom deformations, sizes and orientations of the source, have revealed that all these parameters insignificantly influence wave propagation in the remote zone. We right away note that in the close zone the shape and orientation of the source happen to be decisive parameters.

A very important fact, testifying in favour of the numerical model being adequate, is the good agreement between wave amplitudes, resulting from calculations, and those registered by coastal stations. Thus, for example, the anomalously high values of amplitudes in the remote zone reflect precisely the main directions of wave energy propagation. The coastal stations in Halifax (Canada), Manzanillo (Mexico), Callao (Peru), Arice (Chile) recorded wave heights exceeding 50 cm. Being at a significant distance from the source (over 20,000 km), each of the sites indicated is to be found in an area, related to the end of one of the 'wave rays'.

Numerical calculations, corroborated by in situ data, confirm the assumptions that two main factors influence tsunami propagation: the source configuration (geometry) and the wave-guide properties of mid-oceanic ridges. We recall that the continental shelf can also serve as a wave guide. In many cases waves, captured by the shelf, are the cause of prolonged oscillations of the water level at the coast.

In the nearby zone, the orientation of energy emission was related to the large extension of the tsunami source. The long and narrow region (stretched out in the meridional direction), in which deformations of the ocean bottom, caused by the earthquake of December 26, 2004, were concentrated, formed waves of large amplitude in the perpendicular, i.e. longitudinal, direction. The waves propagating in the meridional direction were of essentially smaller amplitude. This effect not only manifests in simulations, but also follows from analysis of records of mareographs and expedition data. Thus, for example, on the opposite coast of the Indian Ocean, 5,000 km from the source, on the Somalia coast (East Africa) run-up heights from 5 up to 9 m have been observed [Synolakis et al. (2005)]. From numerical calculations of the distribution of maximum amplitudes (Fig. 5.10) it is seen that one of the 'wave rays' ends precisely on this coast.

The main factor, determining the orientation of energy propagation in the remote zone, is now the topography of the bottom of the World Ocean (Fig. 5.11 in the coloured inset). Analysis of the Indonesian tsunami of December 26, 2004, reveals the important role of mid-oceanic ridges in channeling the tsunami energy. From comparison of Figs. 5.10 and 5.11 it is readily seen that the south-west Indian Ridge, as well as the Mid-Atlantic Ridge served as waveguides for propagating the tsunami towards the Atlantic. The Pacific Antarctic Ridge and south-east Indian Ridges, and, also, the East-Pacific Rise contributed to the penetration of waves into the Pacific Ocean. It is interesting that ridges cope well with the role of waveguides until their curvature does not exceed a critical value. Thus, for example, the sharp bend in the Mid-Atlantic Ridge at the parallel of 40° S contributed to the waveguide losing beams. As a result, waves of noticeable amplitude were observed at the Atlantic coast of South America. The numerical model pointed correctly to the significant amplitude ($\sim$1 m) in Rio de Janeiro. Regretfully, no other measurements were carried out at any points of the Atlantic coast of South America.

In the sourthern direction from the source, the waves propagated along the Ninety-East Ridge. In accordance with calculations they could have had a significant height on the Antarctic coast. However, it was practically impossible to check this fact, owing

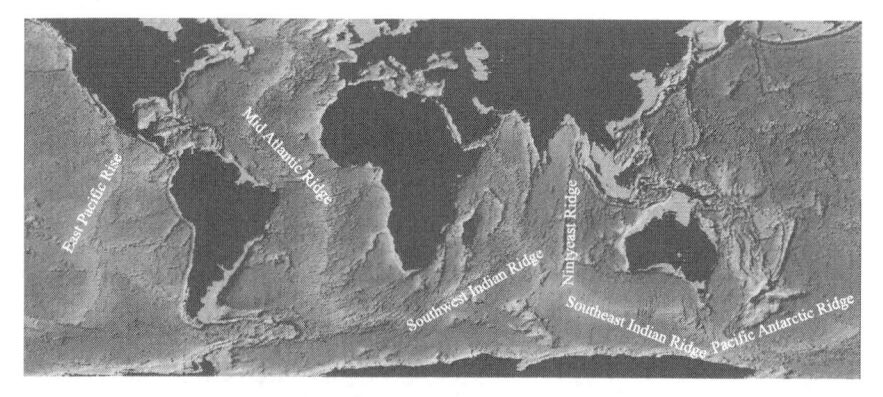

Fig. 5.11 Bottom topography of World Ocean. Mid–oceanic ridges that essentially influenced propagation of the Indonesian tsunami of December 26, 2004 (see also Plate 7 in the Colour Plate Section on page 315)

to the absence of mareograph stations. Two stations (one Japanese, 'Syowa', and the other French, 'Dumont d'Urville'), located approximately 2,000 km to the West and East, respectively, from the point of incidence of the main 'beam', registered moderate wave heights (of amplitude 60–70 cm).

In most cases of tsunami records, obtained in the eastern and central regions of the Indian Ocean (Fig. 5.9), only the first several waves exhibited the maximum amplitude. Further, the amplitude approximately exponentially decreased in time. The duration of anomalously large level oscillations amounted to 12 h. Numerical simulation shows that such a character of level oscillations at the coast corresponds to those cases, when waves of maximum amplitude, being focused by an extended source, travelled directly from the source to the observation point. The pronounced orientation of wave emission, seen well in Fig. 5.10, once more confirms the fact that one of the most important factors determining tsunami propagation in the near zone is the shape of the source.

Tsunami records, obtained in the western part of the Indian Ocean and in other oceans, reveal a significant duration of tsunami 'sounding', while level oscillations of maximum amplitude were observed with an essential delay after the onset of the first wave. This is due to enhancement of the role of waves reflected from the coasts and from irregularities of the ocean bottom, and also to propagation along natural waveguides—submarine ridges. The relatively slow, but energy-saving waveguide propagation provides for a late onset of the largest waves. Numerical simulation has shown that a wave perturbation often consists of two (or more) clearly distinguishable packets. One of them has a relatively small amplitude and propagates straightforwardly with a high velocity, 'taking advantage of' deep areas of the ocean. The second packet has a greater amplitude, but propagates slower along underwater ridges (elevations).

It is interesting that the tsunami penetrated the Pacific Ocean via two routes: directly from the Indian Ocean and through the southern part of the Atlantic Ocean, bypassing the Drake Passage between South America and Antarctica. Numerical simulation reveals the waves that arrived in the Pacific Ocean from the West from the Indian Ocean, and those that came from the East through the Atlantic, to have commensurable amplitudes. For all the Pacific coast, with the exception of southern Chile, the onset of waves arriving from the East occurs later.

To conclude the section we note that no destructions related to the tsunami of December 26, 2004 were reported outside the Indian Ocean. But experience in observations and simulations of the global propagation of tsunamis shows that the penetration of waves into all oceans is possible, in principle. Such a danger can be withstood, if a global system of tsunami warning is created.

5.3 Tsunami Run-up on the Coast

Of all problems relevant to tsunami dynamics, the description of wave transformation in the coastal belt, together with flooding of the coastal zone or uncovering of the ocean bottom, represents one of the most difficult tasks. This is, first of all, due to the problem being non-linear and the boundary, i.e. the shoreline, being movable. The topic of tsunami run-ups on the coast is so vast that it could be the subject of a separate monograph. In this section, we shall only briefly dwell upon some of the main results of and approaches to resolving the tsunami run-up problem and give references to key publications.

The well-known book by J. Stoker [Stoker (1959)] contains the classical formulation of the run-up problem. An extensive bibliography, reflecting development of the issue up to the end of the 1980s of the twentieth century, can be found in [Voltsinger et al. (1989)]. The most significant results achieved before 1995 are presented in [Carrier, Greenspan (1958); Keller et al. (1960); Shen, Meyer (1963); Sielecki, Wurtele (1970); Lyatkher, Militeev (1974); Spielvogel (1975); Hibberd, Peregrine (1979); Pedersen, Gjevik (1983); Kim et al. (1983); Synolakis (1987); Pelinovsky et al. (1993); Tadepalli, Synolakis (1994); Liu et al. (1995); Pelinovsky (1995) and Titov, Synolakis (1995)]. A significant part of monographs written by E. N. Pelinovsky [Pelinovsky (1996)] is devoted to analytical approaches to resolution of the tsunami run-up problem. Publications of the past decade demonstrate significant progress in numerical simulation of the interaction of tsunamis with coasts [Titov, Synolakis (1997), (1998); Fedotova (2002); Chubarov, Fedotova (2003); Titov et al. (2003); Choi et al. (2007), (2008)]; the interest in analytical and experimental studies in this field does also not weaken [Li, Raichlen (2002); Jensen et al. (2003); Chanson et al. (2003); Liu et al. (2003); Carrier et al. (2003); Kanoglu (2004); Tinti, Tonini (2005); Didenkulova et al. (2007)].

There exist different types of tsunami run-ups on a shore. They vary from gradual flooding (like during the tide) to the onslaught on the coast of a vertical wall of turbulent water—a bore. As a rule (in about 75% of events), tsunami waves flood the shore without breaking [Mazova et al. (1983)]. Tsunami run-ups in the form of a wall are quite rare, and usually in the case of waves of significant amplitude.

The three following main types of wave run-ups onto the coast can be identified [Pelinovsky (1996)]:

1. Spilling—crest of wave breaks, foam flows down its frontal slope, peculiar to gently sloping bottom
2. Plunging—crest of wave surpasses foot and curls down, peculiar to inclined bottom slopes
3. Surging—wave floods coast without breaking, peculiar to steep slopes

One and the same tsunami may exhibit different run-up forms at different points of the coast.

The most widespread mathematical model, applied in describing wave dynamics in the coastal zone, makes use of the non-linear equations of long waves, (5.29)–(5.31), in which the Coriolis force is usually neglected. In many cases, for reasons of simplicity, bottom friction is also neglected, although this factor may actually influence the run-up value noticeably. The main ideas of the tsunami run-up process can be understood by considering a one-dimensional problem along the axis perpendicular to the shoreline. Most model studies are performed for a region, representing a slope connected with a smooth horizontal ocean bottom (Fig. 5.12).

5.3 Tsunami Run-up on the Coast

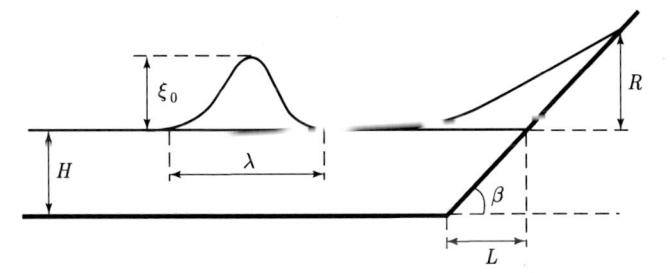

Fig. 5.12 Formulation of the problem of a tsunami run-up on the coast

In determining boundary conditions for practical tsunami calculations the so-called 'vertical wall' approximation has become widespread. A boundary condition of this type provides for total reflection of the wave at a fixed isobath. Note that the vertical wall approximation is not a purely academic abstraction, it imitates quite a type of coast, encountered quite often—a rocky precipice, falling off to the water. In the notation, given in Fig. 5.12, the vertical wall corresponds to $\beta = 90°$ or to $L = 0$.

From elementary theory of linear waves it is known that, if a channel of fixed depth ends in a vertical wall, then the height of the run-up onto the wall is determined as twice the incident wave amplitude, $R_L = 2\xi_0$.

Actually, when approaching the coast, the tsunami amplitude may be commensurable with the depth. Therefore, to determine the run-up height one must, generally speaking, apply non-linear theory. Omitting the details of resolving the non-linear problem of long-wave theory, expounded in the book [Pelinovsky (1996)], we present the resulting analytical formula that relates the run-up onto a vertical wall, R_N, and the wave amplitude far from the coast, ξ_0,

$$R_N = 4H \left(1 + \frac{\xi_0}{H} - \left(1 + \frac{\xi_0}{H} \right)^{1/2} \right), \tag{5.46}$$

where H is the basin depth. It is readily verified that the relation $R_L = 2\xi_0$ is a partial case of formula (5.46) given the condition $\xi_0/H \ll 1$. Comparison of quantities R_L and R_N shows that taking into account non-linearity enhances the run-up amplitude insignificantly. As the non-linearity (of quantity ξ_0/H) increases, the ratio R_N/R_L grows monotonously, but this growth is not without limit,

$$\lim_{\frac{\xi_0}{H} \to \infty} \left(\frac{R_N}{R_L} \right) = 2.$$

This means, the run-up amplitude, calculated with account of non-linearity, cannot be superior to twice the amplitude corresponding to linear theory.

We further consider the one-dimensional problem of a long wave moving along a slope $(0 < \beta < \pi/2)$. We write the non-linear equations for shallow water, taking into account that the basin depth is a linear function of the horizontal coordinate, $H = H_0 - \alpha x$,

$$\frac{\partial U}{\partial t}+U\frac{\partial U}{\partial x}+g\frac{\partial \xi}{\partial x}=0, \quad (5.47)$$

$$\frac{\partial \xi}{\partial t}+\frac{\partial}{\partial x}\left((\xi-\alpha x)U\right)=0, \quad (5.48)$$

Consider the wave, arriving on the shelf, to be characterized by a height ξ_0 and period T. We introduce dimensionless variables (the asterisk '*' will be further dropped)

$$t^* = \frac{t}{T}, \quad x^* = \frac{x\alpha}{\xi_0}, \quad \xi^* = \frac{\xi}{\xi_0}, \quad U^* = \frac{U\alpha T}{\xi_0}.$$

In these variables the systems (5.47) and (5.48) assume the following form [Kaistrenko et al. (1985)]:

$$\frac{\partial U}{\partial t}+U\frac{\partial U}{\partial x}+\frac{1}{Br}\frac{\partial \xi}{\partial x}=0, \quad (5.49)$$

$$\frac{\partial \xi}{\partial t}+\frac{\partial}{\partial x}[(\xi-x)U]=0, \quad (5.50)$$

where $Br = \xi_0/(g\alpha^2 T^2)$ is the only dimensionless parameter, which from a physical point of view represents a criterion for the breaking of a wave running up a plane slope. Note that this criterion is not quite precise, since it does not take into account phase dispersion and bottom friction.

Numerous experimental studies have permitted to introduce the Iribarren number as a criterion for wave breaking [Battjes (1988)],

$$Ir = \frac{\alpha \lambda^{1/2}}{\xi_0^{1/2}},$$

where λ is the deep-water wavelength. We consider the depth along the slope to increase indefinitely, therefore, for waves of any length there exists a region, where they do not 'feel' the bottom. Expressing the wavelength via the period from the dispersion relation for gravitational waves in deep water, $\lambda = gT^2/(2\pi)$, we obtain, that the empirically introduced Iribarren parameter and the quantity Br are uniquely related to each other, $Ir^{-2} = 2\pi Br$. The existence of such a relationship testifies in favour of the correct choice of non-linear long-wave model for describing the tsunami run-up on the shore. Transition from surging to plunging breaker (wave breaking) occurs when $Ir \approx 2$ ($Br \approx 0.04$).

An important step in resolving the run-up problem was the work [Carrier, Greenspan (1958)], in which it was shown that non-linear long-wave equations can be reduced to a linear wave equation, which, unlike the initial system is resolved in semispace with a fixed boundary. We recall that the initial system has an unknown movable boundary—the shoreline. This transformation was subsequently termed the Carrier–Greenspan transformation.

The approach based on the Carrier–Greenspan transformation has permitted to find a whole series of analytical solutions to the problem of tsunami run-up on a plane slope (e.g. [Pelinovsky (1996)]).

5.3 Tsunami Run-up on the Coast

One of the main results of the analysis of non-linear run-up problems consists in the proof that run-up characteristics depend linearly on the wave amplitude far from the coast. This fact provides for the possibility of applying linear theory in calculating the run-up. A rigorous substantiation of such possibility can be found in [Pelinovsky (1982)] and [Kaistrenko et al. (1991)].

Thus, for instance, in the case of the run-up of a monochromatic wave on a plane slope, resolution of the linear problem results in it being possible to construct the following simple approximation for the maximum run-up value:

$$R = \xi_0 \begin{cases} 2, & L < 0.05\lambda; \\ 2\pi \left(\dfrac{2L}{\lambda}\right)^{1/2}, & L > 0.05\lambda. \end{cases} \quad (5.51)$$

If the length of the slope is insignificant as compared with the length of the incident wave, then the run-up process will proceed like in the case of a vertical wall, i.e. the run-up height will turn out to be twice the amplitude of the incident wave. An increase in the slope length L (a decrease of the angle β) will lead to a certain enhancement of the run-up height.

Similar calculations were performed, also, for the run-up of a solitary wave. The maximum run-up in this case, also, is described by a formula identical to (5.51), but with a somewhat different numerical coefficient. It is interesting that oscillations of the shoreline on steep slopes repeat the form of the initial wave. Shoreline oscillations on a gentle slope are related to the form of the incident wave in a more complex manner. Thus, for instance, when a solitary wave (of positive sign) is incident upon a slope, shoreline oscillations turn out to alternate in sign.

The relation between the wave height far from the coast and its run-up height on a plane slope being linear is confirmed by results of laboratory and numerical experiments. Figure 5.13 presents such a relation, obtained for the run-up of solitary

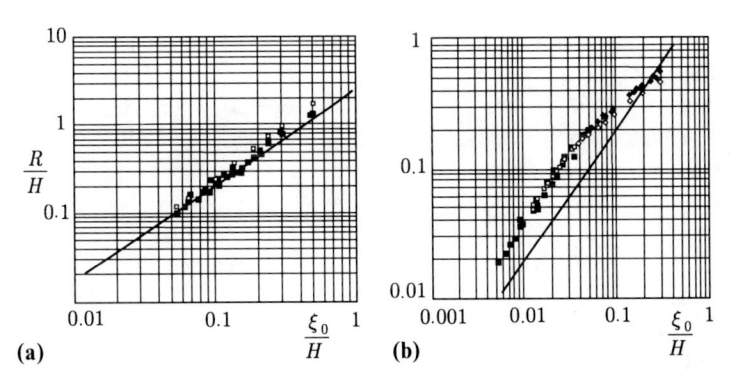

Fig. 5.13 Normalized maximum run-up of solitary waves on plane slope (a—1:1, b—1:19.85) versus normalized height of incident wave. Squares—nonbreaking data, rhombs—breaking data. Full and empty symbols correspond to laboratory and numerical experiments, respectively. Solid line—run-up height on vertical wall, calculated by formula (5.46) (Adapted from [Titov, Synolakis (1995)])

waves on steep (1:1) and gentle (1:19.85) slopes. The solid line in Fig. 5.13 shows the dependence corresponding to the run-up height on a vertical wall, calculated by formula (5.46). In the case of a run-up on a steep slope, the dependence of $R(\xi_0)$ is actually very close to linear. But in the case of the run-up of waves of large amplitude on a gentle slope deviation is seen of the dependence from linearity towards a decrease in the run-up height. The bend in the dependence, observed at $\xi_0/H \sim 0.03$ (Fig. 5.13b) corresponds to transition to run-ups involving wave breaking.

In Fig 5.13a. it is seen that the run-up on a steep slope is approximately twice the wave amplitude far from the coast, which complies with the theoretical result for the run-up on a vertical wall (5.46). In the case of a run-up on a gentle slope (Fig. 5.13b) the height increases noticeably; here, it is three to four times higher than the wave amplitude far from the coast. In any case, the energy losses, related to wave breaking, result in a reduction of the run-up height. In certain conditions the run-up height on a gentle slope, involving wave breaking, may even turn out to be smaller than the same quantity in the case of a vertical wall.

The data of numerical simulation, presented in Fig. 5.13, are in good accordance with the results of laboratory experiments. Details of the numerical algorithm are described in [Titov, Synolakis (1995)].

In conclusion of this section we shall dwell upon certain difficulties arising in numerical simulation of a tsunami run-up with account of the real relief of the coastal area. The *first difficulty* is related to the absence of or insufficiently detailed bathymetric and topographical data. For modelling tsunamis in the open ocean, where wavelengths are significant, of the order of 100 km, the existing global data, for example, ETOPO-2 with a resolution of 2 angular minutes (~ 4 km) are quite sufficient. But for reliable numerical simulation of the tsunami dynamics in the coastal zone it is necessary to have data on the reliefs of the bottom and of the coastal area with a space resolution hundreds of times better (~ 10 m). This requirement is related not only to the significant reduction of wavelengths in shallow water. The quality of topographical data directly influences the precision in resolving the practical problem—determination of the run-up boundaries. Here, it must also be noted, that for resolving the run-up problem accurately it is also necessary to have at one's disposal information on tidal-level oscillations.

The most reliable criterion of applicability of one or another tsunami run-up model consists in practical tests. Laboratory experiments, naturally, permit to judge the efficiency of numerical models, but in any case the most reliable test consists in comparison of the results of simulation with data on real tsunamis. Here, we encounter the ***second difficulty***, related to the existence and quality of results of in situ measurements. To test a model detailed measurements of the run-up area are required, desirably supplemented with information on the water flow parameters on the coast. In recent years, the database of run-up parameters is regularly upgraded with high-precision measurements performed by international expeditions, for which the investigation of coastal areas hit by tsunamis is mandatory. Contributions to the resolution of this problem are also provided by high-quality satellite photographs, permitting to determine the run-up area. But the velocities of water flows usually have to be estimated from indirect data.

The ***third difficulty*** is related to the fact that strong tsunami waves are capable of changing the initial aspect of a coast, including the topography of the coastal belt (erosion, demolition of buildings and destruction of vegetation). Thus, subsequent waves will interact with a coast, the properties of which (topography, irregularities) were altered by the preceding wave. High-precision run-up simulation will inevitably encounter the necessity of taking these effects into account.

References

Abe K. (1978): A dislocation model of the 1933 Sanriku earthquake consistent with tsunami waves. J. Phys. Earth **26**(4) 381–396

Abe K. (1979): Size of great earthquakes of 1837–1974 inferred from tsunami data. J. Geophys. Res. **84** 1561–1568

Aida I. (1969): Numerical experiments for the tsunami propagation the 1964 Niigata tsunami and 1968 Tokachi–Oki tsunami. Bull. Earthquake Res. Inst. Univ. Tokyo **47**(4) 673–700

Aida I. (1974): Numerical computation of a tsunami based on a fault origin model of an earthquake. J. Seismol. Soc. Jpn **27**(2) 141–154

Battjes J. A. (1988): Surf-zone dynamics. Annn. Rev. Fluid Mech. **20** 257–293

Carrier G. F., Wu T. T., Yeh H. (2003): Tsunami run-up and drawdown on a plane beach. J. Fluid Mech. **475** 449–461

Carrier G. F., Greenspan H. P. (1958): Water waves of finite amplitude on a sloping beach. J. Fluid Mech. **4** 97–109

Chanson H., Aoki S., Maruyama M. (2003): An experimental study of tsunami runup on dry and wet horizontal coastlines. Science of Tsunami Hazards **20**(5) 278–293

Choi B. H., Pelinovsky E., Kim K. O., Lee J. S. (2003): Simulation of the trans-oceanic tsunami propagation due to the 1883 Krakatau volcanic eruption. Nat. Hazards Earth Sys. Sci. **3** 321–332.

Choi B. H., Kim D. C., Pelinovsky E., Woo S. B. (2007): Three-dimensional simulation of tsunami run-up around conical island. Coastal Engineering 54 618629

Choi B. H., Pelinovsky E., Kim D. C., Didenkulova I., Woo S. B. (2008): Two- and three-dimensional computation of solitary wave runup on non-plane beach. Nonlin. Processes Geophys. 15 489–502

Chubarov L. B., Fedotova Z. I. (2003): An Effective High Accuracy Method for Tsunami Run-up Numerical Modeling. In: Submarine Landslides and Tsunamis. Book Series: NATO SCIENCE SERIES: IV. Eds. Ahmet C. et al. Kluwer Dordrecht, pp. 203–216

Chubarov L. B., Shokin Yu. I., Gusiakov V. K. (1984): Numerical modeling of the 1973 Shikotan (Nemuro–Oki) tsunami. Comput. Fluid. **122** 123–132

Didenkulova I. I., Kurkin A. A., Pelinovsky E. N. (2007): Run-up of solitary waves on slopes with different profiles. Izvestiya RAN, Atmos. Ocean. Phys. **43**(3)

Fedotova Z. I. (2002): Substantiation of the numerical method for simulating the run-up of long waves on a shore (in Russian). Comput. Technol. **7**(5) 58–76

Gonzalez F. I., Bernard S. N., Milbern H. B., et al. (1987): The Pacific Tsunami Observation Program (PacTOP), In: Proc. IUGG/IOC, Intern. Tsunami Symp. 3–19

Goto C., Ogawa Y., Shuto N., Imamura N. (1997): Numerical method of tsunami simulation with the leap-frog scheme (IUGG/IOC Time Project), IOC Manual, UNESCO No 35

Gusyakov B. K., Chubarov L. B. (1982): Numerical simulation of the Shikotan (Nemuro-oki) tsunami of June 17, 1973 (in Russian), In: Tsunami evolution from the source to the coast run-up, pp. 16–24. Radio i svyaz', Moscow

Gusyakov B. K., Chubarov L. B. (1987): Numerical simulation of tsunami excitation and propagation in the coastal zone. Earth Phys. (in Russian) (1) 53–64

Hibberd S., Peregrine D. H. (1979): Surf and run-up on a beach: a uniform bore. J. Fluid Mech. **95** (Part 2) 323–345

Horrillo J., Kowalik Z., Shigihara Y. (2006): Wave dispersion study in the Indian Ocean tsunami of December 26, 2004. Science of Tsunami Hazards **25**(1) 42–63

Israeli M., Orzag S. A. (1981): Approximation of radiation boundary conditions. J. Comput. Phys. **41** 115–135

Jacques V. M., Soloviev S. L. (1971): Remote registration of weak tsunami-type waves on the shelf of the Kuril islands. DAN SSSR (in Russian) **198**(4) 816–817

Jensen A., Pedersen G. K., Wood D. J. (2003): An experimental study of wave run-up at a steep beach. J. Fluid Mech. 486 161–188 DOI: 10.1017/S0022112003004543

Kaistrenko V. M., Pelinovsky E. N., Simonov K. V. (1985): Wave run-up and transformation in shallow water (in Russian). Meteorol. Hydrol. (10) 68–75

Kaistrenko V. M., Mazova R. Kh., Pelinovsky E. N., Simonov K. V. (1991): Analitical theory for tsunami run-up on a smooth slope. J. Tsunami Soc. **9**(2) 115–127

Kanoglu U. (2004): Non-linear evolution and run-up–rundown of long waves over a sloping beach. J. Fluid Mech. **513** 363–372

Keller H. B., Levine D. A., Whitham G. H. (1960): Motion of a bore over sloping beach. J. Fluid Mech. **7** 302–316

Kim S. K., Liu Ph. L.-F., Liggett J. A. (1983): Boundary integral equation solutions for solitary wave generation, propagation and run-up. Coastal Engng. **7** 299–317

Kosloff R., Kosloff D. (1986): Absorbing boundaries for the wave propagation problem. J. Comput. Phys. **63** 363–376

Kowalik Z., Knight W., Logan T., Whitmore P. (2005): Numerical modelling of the global tsunami: indonesian tsunami of December 26 2004. Sci. Tsunami Hazard, **23**(1) 40–56

Kulikov E. A., Gonzalez F. I. (1995): Reconstruction of the shape of a tsunami signal at the source by measurements of hydrostatic pressure oscillations using a remote bottom sensor (in Russian). DAN RF **344**(6) 814–818

Kulikov E. A., Rabinovich A. B., Thomson R. E., Bornhold B. D. (1996): The landslide tsunami of November 3, 1994, Skagway Harbor, Alaska. J. Geophys. Res. **101**(C3) 6609–6615

Kulikov E. A., Medvedev P. P., Lappo S. S. (2005): Registration from outer space of the December 26, 2004, tsunami in the Indian Ocean (in Russian). DAN RF **401**(4) 537–542

Lamb H. (1932): Hydrodynamics. Cambridge University Press, Cambridge

Landau L. D., Lifshitz E. M. (1987): Fluid Mechanics, V.6 of Course of Theoretical Physics, 2nd English edition. Revised. Pergamon Press, Oxford-New York-Beijing-Frankfurt-San Paulo-Sydney-Tokyo-Toronto

Li Y., Raichlen F. (2002): Non-breaking and breaking solitary wave run-up. J. Fluid Mech. 456 295–318 DOI: 10.1017/S0022112001007625

Lighthill J. (1978): Waves in fluids. Cambridge University Press, Cambridge

Liu P. L.-F., Lynett P., Synolakis C. E. (2003): Analytical solutions for forced long waves on a sloping beach. J. Fluid Mech. **478** 101–109

Liu Ph. L.-F., Cho Yo-S., Briggs M. J., Kanoglu U., Synolakis C. E. (1995): Run-up of solitary waves on a circular island. J. Fluid Mech. **302** 259–285

Lyakhter V. M., Militeev A. N. (1974): Calculation of run-up on slope for long gravitational waves (in Russian). Oceanology (1) 37–43

Marchuk An. G., Chubarov L. B., Shokin Yu. I. (1983): Numerical simulation of tsunami waves (in Russian). Nauka, Siberian Branch. Novosibirsk

Mazova R. Kh., Pelinovsky E. N., Soloviev S. L. (1983): Statistical data on the character of the run-up of tsunami waves (in Russian). Oceanology **23** (6) 932–937

Milburn, H. B., Nakamura A. I., González F. I. (1996): Real-time tsunami reporting from the deep ocean. In: Proceedings of the Oceans 96 MTS/IEEE Conference, 23–26 September 1996, Fort Lauderdale, FL, pp. 390–394

Mofjeld H. O., Titov V. V., González F. I., Newman J. C. (2000): Analytic Theory of Tsunami Wave Scattering in the Open Ocean with Application to the North Pacific. NOAA Technical Memorandum OAR PMEL-116. PMEL, Seattle, Washington

Murty T. S. (1984): Storm surges. Meteorological ocean tides. Department of Fisheries and Oceans, Bulletin 212, Ottawa

Myers, E. P., Baptista A. M. (1995): Finite Element Modeling of the July 12, 1993 Hokkaido Nansei–Oki Tsunami. Pure Appl. Geophys. 144(3/4) 769–802

Nakoulima O., Zahibo N, Pelinovsky E., Talipova T., Kurkin A. (2005): Solitary wave dynamics in shallow water above periodic bottom. Chaos 15(3) 037107

Nekrasov A. V. (1973): On the Reflection of tidal waves from the shelf zone (in Russian). Oceanology 13(2) 210–215

Okada Y. (1985): Surface deformation due to shear and tensile faults in a half-space. Bull. Seismol. Soc. Am. 75(4) 1135–1154

Okal E. A., Piatanesi A. Heinrich P. (1999): Tsunami detection by satellite altimetry. J. Geophys. Res. 104 599–615

Pedersen G., Gjevik B. (1983): Run-up of solitary waves. J. Fluid Mech. 135 283–290

Pelinovsky E. (1995): Non-linear hyperbolic equations and run-up of Huge sea waves. Appl. Anal. 57 63–84

Pelinovsky E. N., Stepanyants Yu., Talipova T. (1993): Non-linear dispersion model of sea waves in the coastal zone. J. Korean Soc. Coastal Ocean Eng. 5(4) 307–317

Pelinovsky E. N. (1996): Hydrodynamics of tsunami waves (in Russian). Institute of Applied Physics, RAS, Nizhnii Novgorod

Pelinovsky E. N. (1982): Non-linear dynamics of tsunami waves (in Russian), Institute of Applied Physics, USSR AS.Gorky

Rabinovich A. B., Lobkovsky L. I., Fine I. V., et al. (2008): Near-source observations and modeling of the Kuril Islands tsunamis of November 15, 2006 and January 13, 2007. Adv. Geosci. 14 105–116

Rivera P. C. (2006): Modeling the Asian tsunami evolution and propagation with a new generation mechanism and a non-linear dispersive wave model. Science of Tsunami Hazards 25(1) 18–33

Satake K. (1988): Effects of bathymetry on tsunami propagation: application of ray tracing to tsunamis, PAGEOPH 126 27–36

Satake K., Imamura F. (1995): Tsunamis: seismological and disaster prevention studies. J. Phys. Earth 43(3) 259–277

Shen M. C., Meyer R. E. (1963): Climb of a bor on a beach. Part 3. Run-up. J. Fluid Mech. 16 113–125

Sielecki A., Wurtele M. (1970): The numerical integration of the non-linear shallow water equations with sloping boundaries. J. Comput. Phys. 6 219–236

Spielvogel L. Q. (1975): Single-wave run-up on sloping beaches. J. Fluid Mech. 74 685–694

Suleimani E., Hansen R., Kowalik Z. (2003): Inundation modeling of the 1964 tsunami in Kodiak Island, Alaska. In: Submarine Landslides and Tsunamis (edited by Yalciner A. C., Pelinovsky E. N., Okal E. Synolakis C. E.), 21 191–201. Kluwer, Dordreecht

Synolakis C. E. (1987): The run-up of solitary waves. J. Fluid Mech. 185 523–545

Synolakis C. E., Fritz M. H., Borrero C. J. (2005): Far Field Surveys of the Indian Ocean Tsunami: Sri Lanka, Maldives and Somalia. In: 22nd International Tsunami Symposium, Chania, Crete isl. Greece, June 27–29, 2005 (edited by Papadopoulos G. A., Satake K.), pp. 57–64

Stoker J. J. (1959): Water waves, Interscience, New York

Tadepalli S., Synolakis C. E. (1994): The run-up of N-waves on sloping beaches. Proc. R. Soc. Lond. A. 445 99–112

Takahashi To., Takahashi Ta., Shuto N., Imamura F., Ortiz M. (1995): Source models for the 1993 Hokkaido–Nansei–Oki earthquake tsunami. Pure Appl. Geophys. 144(3/4) 747–768

Tanioka Y., Seno T. (2001): Sediment effect on tsunami generation of the 1896 Sanriku tsunami earthquake. Geophys. Res. Lett. 28(17) 3389–3392

Tinti S., Tonini R. (2005): Analytical evolution of tsunamis induced by near-shore earthquakes on a constant-slope ocean. J. Fluid Mech. 535 33–64

Titov V. V., Gonzalez F. I. (1997): Implementation and testing of the Method of Splitting Tsunami (MOST) model. NOAA Technical Memorandum ERL PMEL-112

Titov V. V., Synolakis C. E. (1998): Numerical modeling of tidal wave run-up. J. Waterway, Port, Coastal Ocean Eng. **124**(4) 157–171

Titov V. V., Gonzalez F. I., Mofjeld H. O., Venturato A. J. (2003): NOAA Time Seattle Tsunami Mapping Project: Procedures, Data Sources, and Products. NOAA Technical Memorandum OAR PMEL-124

Titov V. V., Synolakis C. E. (1995): Modelling of breaking and nonbreaking long wave evolution and run-up using VTCS-2. J. Waterway, Port, Coastal Ocean Eng. **121**(6) 308–316

Titov V. V., Synolakis C. E. (1997): Extreme inundation flows during the Hokkaido-Nansei-Oki tsunami. Geophys. Res. Lett. **24**(11) 1315–1318

Titov V. V., Gonzalez F. I., Bernard E. N., et al. (2005): In: Real-Time Tsunami Forecasting: Challenges and Solutions. Nat. Hazards, **35**(1), Special Issue, pp. 41–58. U.S. National Tsunami Hazard Mitigation Program

Titov, V. V., Rabinovich A. B., Mofjeld H. O., Thomson R. I., Gonzalez F. I. (2005): The Global Reach of the 26 December 2004 Sumatra Tsunami. Science, 309 (23 Sep 2005) 2045–2048

Voltsinger N. E., Klevanny K. A., Pelinovsky E. N. (1989): Longwave dynamics of coastal zone (in Russian). Gidrometeoizdat, Leningrad

Yeh H., Liu Ph., Briggs M., Synolakis C. (1994): Propagation and amplification of tsunamis at coastal boundaries. Lett. Nat. **372** 353–355

Zaitsev A. I., Kurkin A. A., Levin B. W., et al. (2005): Numerical Simulation of Catastrophic Tsunami Propagation in the Indian Ocean (December 26, 2004). Doklady Earth Sci. **402**(4) 614

Chapter 6
Methods of Tsunami Wave Registration

Abstract The traditional method for tsunami wave registration by coastal mareographs including modern telemetric complexes, is described. The technique is described for measuring tsunami waves in the open ocean with the aid of bottom pressure sensors. The advantages of this technique are discussed. The technique for studying and documenting effects of tsunami influence on the coast are briefly expounded. The significance of searching for and identifying paleotsunami sediments is discussed. The application is described of satellite altimeters for registering tsunamis in the open ocean, and data are also given concerning registration of the 26 December 2004 tsunami in the Indian Ocean by the JASON-1 satellite.

Keywords Tsunami registration · mareograph· sea level · bottom pressure · telemetric complex · satellite communications · tide · DART · JAMSTEC · tsunami sensor · tsunami manifestation · tsunami run-up · run-up heights · tsunami deposit · erosion · abrasion · inundation · topographic profile· paleotsunami · satellite altimetry · dispersion · wave front

The first information on the potential generation of a tsunami comes from the World Seismic Network. Data on the time, epicentre coordinates and energy (magnitude) of an underwater earthquake permit to estimate the location of the source, the probability for the tsunami to originate and the time the wave will arrive at the coast. But tsunami waves are not related in a unique manner to a seismic event. A strong earthquake is sometimes accompanied by an insignificant tsunami, while, contrariwise, a weak earthquake in a number of cases causes the formation of catastrophic waves. For a more accurate estimation of the tsunami hazard, information is required on the actual development of the wave process with time. In absence of such information the existence of a great number of false alert signals and neglected tsunami events is practically inevitable.

Tsunami registration is performed by various methods, including traditional measurements of the sea level close to the coast (mareographs), measurements with the aid of bottom pressure sensors in the open ocean, as well as methods, recently undergoing development, of distance measurements, which primarily involve satellite altimetry. Investigation of tsunami manifestations on the coast is performed by in situ expeditions immediately after the event. Some traces of the influence of tsunamis on the coast are conserved for many thousands of years. Searching for and analysing such traces of prehistoric tsunamis, or paleotsunamis, permits to significantly supplement tsunami catalogues.

6.1 Coastal and Deep-water Measurements of Sea Level

The first instrumental registration of a tsunami was obtained by a coastal mareograph, a device intended for measuring low-frequency variations of the ocean level (primarily, of tides). The scheme of a traditional mareograph is presented in Fig. 6.1.

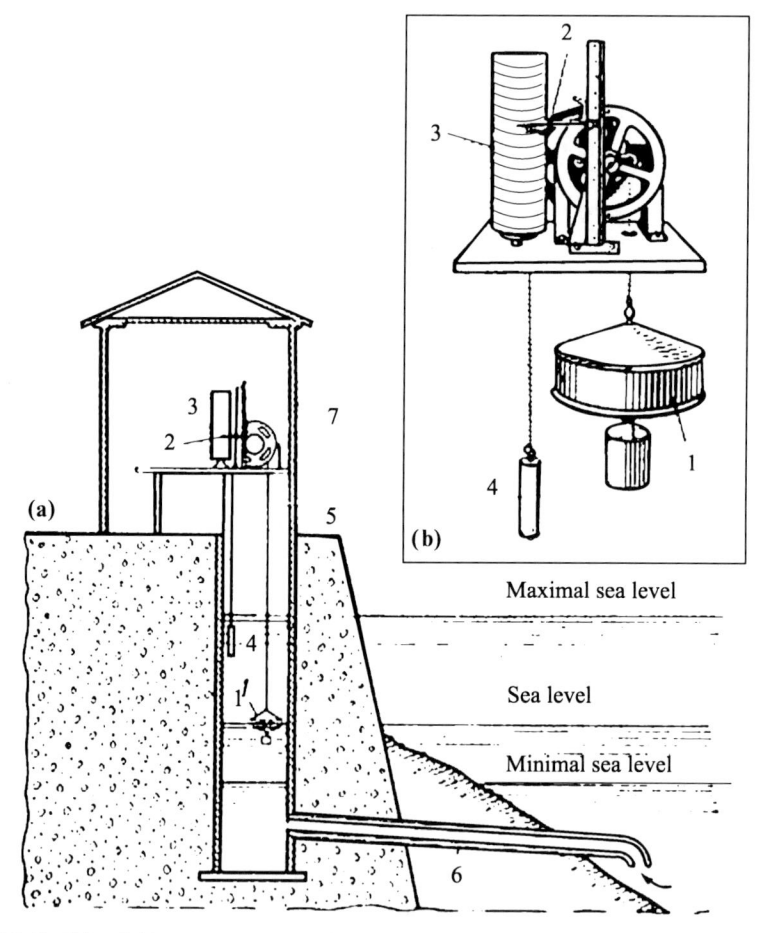

Fig. 6.1 Traditional tide gauge (mareograph) in a shaft (**a**) and its layout (**b**): 1—float; 2—pen; 3—drum with chart strip; 4—counterweight; 5—shaft; 6—pipe connecting shaft with sea; 7—cabin housing the device

6.1 Coastal and Deep-water Measurements of Sea Level

Level measurements are carried out in a shaft connected with the ocean by a relatively thin pipe. Such a scheme permits to automatically filter out high-frequency-level oscillations, related to wind waves.

A standard modern telemetric complex for controlling the ocean level (and for tsunami registration) is composed of a traditional registrator (mareograph), a footstock, and, also, means for collecting and transmitting data (see Fig. 6.2). About 70 such complexes have been established on the coasts of many countries, joined in the International Pacific Tsunami Warning System, for instance, in the harbours of Severo-Kurilsk and Ust-Kamchatsk of the Far-East region of Russia.

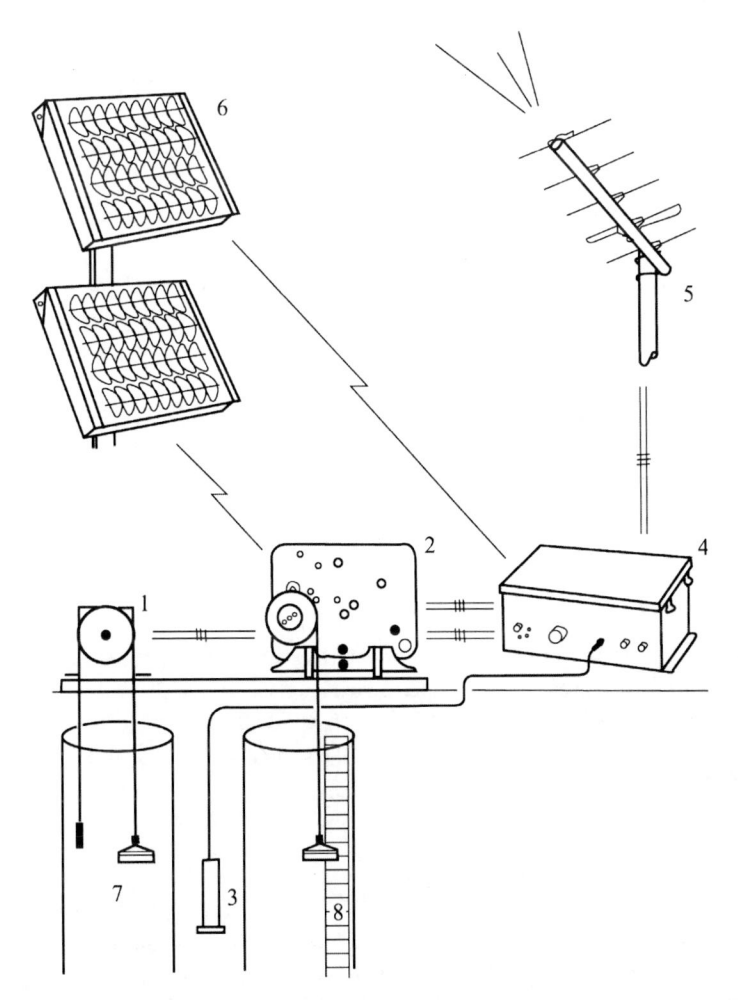

Fig. 6.2 Layout of modern standard complex of Pacific Tsunami Warning Service. 1—differential (incremental) coding device (Handar), 2—Leopold–Stevens device with analog-to-digital registrator, 3—pressure transformer, 4—satellite platform for data collection (Handar), 5—satellite antenna (GOES), 6—solar batteries, 7—wave damper (mareographic shaft), 8—footstock

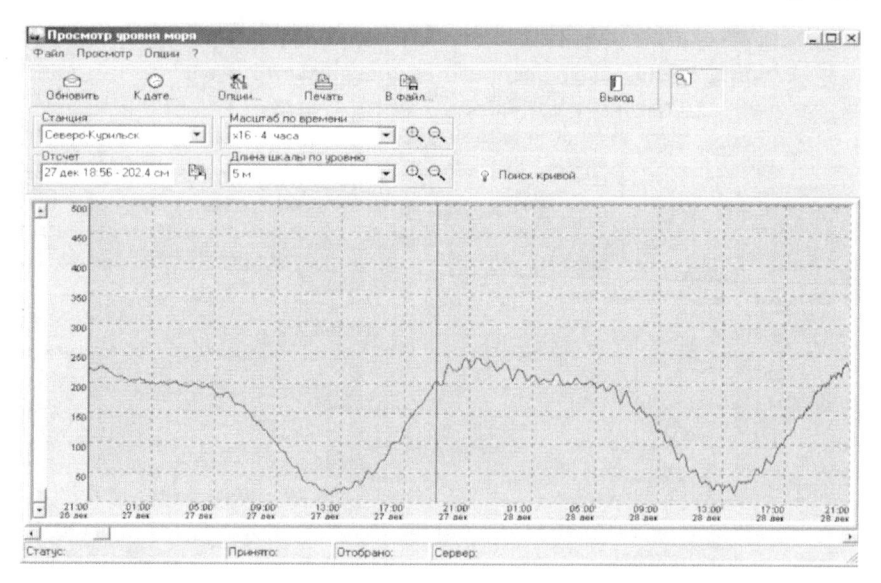

Fig. 6.3 Example of telemetric data presentation at the Sakhalin Tsunami Center, data were received from the registrator in Severo-Kurilsk. The mark indicates the onset of the front of the Indonesian tsunami of December 26, 2004

The telemetric complex comprises a pressure gauge PTX160 of the Druck company (pressure measurement precision: ±0.1%), a universal device for data collection and transformation, 555 DCP of the Handar company, including a radio transmitter of the satellite data collection system via GOES or GMS-5 and a power supply. The complex is also equipped with an antenna for satellite communications. The distance of the pressure gauge from the coastal device can amount to 1.5 km. Usually, the gauge is put in a pipe with draining openings, or inside a mareograph shaft at a depth of about 3 m.

Telemetric information on the sea level is received and processed in national tsunami warning centres. An example of such registration at the Sakhalin Tsunami Center (Yuzhno-Sakhalinsk) is shown in Fig. 6.3. The tsunami wave due to the earthquake of December 24, 2004, in the Indian Ocean reached Severo-Kurilsk in 41 h 17 min, its amplitude amounted to 29 cm.

Until recently all ideas of the character of tsunami wave evolution in the open ocean were based exclusively on coastal measurements. In the 1960s–1970s S. L. Soloviev proposed an essentially new method for effective prognosis, based on the registration of waves far from the coast [Soloviev (1968); Jacques, Soloviev (1971)]. Upon having undergone significant technological development [UNESCO Tech. Pap. (1975); Cartwright et al. (1979); Gonzalez et al. (1987), (2005)], this method is at present applied in the USA (DART) and Japan (JAMSTEC).

The creation of an automatized system of level observation in the open ocean represents a promising way for providing reliable and timely tsunami warning. Moreover, deep-water-level measurements are important for developing an understanding of the processes of wave excitation and propagation. Compared to coastal measurements, deep-water tsunami registration displays a whole number of important advantages [Titov et al. (2005)]. First, owing to the tsunami velocity depending on the ocean depth, the deep-water sensor registers a wave faster than a coastal mareograph, located at the same distance from the source. Second, a tsunami wave approaching the coast is strongly distorted (for instance, owing to resonances in bays) and it 'forgets' the properties of the source that generated it. Therefore, coastal mareographs are not sensitive to the true frequency spectrum of a tsunami. At the same time, a tsunami signal in the open ocean is not distorted or filtered and contains all the components of the original spectrum. Third, the frequency response function of bottom pressure sensors is totally flat within the range of tsunami waves, while the response function, peculiar to many coastal mareographs, is complex and not constant. Most mareographs are, generally speaking, not intended for tsunami measurements, since they were created for observing relatively low-frequency tidal-level oscillations. Fourth, the amplitude of a tsunami in the open ocean is small compared to the ocean depth, therefore, wave propagation is described with a very good accuracy by simple linear models. For this reason, the results of deep-water measurements can be applied effectively in resolving inverse problems (reconstruction of perturbation forms at sources etc.).

Level registration in the open ocean (at large depths) is a difficult technical task that has been accomplished only in recent decades. Of the various numerous systems, quartz sensors provide the best measurement precision and stability.

Variations of pressure at the ocean bottom exhibit a broad frequency spectrum and are due to a whole complex of processes in the atmosphere, ocean and lithosphere. Surface, internal and elastic waves in the water column, as well as seismic surface waves and changes in the atmospheric pressure all contribute to variation of the bottom pressure.

Pressure variations, related to surface waves, are known to be felt at the bottom only in the case of long waves ($\lambda \gtrsim H$), so for sensors established at large depths high-frequency sea-level oscillations (wind waves) do not influence the bottom pressure. But in the case of tsunamis or tidal waves changes in the bottom pressure Δp_{bot} are mostly determined by displacements of the ocean-free surface, $\Delta \xi$, with account of changes in the atmospheric pressure $\Delta p_{\text{bot}} = \rho g \Delta \xi + \Delta p_{\text{atm}}$, where ρ is the density of water and g is the acceleration of gravity. Estimation [Wunsch (1972)] shows that the influence of baroclinity on the ocean bottom pressure can be neglected. Elastic oscillations of the water column, caused by seismic movements of the bottom (for instance, due to surface seismic waves), may provide a significant, and even definitive, contribution to the bottom pressure $\Delta p = \rho c U^*$, where c is the velocity of sound in water, U is the vertical velocity of movement of the bottom. But, owing to the difference in propagation velocity between tsunamis and seismic waves, the arrival times of 'seismic noise' and of the tsunami signal may differ. In those cases, when the pressure sensor is located near the epicentre, the tsunami signal can be singled out by filtration of the high-frequency seismic component.

---

* On condition that $T > 4H/c$ (incompressible ocean): $\Delta p = \rho H a$, where a is the vertical acceleration of movement of the bottom, T is the period of movement of the bottom.

Tidal oscillations of the ocean level usually provide for 85–90% of the total energy of long-wave 'noise', the spectral level, corresponding to 24-h and 12-h tides, is of the order of 10^3 cm^2. The tidal component, unlike pressure variations, related to the random influence of atmospheric and other processes, is strictly deterministic—it is determined by changeless astronomical constants.

An unavoidable task, related to singling out and analysing the tsunami signal in a record obtained by a sensor of bottom pressure, consists in removal of the tidal component. Tsunami wave frequencies usually exceed tidal frequencies by more than an order of magnitude, and, moreover, tides are not stochastic processes. Therefore, there exist no essential difficulties in singling out a tsunami signal. Certain technical difficulties may arise during operational tsunami identification, since tuning the filter, that cuts off low frequencies, requires significant time. But, since tidal-level oscillations can be calculated aforehand from known constants, they can just be subtracted from the record. The main drawback in this approach is that it cannot be applied immediately after the sensor has been established on the ocean bottom. For calculation of the constants with the necessary precision the level must be recorded during at least 1 month.

Within the National Tsunami Hazard Mitigation Program (NTHMP, USA) there exists a project aimed at revealing tsunami waves in the open ocean and estimating their danger (Deep-ocean Assessment and Reporting of Tsunami, DART) [Bernard et al. (2001); Eble et al. (2001); Kong (2002); Meinig et al. (2005); Green (2006)]. DART buoy stations are established near the regions, within which the formation of destructive tsunamis is possible (Fig. 6.4). Information from the stations is sent to tsunami warning centres in a real-time mode via the satellite communication channel. The choice of sites for the stations guarantees, that a tsunami arising in these regions will be revealed in 30 min after having originated.

The system is based on deep-water pressure sensors, developed in one of the NOAA divisions (Pacific Marine Environmental Laboratory, PMEL). A DART

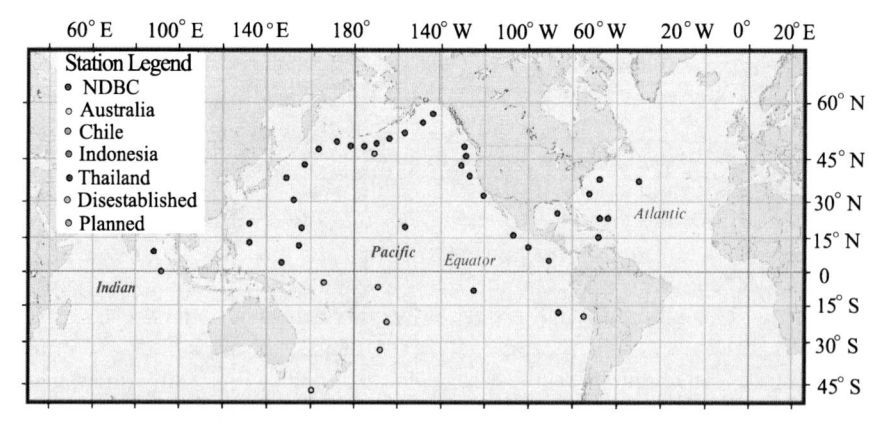

Fig. 6.4 Location of buoy stations, for registering tsunamis in the open ocean (as of November 2007). The figure is taken from http://www.ndbc.noaa.gov/dart.shtml (see also Plate 8 in the Colour Plate Section on page 316)

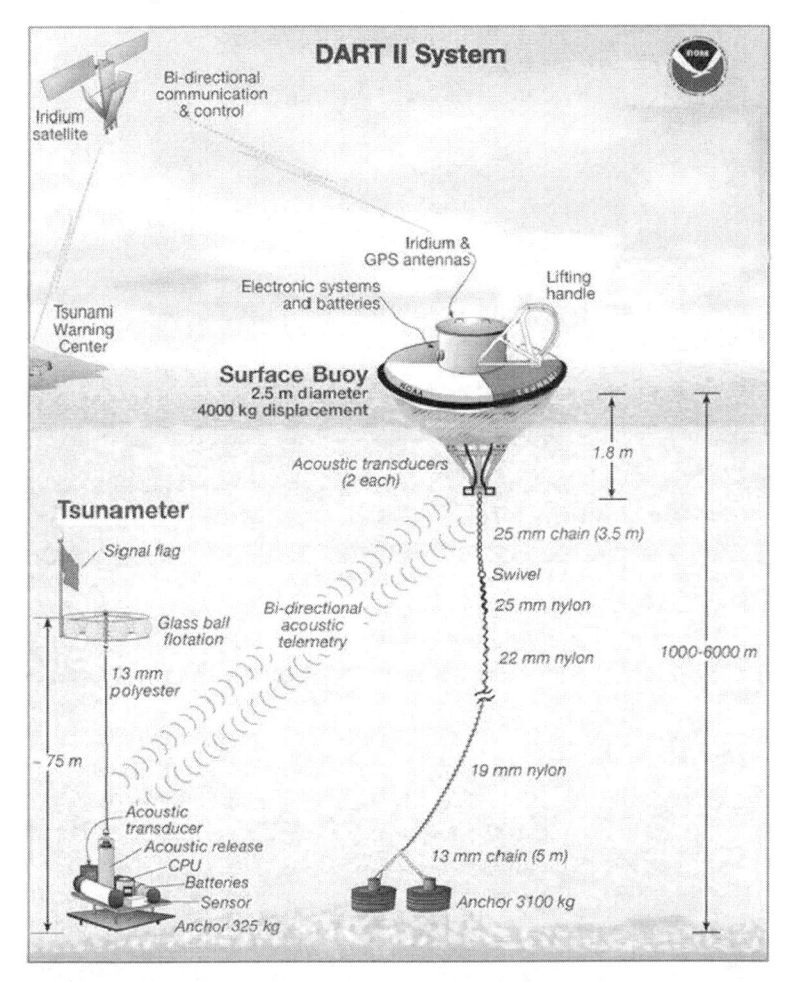

6.1 Coastal and Deep-water Measurements of Sea Level

Fig. 6.5 Layout of station (DART II) for registering tsunamis in the open ocean. The figure is taken from http://www.ndbc.noaa.gov/dart.shtml (see also Plate 9 in the Colour Plate Section on page 317)

station consists of an anchored surface buoy and a bottom platform; information is exchanged between them via an acoustic communication channel (Fig. 6.5). The sensor on the platform measures the pressure, averaged over 15 s, with an accuracy, corresponding to a 1 mm water column. The level of the natural long-wave noise (without account of tides) in the deep-water part of the ocean is not high (its root-mean-square value is ∼0.2 cm). This makes it possible to reliably single out tsunamis of heights of merely 1 cm, when the ocean depth at the site of the sensor is about 6,000 m. The surface buoy is equipped with a satellite system for data transmission (GOES) to the respective Tsunami Warning Centers for Hawaii and Alaska and, also to PMEL. At present, 43 DART stations are in operation. Information can be found at the freely accessible site http://www.ndbc.noaa.gov/dart.shtml.

Fig. 6.6 Location of JAMSTEC bottom stations: 'Hatsushima' (**A**), 'Muroto' (**B**), 'Kushiro-tokachi' (**C**)

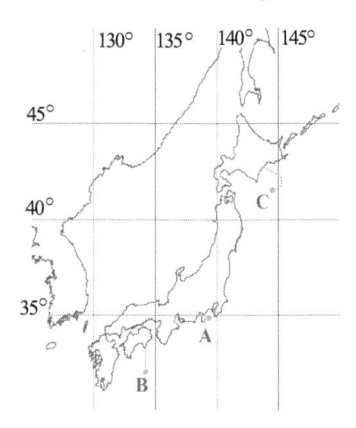

Since 1997, the Japanese agency JAMSTEC has also organized a network of deep-water bottom stations, equipped with seismometers, hydrophones and pressure sensors. The latter are intended for operative tsunami forecasting and are even called 'tsunami sensors'. Unlike the American system DART, described above, the Japanese stations do not make use of a satellite communication channel, but are connected to the shore by cable lines. The resolution of the pressure sensors is ∼0.3 mm of the water column. The frequency of data discretization is essentially higher than in the American system; it amounts to 1 Hz. At present, 5 bottom stations are functioning: Hatsushima (1 sensor), Muroto (2 sensors) and Kushirotokachi (3 sensors). Their locations are shown in Fig. 6.6. Measurement data are available at the address http://www.jamstec.go.jp/scdc/top_e.html. Detailed descriptions of the technical features of these systems can be found, for example, in [Hirata et al. (2002)] or at the aforementioned network site.

Since installation, the system has registered numerous seismic events and tsunamis of small amplitudes (microtsunamis). The exceptionally strong 2003 earthquake ($M_w = 8.3$), that occurred in the vicinity of Hokkaido Island was also successfully registered by the system. The earthquake turned out to be tsunamigenic, it caused a wave 4 m high on the coast of Hokkaido. Two bottom pressure sensors happened to be in the immediate vicinity of the tsunami source area. This was the first direct measurement in history, made at the source at the moment of the wave origination. It permitted not only to measure the residual vertical deformations of the bottom at the tsunami source [Watanabe et al. (2004)], to estimate the velocity and duration of these deformations [Nosov (2005)], but also to reveal low-frequency elastic oscillations of the water column. These results were analysed in detail in Sect. 3.1.6.

The catastrophe that took place in the Indian Ocean on December 26, 2004, once more demonstrated the necessity of developing a global system for controlling the ocean level. In spite of the high costs (a DART station costs 250,000$ US), bottom measurements of the ocean level represent an extremely promising and reliable means of operative tsunami forecasting.

6.2 Geomorphological Consequences of Tsunami: Deposits of Paleotsunamis

Detailed investigation and documentation of the effects of the tsunami influence on the coast is a very important task, which permits to study more profoundly the nature of this phenomenon, to develop necessary recommendations for tsunami zoning, to improve schemes for evacuating the population, to determine the local influence of the coast morphology on the tsunami effects and much more.

A tsunami drags along a whole number of geomorphological consequences. The most widespread of them consists in the deposition of sea sand, silt and uprooted material far from the shoreline. The most strong tsunamis move material over distances up to 5 km, and, when the water draws back, the territory, that was flooded, turns out to be covered with a layer of 'tsunamigenic deposits' of thicknesses from centimeters up to meters. Tsunamigenic deposits, as a rule, consist of the material making up beaches and the shallow-water part of the coastal zone. Moreover, tsunamis shift a large amount of torn-up and broken trees and bushes, if there were any on the shore. In inhabited places a tsunami carries away a lot of garbage and fragments of destroyed buildings. Thus, for example, the front of the Indonesian tsunami of December 26, 2004 carried so much material with it in the region of Banda Aceh city that it looked more like a mud flow or lahar (Fig. 6.7).

One more, very widespread, effect, caused by tsunamis on shores, consists in soil erosion and abrasion of hillsides and terraces. As a rule, these effects accompany strong tsunamis with high flow velocities. In such cases, a tsunami strips the layer of soil and vegetation away from the earth's surface, washes away coastal swells, river sandbars, ledges of terraces and washes out niches along slopes, made up of sedimental or weakly consolidated rock (Fig. 6.8). Investigation of the consequences of

Fig. 6.7 City Banda Aceh (North of island Sumatra) destroyed by the tsunami of December 26, 2004. The photograph was taken at a distance of 3 km from the shore, in the very beginning of the zone, where material and fragments of buildings displaced from the shore were deposited (Photo by T. K. Pinegina) (see also Plate 10 in the Colour Plate Section on page 318)

Tsunami height determined
from traces on the trees

Fig. 6.8 Segment of shore 15 km south-west of the northern end of island Sumatra. The tsunami height amounted to 15–20 m, here, while the heights of individual run-ups were up to 35 m above the sea level. In this region the tsunami exerted a strong erosive and abrasive influence on the shore and the nearby hills. In the background of the photograph one can see an abrasion niche, formed after the tsunami. The distance from the shoreline is 300 m (Photo by T. K. Pinegina) (see also Plate 11 in the Colour Plate Section on page 319)

a whole number of historical tsunamis in various regions over the world has shown that tsunamis do not form new shapes of the coastal relief (for instance, like new swells), but are capable of very strongly washing away the existing shapes. The width of the zone subjected to erosive activity depends, first of all, on the tsunami intensity and on the flow velocity. Thus, for example, in the Aceh province, in the North of Sumatra, the width of the 'erosion' zone of the December 26, 2004, earthquake impact amounted to 2–3 km and, farther, at a distance of 5–10 km was the zone of accumulation of tsunamigenic material.

6.2 Geomorphological Consequences of Tsunami: Deposits of Paleotsunamis

Tsunamis approach a coast either in the form of gradual inundation or a rapid tide. The wave front, when running up the land, is rendered complex by various turbulent flows, therefore, the direction of motion of the tsunami and its height may vary very strongly even within hundreds of meters. In this connection, tsunami effects on the shore may differ even in neighbouring regions. The width of the tsunami inundation zone depends on the relief topography, the tsunami height and its wavelength.

Investigation of the consequences due to a tsunami traditionally includes the documentation of its run-up height, inundation height and the distance covered by the run-up. It is also necessary to pay attention to the direction of fallen trees, pillars and other constructions; to photograph regions of eroded land and of accumulation of tsunamigenic deposits. The maximum inundation level often leaves a muddy sign on the walls of houses, which also has to be recorded. The tsunami height can be estimated from the garbage left on the branches of surviving trees and, also, by the height, at which there are broken twigs and peeled off bark. It is necessary to note the variation of water level with the distance from the shore, since this provides a basis for determining the wavelength. On surviving houses it is necessary to measure the height of water at the wall open to the sea and at the opposite side of the house, since the difference in levels permits to determine the flow velocity. Besides the above, it is necessary to make bores at various distances from the shore and to determine the thickness of the tsunamigenic layer and, also, to take samples (100–200 g) from this layer for a granulometric analysis. Such analysis gives an idea of changes in the flow velocity. If possible, it is necessary to take note of the number of strata—this provides information on the number of waves.

The main method of measuring the heights and depths of tsunami inundation consists in gaging topographic profiles on the shore—from the sea level (perpendicular to the shoreline) up to the height of maximum inundation at the maximum distance covered by the run-up (Fig. 6.9). Gaging the topographic profile involves registration of the time the height of the first point on the shore above the water level was determined. Then, the table of tides and ebb-tides is applied in recalculating the profile height relative to the average sea level, and, also, with respect to the sea level at the tsunami arrival time.

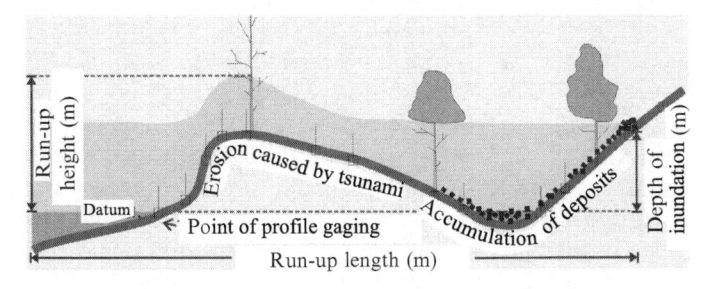

Fig. 6.9 Scheme for gaging the main characteristics of tsunami manifestations along the topographic profile

If the area investigated is inhabited, inquiries must be made among the population to clarify the beginning time of the tsunami, its period, the number of waves and so on, and, also, to analyse thoroughly the destruction of buildings.

It is necessary to make use of the results of in situ studies, aerial surveys and cosmic photography to compile the map showing the distribution of run-up heights and distances over the entire territory, subjected to the influence of a tsunami.

Although tsunamis occur every year, strong and catastrophic tsunamis are quite rare phenomena, and for most tsunamidangerous coasts the catalogue of these events is insufficiently representative for statistical estimations. Therefore, studies of *tsunami deposits*, both historical and prehistorical (*paleotsunami deposits*), are carried out throughout the entire world since the end of the 1980s.

The first studies of tsunami deposits were performed in Japan in the middle of the 1980s [Atwater (1987); Minoura et al. (1996)]. Subsequently, similar investigations were carried out on the West coast of the USA and Canada, in Chile, Australia, New Zealand and in a number of other countries—in most of the regions subject to tsunami. In Russia detailed work in this direction started in the middle of the 1990s—on Sakhalin, Kamchatka, the Kuril islands—in the most tsunamidangerous areas of the Russian Far East [Pinegina et al. (1997), (2000); Pinegina/Bourgeois (2001)].

The following can be considered the main peculiarities of tsunami deposits:

1. They are related to the coastal belt outside the reach of storms and to various hypsometric levels (approximately up to 30 m above the sea level).
2. The presence in deposits of sea sand and of smoothed out pebbles.
3. Insignificant thickness of deposits (from several millimeters up to several tens of centimeters, rarely down to a few meters).
4. Periodicity of deposit formation (few tens–few hundreds of years).

Preliminary sites of searches for, and investigation of, tsunami deposits are chosen after a thorough analysis of aerial-survey and cosmic photographs and topographic materials. They are used for identifying key areas on the coasts, where tsunamigenic deposits may have remained intact for a long time. These areas must not be within zones subject to the influence of alluvial and slope processes (e.g. freshets, rock slides, etc.). It is desirable for the coastal relief to have various height levels, and for the coastal configuration not to hinder free tsunami penetration. Moreover, the descriptions of historical tsunamis in the proposed search area are collected. Deposits of historical tsunamis serve as benchmarks for revealing the intensities of more ancient events.

During in situ work at the chosen coastal area levelling photography of the topographical profiles is carried out, as shown in Fig. 6.9, from the shoreline across the beach and coastal swells up to the distance of maximum tsunami run-ups. The profiles are usually determined for several kilometers along the shore. Along the profiles geological bores are made, in terms of which geological cuts are described. Usually from 5 to 20 bores are made along each profile. The average depths of the bores are between 1 and 4 m—depending on the age of the surface and on the growth rate of the soil. During geological description of the bores, strata of soils

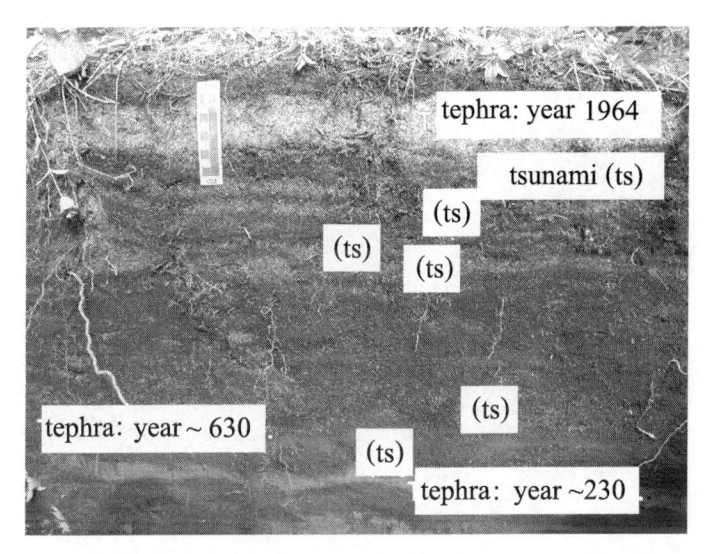

6.2 Geomorphological Consequences of Tsunami: Deposits of Paleotsunamis 245

Fig. 6.10 Photograph of wall of bore with strata of tephra and tsunamigenic horizons in the soil. The bore is located at the 20th terrace 500 m from the shoreline of today (peninsula of the Kamchatka cape, Kamchatka) (see also Plate 12 in the Colour Plate Section on page 320)

and peat, as well as tsunamigenic and other deposits are identified. Samples are selected from these prolayers for mineralogical, radio-carbonic, diatomic, spore-pollen, granulometric and other sorts of analysis—for determining the genesis and age of the prolayers.

In the Far East, for instance, on the Kamchatka, the investigation of tsunami deposits is closely related to the possibility of applying the method of tephrochronology. The method is based on investigating and correlating the marking horizons of volcanic ash (tephra), each of which exhibits a characteristic outlook, chemical and mineralogical composition and abundance over a large territory. In studies of tsunami deposits on the Kamchatka the tephrochronological method is applied as the base method for correlating and dating deposits (Fig. 6.10).

When the description of geological cuts, the identification of deposits and the correlation of tsunamigenic prolayers are completed, a synoptical geological column and a geochronological (time) cut are compiled, from which it is possible to calculate the tsunami repetition frequency on the coast. Such work has resulted in significant success, for instance, in Russia. Thus, for many regions of the Kamchatka and Kuril coasts traces were revealed of deposits due to tens of ancient tsunamis that occurred in the past 2,000–7,000 thousand years.

The data on the tsunami repetition rates at individual sites of the coast are insufficient for estimation of the intensities of ancient tsunamis and of the earthquakes that caused them. To resolve such problems it is necessary to know the run-up or inundation height, the penetration distance and the length of the coast, subjected to the influence of each of the tsunamis identified. This task is very complicated and complex. For each of the regions investigated it is necessary to perform

Fig. 6.11 Example of coseismic deformation on the island Simeulue, village of Busung, due to strong aftershock on March 28, 2005 at the coast of Indonesia (Photo by V. Kaistrenko) (see also Plate 13 in the Colour Plate Section on page 321)

reconstruction of the positions of the ancient shoreline and of the relief height for different moments of time. And only after this has been done it becomes possible to determine the parameters of ancient tsunamis. In the conditions of rapid and large uplift or subsidence of the coasts (which is peculiar to subduction zones) it is especially important to perform reconstructions. The difficulty of such work is enhanced, also, by the coast along the subduction zones being subject to sharp coseismic elevations and depressions, which often make it difficult to determine the position of the shoreline for the moment of time investigated (see Fig. 6.11).

In spite of the difficulties encountered in geological interpretation of tsunami deposits, such work represents the only possibility for obtaining objective data both for tsunami zoning and for estimating distribution in space–time of the sources of past strong tsunamigenic earthquakes.

6.3 Tsunami Detection in the Open Ocean by Satellite Altimetry

Revelation of the place and time a tsunami wave originates is based on seismic information obtained immediately after the earthquake. The absence of observational data on the tsunami parameters at the source leads to a low efficiency of the computational models, determining the arrival time and amplitude of a wave at each concrete point. As a result, the level of false tsunami warnings increases.

Thus, for example, during the Shikotan tsunami of October 4, 1994, which happened to be catastrophic for the Southern Kurils and the Hokkaido island, the international service sent a warning to the Hawaii islands about the possible approach of a tsunami several meters high. Significant financial means (up to US$ 30 million) were applied in evacuating thousands of people, although the height of the tsunami turned out to be about half a metre. In these conditions the application of remote satellite methods for registering tsunami waves would have permitted to obtain the lacking information at the very moment of the earthquake or immediately afterwards.

Many thousands of human lives could have been saved during the tsunami of December 26, 2004, if the system for satellite registration of tsunami waves had already been functioning within the Indian Ocean.

Direct tsunami measurements, received by coastal-level recorders, contain oscillations strongly distorting the initial record of a wave in the open ocean. The arrival of a wave in shallow water and its reflection from the coast leads to enhancement of its amplitude, but the spectrum (shape) of the signal is distorted by the resonance properties of the shelf, bays and straits. The best in quality records of tsunamis in the open ocean are provided by sensors of the bottom hydrostatic pressure [Kulikov, Gonzalez (1995)]. However, such systems are very expensive and point-like measurements do not provide for total coverage of probable zones of tsunami wave origination. The rapid development of remote (satellite) methods opens up new possibilities for resolving problems of efficient tsunami forecasting.

At present, a cardinal way for resolving the problem of sea-level investigation not only near, but also at a significant distance from the coast with clear connection to a unique geodetic reference system consists in the application of satellite altimetry and, for example, high-precision radio-altimetry measurements using the Earth's artificial satellites (EAS) GEOSAT, TOPEX/POSEIDON, ERS-1,2, JASON-1 and ENVISAT. For this purpose, in the future, measurements can be made use of, that were made by the Russian geodetic EAS 'Musson-2' and other satellites with altimeters, designed in other countries. The accuracy, with which the data on the sea level are correlated to the common system of heights is provided for by the receivers of one of the navigational systems, GLONASS, GPS or DORIS, established on board the satellite. At present, the data of satellite altimetry are widely applied in investigations of mesoscale variations of flows, tides, etc.

Modern systems and means of satellite altimetry are successfully applied for studying properties of the oceanic lithosphere, determination the parameters of lunar–solar tides in the ocean. The existence of a correlation has been established between the level topography of the World Ocean and circulation of water masses and meteorological phenomena. Moreover, satellite altimetry has been shown and confirmed experimentally to serve as an effective means for studying deviations of the vertical slope, the ocean bottom relief and the dynamics of the World Ocean surface.

Taking into account the prospects of applying satellite altimetry for resolving the above-indicated broad class of problems, intensive work was initiated during the period after 1980 for creating a new class of high-precision radio-altimeters and EAS, to be equipped with them. From the middle of the 1980s till the present day, seven EAS with high-precision radio-altimeters on board were put into orbit. During this period the precision in determining the orbit improved from 45 to 5 cm, while the measurement precision was improved from 1.5 m to 3–6 cm. The latest models of EAS (JASON-1, ENVISAT-1), put into orbit in 2001 and 2002 provide for a root-mean-square measurement error within 2 cm.

Original satellite altimetry databases have been created and are regularly updated, which are available for scientific purposes in the Distributed Archive of physical oceanography of the Jet Propulsion Laboratory of the California Institute of Technology (PO.DAAC) in the USA and the Center for satellite oceanographic data storage, control and interpretation (AVISO) in Europe.

An integrated database for satellite altimetry (IDBSA) and a System for automatized satellite altimetry data processing have also been created in the RAS geophysical center (RAS GC) with support of the RFBR. The database includes altimetry data from satellites GEOIC, GEOSAT, TOPEX/POSEIDON, ERS 1,2, GFO and JASON for the period from 1985 to 2003. The System for automized data processing, designed in the RAS GC, supports all data formats adopted in foreign centres, and IDBSA.

The first attempts at applying satellite altimetry data in searching for tsunamis in the open ocean were evidently made by American specialists in 1994–1999 [Okal et al. (1999)]. They analysed satellite altimetry data, obtained in the experiments TOPEX/POSEIDON, related to several strongest tsunamigenic earthquakes: the Nicaragua tsunami of September 2, 1992, the Okushiri tsunami of July 12, 1993, the tsunami of June 2, 1994, on the island Java and the Shikotan tsunami of October 4, 1994. Spectral analysis has only permitted to identify definitively the wave of the 1992 Nicaragua tsunami.

Tsunami researchers recently developed an improved procedure [Zaichenko et al. (2005)] of satellite information processing that comprised several stages for the registration of tsunami waves in the open ocean. At the first stage, the satellite cycle, corresponding to the tsunamigenic earthquake, was chosen. Then, all the routes of this cycle covering the Pacific Ocean were selected from the database of the RAS GC. The satellite circuits preceding the earthquake were further discarded. At the same time the model, chosen for calculation of the arrival time of the waves, was applied to calculate the position of the wave front in the case of each tsunami source. Calculation of the arrival times was based on bathymetric data with a space resolution of 2 min. Further, for each point of the satellite route comparison was performed of the satellite time, counted from the earthquake origination moment, and

6.3 Tsunami Detection in the Open Ocean by Satellite Altimetry

the calculated time for the position of the wave front. This resulted in determination of the position of the intersection point of the satellite's route and the tsunami front.

The position of the wave front is to a certain extent conventional. Calculations are performed assuming the maximum of the wave propagation velocity to be $c = \sqrt{gH}$, where c is the wave propagation velocity, g is the acceleration of gravity and H is the ocean depth. In reality, the wave velocity also depends on its length, $c = \sqrt{g\tanh(kH)/k}$, where $k = 2\pi/\lambda$, λ is the wavelength. The shorter the wavelength, the slower it propagates. Owing to this dispersion effect, the real front, as a rule, lags behind this theoretical estimate. The longer the propagation path, the stronger is the dispersion effect. The wave front can be distorted even stronger in shallow water and during tides. Nevertheless, the estimate of the tsunami wave front position, calculated by formula $c = \sqrt{gH}$, is quite accurate and its error does not exceed the size of the tsunami source $\sim$50–100 km.

The evolution of a wave packet is due to the dependence of the group velocity on the frequency:

$$c_g = \frac{d\omega}{dk} = \frac{\omega}{2k}\left(1 + \frac{2kH}{\sinh(2kH)}\right), \tag{6.1}$$

where ω is the angular wave frequency, k is the wave number and H is the ocean depth.

A detailed analysis of altimetry data has been performed in [Zaichenko et al. (2004)] for six strongest tsunamis: the Shikotan tsunami of 1994, the 1996 tsunami near the island Irian Jaya, the Okushiri tsunami of 1993, the 1998 tsunami near the coast of Papua New Guinea, the tsunamis near Island Java (1994) and the coast of Peru (2001). In the first four cases it turned out to be possible to reveal specific perturbations of the ocean level, which appear in the records within the time range close to the moment, when the calculated tsunami front passes. Such perturbations were most strikingly revealed in the records obtained on July 17, 1998 (Papua New Guinea). In the last two cases, searches for traces of tsunamis in the records were not met with success.

Figures 6.12 and 6.13 present a map of the investigated region of the Pacific Ocean with isolines of tsunami arrival times and oceanic-level profiles at the coast of Papua New Guinea, on which the perturbation, supposedly corresponding to the 1998 tsunami passage time, is indicated.

Extremely interesting results were obtained from the analysis of radioaltimetry observations of the catastrophic tsunami of December 26, 2004, in the Indian Ocean [Kulikov et al. (2005)]. All available altimetry data from TOPEX/POSEIDON, ENVISAT and JASON-1 were analysed for the period immediately after the seismic shock. Individual routes revealed anomalous-level variations, probably related to the passage of tsunami waves. The best-quality record JASON-1 (cycle 109, circuit 129) was chosen for further calculations.

Figure 6.14 presents the map of the north-eastern part of the Indian Ocean with epicentres of the main earthquake (the black square) and the main aftershocks (circles). The isochrones, showing the calculated position of the tsunami front, are constructed with an interval of 0.5 an hour. The figure also shows the route of the satellite JASON-1 (cycle 109, circuit 129) and the respective profile of the ocean

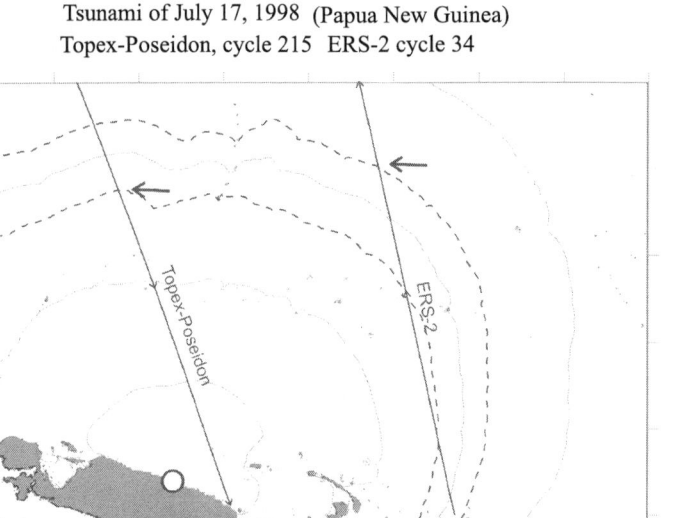

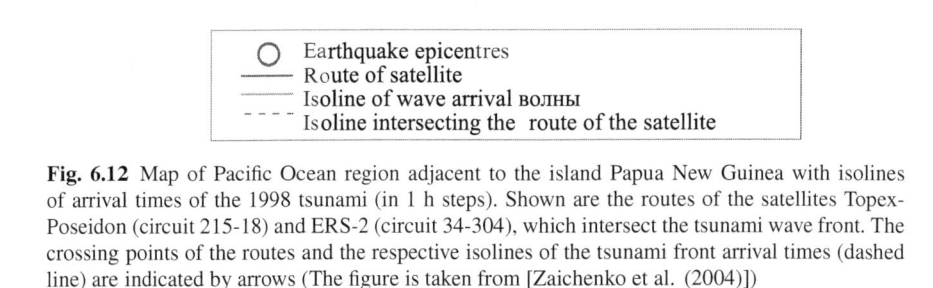

Fig. 6.12 Map of Pacific Ocean region adjacent to the island Papua New Guinea with isolines of arrival times of the 1998 tsunami (in 1 h steps). Shown are the routes of the satellites Topex-Poseidon (circuit 215-18) and ERS-2 (circuit 34-304), which intersect the tsunami wave front. The crossing points of the routes and the respective isolines of the tsunami front arrival times (dashed line) are indicated by arrows (The figure is taken from [Zaichenko et al. (2004)])

level, measured by the altimeter. In Fig. 6.15b (see colour section) the profile is depicted in an enhanced scale. For comparison, the level profile, obtained 10 days before the tsunami (preceding cycle 108), is also plotted. The time it took the satellite to cross the Indian Ocean corresponds to the period between 2 h 51 min (12° of southern latitude) and 3 h 02 min (20° of northern latitude), i.e. approximately 2 h after formation of the tsunami wave. The wave front is seen well at approximately 6° of southern latitude. The maximum wave amplitude amounts to 80 cm.

From Fig. 6.14 the intersection angle of the satellite's route and the wave front is seen to be approximately 45°. Therefore, the real horizontal scales of ocean-level variations must be (for geometrical reasons) divided by 1.4. The main wave length in

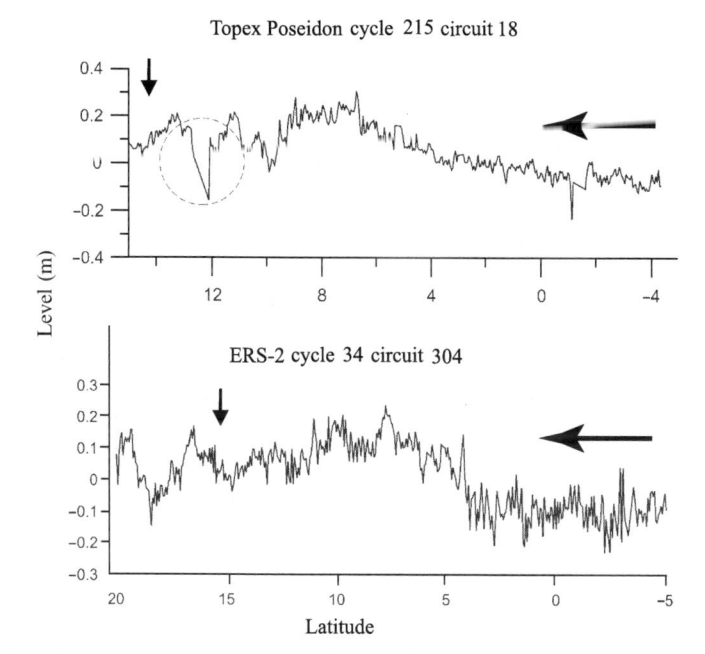

6.3 Tsunami Detection in the Open Ocean by Satellite Altimetry

Fig. 6.13 Profiles of the ocean level with indication of the direction of wave motion (horizontal arrows) and the crossing point of the theoretical tsunami front and the route of the satellite (vertical arrows) (The figure is taken from [Zaichenko et al. (2004)])

the record of an altimeter amounts to about 700 km, which corresponds to 500 km for the length of a real tsunami wave. On the basis of the average wave velocity in the open ocean, equal to ∼200 m/s, it is possible to calculate the main tsunami period $T \approx 40$ min. Attention should be drawn to manifestations of higher-frequency oscillations of the level being observed noticeably more to the north, i.e. closer to the source. This actually reflects linear tsunami wave dispersion, when short-period waves exhibit propagation velocities inferior to those of long-period components.

To analyse the effect of linear tsunami wave dispersion the dependence was calculated of the spectral amplitude of the signal on time and the wave number (wavelet diagram). The result of calculations is presented in Fig. 6.15b. The wave front is clearly seen to be related to the onset moment of the low-frequency components ($k < 0.05$ km^{-1}). High-frequency components appear to the north of the front. The dispersion curve $c_g(k)$, corresponding to formula (6.1), is also shown in the figure. It was calculated taking into account the intersection angle between the satellite's route and the front. Good accordance is observed between 'onsets' of the signal spectral components and the theoretical curve.

The revealed effect of tsunami wave dispersion demonstrates the restrictions of the long-wave approximation, widely applied in numerical models of tsunami wave propagation. Owing to the 'lag' of the high-frequency components in the wave spectrum, the tsunami amplitude decreases more rapidly, than in the 'shallow-water'

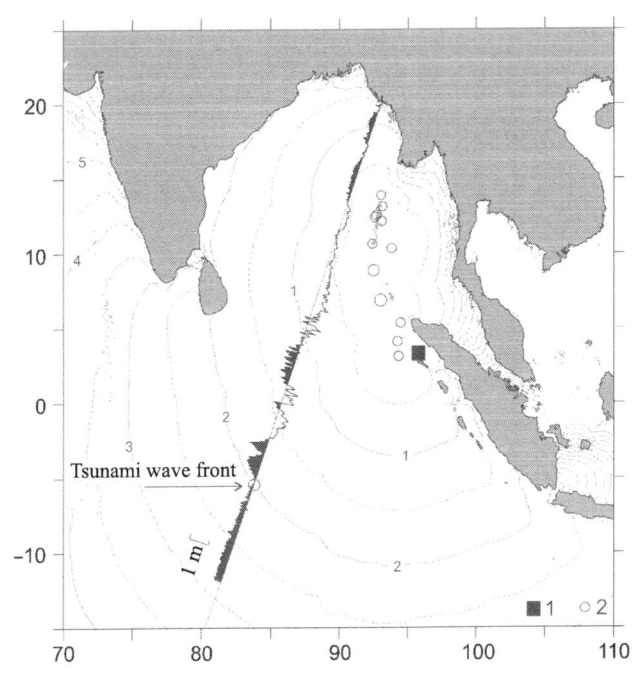

Fig. 6.14 Map of north-eastern part of the Indian Ocean with isochrones, showing the calculated front position of the tsunami of December 24, 2004 (with an interval of 0.5 an hour). 1—epicentre of the main earthquake, 2—epicentres of main aftershocks. Route of JASON-1 satellite (circuit 109-129). The profile of the ocean level determined from altimetry data is shown along the route (Adapted from [Kulikov et al. (2005)]) (see also Plate 14 in the Colour Plate Section on page 322)

model. This error is especially noticeable in calculations of the wave field at significant distances from the source. In [Kulikov et al. (1995)] it was shown that the effect of linear dispersion can actually totally distort the form of the tsunami signal in the open ocean. In this case, the main energy is concentrated in the region of periods around 30–50 min, and at a distance of about 1,000 km from the source the distortion is not so significant.

The results described above, in principle, demonstrate the possibility of timely registration of a dangerous tsunami wave in the open ocean with the aid of modern systems, permitting to observe the ocean from outer space. Such methods of effective tsunami forecasting will obviously develop in the direction of creating a technology for continuous monitoring of the ocean surface both with the aid of sensors of the open ocean level, equipped with telemetric connection to the processing centers, and making use of satellite altimetry measurements.

Note that the sensor of bottom pressure (see Sect. 6.1) used for measuring the level of the open ocean possesses an essential advantage as compared to the satellite altimeter. The point is that the frequency range of tsunami waves is practically free of irrelevant signals, while the corresponding range of wavelengths is quite noisy (e.g. owing to vortical formations). Therefore, a tsunami wave can be

6.3 Tsunami Detection in the Open Ocean by Satellite Altimetry

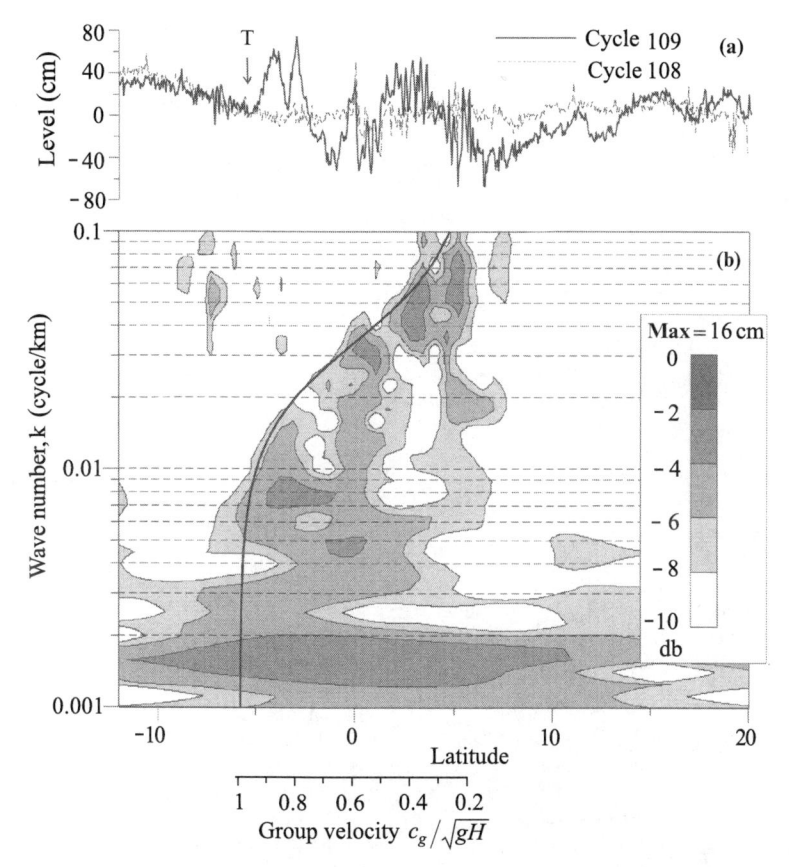

Fig. 6.15 (a) Altimetry sea-level taken along track 129 of the Jason-1 satellite for Cycle 109 and along the same track 10 days earlier for Cycle 108; and (b) wavelet analysis of the sea level profile in (a). The theoretical curve, calculated in accordance with the linear dispersion law for surface gravitational waves, shows the calculated 'onset' moments of the respective spectral components. Letter "T" indicates the tsunami wavefront (Courtesy of E.A. Kulikov) (see also Plate 15 in the Colour Plate Section on page 323)

readily singled out in the record of ocean-level variations in time, but not in space. A sole 'instantaneous shot' of the ocean level along the track is insufficient for reliable identification of a tsunami wave—it will just be invisible against the background of other processes. But, by comparing the data obtained from two or more satellites travelling along the same track with a certain time delay between them, it is possible to single out a tsunami wave against a noisy background. Anyhow, such an approach will require significant enhancement of the number of satellites, equipped with altimeters.

References

Atwater B. F. (1987): Evidence for Great Holocene Earthquakes along the Outer Coast of Washington State. Science **236** 942–944

Bernard E. N., Gonzalez F. I., Meining C., Milburn H. B. (2001): Early Detection and Real-Time Reporting of Deep-Ocean Tsunamis. In: International Tsunami Symposium 2001. Proceedings, NTHMP Review Session, Paper R-6

Cartwright D. E., Zettler B. D., Hamon B. V. (1979): Pelagic tidal constants. Int. Assoc. Phys. Sci. Oceans, Publ. Sci. **30**

Eble M. C., Stalin S. E., Burger E. F. (2001): Acquisition and Quality Assurance of DART data. In: International Tsunami Symposium 2001 Proceedings, Session 5, Papers 5–9

Gonzalez F. I., Bernard S. N., Milbern H. B., et al. (1987): The Pacific Tsunami Observation Program (PacTOP), In: Proc. IUGG/IOC, International Tsunami Symposium 3–19

Green D. (2006): Transitioning NOAA Moored Buoy Systems From Research to Operations. In: Proceedings of OCEANS'06 MTS/IEEE Conference, 18–21 September 2006, Boston, MA, CD-ROM

Gonzalez F. I., Bernard E. N., Meinig Ch., Eble M. C., Mofjeldand H. O., Stalin S. (2005): The NTHMP Tsunameter Network. Nat. Hazards, **35** 25–39

Hirata K., Aoyagi M., Mikada H., et al. (2002): Real-time geophysical measurements on the deep seafloor using submarine cable in the southern Kurile subduction zone. IEEE J. Ocean Eng. **27**(2)

Jacques V. M., Soloviev S. L. (1971): Remote registration of weak tsunami-type waves on the shelf of the Kuril islands. DAN SSSR (in Russian), **198**(4) 816–817

Kong L. (2002): DART buoys provide real-time reporting of tsunami. Tsunami Newsletter **XXXIV**(2) 3–8

Kulikov E. A., Medvedev P. P., Lappo S. S. (2005): Satellite recording of the Indian Ocean tsunami on December 26, 2004. Doklady Earth Sciences **401**(3) 444–448

Kulikov E. A., Gonzalez F. I. (1995): Reconstruction of the shape of a tsunami signal at the source by measurements of hydrostatic pressure oscillations using a remote bottom sensor (in Russian). DAN RF. **344**(6) 814–818

Meinig C., Stalin S. E., Nakamura A. I., Milburn H. B. (2005): Real-Time Deep-Ocean Tsunami Measuring, Monitoring, and Reporting System: The NOAA DART II Description and Disclosure

Minoura K., Gusiakov V. K., Kurbatov A. V. (1996): Tsunami sedimentation associated with the 1923 Kamchatka earthquake Sediment. Geol. **106**(1–2) 145–154

Nosov M. A. (2005): Elastic oscillations of water layer in the 2003 Tokachi-Oki tsunami source. In: Proceedings of the 22nd International Tsunami Symposium, Chania, Crete isl. Greece, 27–29 June, 2005. Edited by Papadopoulos G. A., Satake K. P. 168–172

Okal E. A., Piatanesi A. Heinrich P. (1999): Tsunami detection by satellite altimetry. J. Geophys. Res. **104** 599–615

Pinegina T. K., Bourgeois J. (2001): Historical and paleo-tsunami deposits on Kamchatka, Russia: long-term chronologies and long-distance correlations. Nat. Hazards Earth Syst. Sci. **1**(4) 177–185

Pinegina, T., Bourgeois J., Bazanova L., et al. (2003): A millennial–scale record of holocene tsunamis on the Kronotskiy Bay coast, Kamchatka, Russia. Quat. Res. **59** 36–47

Pinegina T. K., Bazanova L. I., Melekestsev I. V., et al. (2000): Prehistorical tsuanmis in Kronotskii gulf, Kamchatka, Russia: A progress report. Volcanology and Seismology **22**(2) 213–226

Pinegina T. K., Melekestsev I. V., Braitseva O. A., et al. (1997): Traces of prehistoric tsunamis on the eastern coast of Kamchatka (in Russian). Priroda (4) 102–106

Soloviev S. L. (1968): The tsunami problem and its significance for the Kamchatka and the Kuril islands (in Russian). In: The tsunami problem, pp. 7–50. Nauka

References

Titov V. V., Gonzalez F. I., Bernard E. N., et al. (2005): In: Real-time tsunami forecasting: challenges and solutions. Nat. Hazards **35**(1), Special Issue, pp. 41–58. US National Tsunami Hazard Mitigation Program

UNESCO Tech. Papers (1975): An intercomparison of open sea tidal pressure sensors. UNESCO Tech. Papers in Mar. Sci. **21**

Watanabe T., Matsumoto H., Sugioka H., et al. (2004): Offshore monitoring system records recent earthquake off Japan's northernmost island. Eos. **85**(2) 13 January

Wunsch, C. (1972): Bermuda sea-level in relation to tides, weather and baroclinic fluctuations. Rev. Geophys. Space Phys. **10**(1)

Zaichenko M. Yu., Kulikov E. A., Levin B. W., Medvedev P. P. (2004): Examples of tsunami registration in the open ocean based on data from a satellite altimeter (1993–2001). Preprint IORAH (in Russian), Tsunami laboratory. Yanus-K, Moscow

Zaichenko M. Yu., Kulikov E. A., Levin B. V., Medvedev P. P. (2005): On the possibility of registration of tsunami waves in the open ocean with the use of a satellite altimeter. Oceanology **45**(2) 194–201

Chapter 7
Seaquakes: Analysis of Phenomena and Modelling

Abstract The main manifestations of seaquakes and possible consequences of this phenomenon are described. The seaquake intensity scale is given. Historical evidence of seaquakes, originating in the Pacific Ocean and in the Mediterranean Sea, has been collected and analysed. Information is presented on instrumental observations of variations in the temperature fields in the ocean after submarine earthquakes. The possibility is estimated for a submarine earthquake to result in destruction of the temperature stratification in the ocean. Theoretical ideas of the parametric generation of surface waves, due to an underwater earthquake, are expounded. The results are described of laboratory experiments, devoted to the investigation of wave structures on the surface of a liquid and of the transformation of stable stratification in its column, in the case of bottom oscillations.

Keywords Underwater earthquake · seaquake · Faraday ripples · stratification · vertical exchange · SST anomaly · internal waves · biogenes · upwelling · sea colour · historical testimony · turbulence · mixing · parametric resonance · non-linear currents · weather phenomena · turbulence energy · earthquake energy · Mathieu equation · instability · increment · experimental set-up · bottom oscillations · dynamic modes · dissipative structures · temperature profile · turbulent exchange coefficient · shadow method

Every year over 100 strong ($M > 6$) earthquakes, most of which are under water, take place on the Earth. The enormous energy released during an earthquake, even if an insignificant part is transferred to the ocean near the source, can lead to strong and even catastrophic dynamic perturbations of the water mass, which is known as a seaquake.

A seaquake usually represents a strong perturbation of a large area (of diameter of the order of 10–100 km) of the surface layer of water, which continues several minutes, and is manifested in the generation of a system of very steep standing waves of large amplitudes (up to 10 m), the sudden ejection of vertical water jets, the development of a foaming cavitation zone over the entire area, generation of local eddies and splashes of water, strong blows of compression waves against the boat's bottom and the appearance of powerful low-frequency acoustic effects in the atmosphere [Levin (1996)].

This phenomenon is considered quite rare, in spite of the fact that tsunami catalogues, scientific literature and maritime navigation charts refer to the descriptions of more than 250 such events during the written history of mankind. A synthesized description of a seaquake is presented in Sect. 1.6. From the point of view of the influence upon a human being and the danger for seafaring, a seaquake can quite compete with tsunamis and killer waves. Seaquakes lead to the destruction of structures on board ships, demoralization of the crew, the rise of critical and emergency situations on vessels and the mass death of fish and other inhabitants of the ocean. The physics of this phenomenon is not quite clear yet, and it requires further investigation.

From general arguments it is clear that a seaquake results from the influence of seismic oscillations of the ocean bottom on the water column, since this process undergoes development in the epicentral zone and terminates, when the action of the earthquake finishes. The effects of a seaquake, as a natural phenomenon, and their influence on ship constructions were studied by the renown geophysicists B. Gutenberg and A. Zieberg [Richter (1963)].

Below, we present the intensity scale for seaquakes, developed by A. Zieberg and modified in [Levin, Soloviev (1985)].

I   A vibration, a light crackling of the deck.
II   A clear crackle, like a light scratching.
III   A strong jolt, as if running aground in shallow water, or on rocky bottom, or onto reef. Loud crack, vibration of objects.
IV   The vessel cracks and is shaken, unstable subjects fall.
V   People cannot stand up, large objects turn over and fall out of supports, vessel loses speed, constructions creak painfully.
VI   The vessel may be thrown out of the water, masts and deck constructions are broken, emergency situation.

In the opinion of C.F. Richter, one of the founding fathers of seismology, 'although the problem of seaquakes contains no unresolvable riddles, to the best of our knowledge, it was neglected for so long, that at present it represents a promising field of studies for new individuals with new ideas'.

The first physical studies of wave structures, arising on the surface of an oscillating liquid, were performed by Michael Faraday back in 1831. In M. Faraday's experiments a flat cavity (saucer) with the liquid was put on a vibrating membrane or console, and the originating sets of standing waves were studied for various vibration modes. The liquids used comprised water, glycerine, vegetable oil and even egg yolk. For the first time, a stable set of standing waves with wavelengths between 3 and 30 mm was found to form on the surface of a liquid vibrating with acoustic frequencies. The wave cells were shaped like squares or hexagons, and their size increased, as the vibration frequency decreased. The phenomenon described of a wave structure arising on the surface of a vibrating liquid was subsequently termed 'Faraday ripples'. Various aspects of this phenomenon were investigated in [Ezersky et al. (1985); Levin, Trubnikov (1986); Alexandrov et al. (1986); Levin (1996); Vega et al. (2001)] and [Higuera et al. (2000)].

Most regions of the World Ocean are known to be characterized by a clearly pronounced stable temperature stratification—the cold lower column several kilometers thick is separated from the atmosphere by a relatively thin (measured by hundreds of meters) warm 'film', comprising a thermocline and a mixed layer. The possibility of cold deep-water masses being transferred up to the surface layer of the ocean by non-linear flows in the seaquake zone was first mentioned in [Levin et al. (1993)] in relation to the identification of a tsunami source from outer space. Later it was pointed out in [Nosov, Ivanov (1994)] that a seaquake can cause the development of such powerful turbulence as to result in the warm film destruction and in formation on the surface of the ocean of a cold 'spot' with an area exceeding $1,000\,\text{km}^2$; it was noted that such a 'spot' is capable of exerting significant influence on the structure of the temperature field of near-the-water layer of the atmosphere and to lead to weather anomalies.

7.1 Manifestations of Seaquakes: Descriptions by Witnesses and Instrumental Observations

7.1.1 Historical Evidence

In this section we shall mostly deal with data to be found in tsunami catalogues for the Mediterranean sea and the Pacific Ocean [Soloviev et al. (1997); Soloviev, Go (1974), (1975)]. About 300 events of tsunamis and of similar phenomena are known in the case of the Mediterranean sea (2000 BC–AD 1991), and for the Pacific Ocean there are about 1,000 events (years 173–1968). Among all the cases we have singled out 25 events in the Mediterranean sea and 65 events in the Pacific Ocean, the descriptions of which refer directly or indirectly to an intensification of the vertical exchange in the ocean resulting from seismic movements of the ocean bottom. It is important to note, here, that the information we are interested in is not the main information for tsunami catalogues.

Most cases, recorded in catalogues, are based on evidence presented by witnesses, who once in a while happened to see phenomena proceeding in the open ocean or at uninhabited coasts. When interpreting testimonies of catastrophic earthquakes and tsunamis it is necessary to take into account the extremely strong shock felt by the witnesses—the people could have not noticed a certain phenomenon, or, contrariwise, might describe something that actually had not taken place. Thus, it would not be correct to do any probability estimations on the basis of the number of singled out events (90) and the total number of cases ($\sim$1,300), described in tsunami catalogues.

The technique for singling out an event among the entire amount was based on the assumption of consequences of short and sharp enhancement of the vertical exchange in the ocean, which from our point of view may be the following:

1. Transfer of bottom sediments, sand, and so on up to the surface and, as a consequence, the water becoming turbid or changing colour—similar phenomena take place in shallow water during a storm.

2. A change in the sea surface temperature (the formation of an SST anomaly), to which the atmosphere should react, i.e. weather anomalies can also be expected. Since, as a rule, the temperature of the water decreases, as the depth increases, the formation of precisely cold SST anomalies is to be expected, in most cases.
3. Transfer of biogenes up to the surface layer, usually impoverished of these substances, and, as a consequence—similar to what happens in upwelling zones, a short-lived, but noticeable enhancement in the concentration of phytoplankton (certain species are capable of reproducing at a rate of 2–3 multiplications per day). Taking into account that the phytoplankton, being the primary link in the trophic chain, determines the bioproductivity of water, the migration of fish and sea animals is possible.
4. Evolution of the zone of violated and, therefore, unstable stratification should create a powerful system of internal waves with amplitudes many times greater than the amplitude of internal tsunami waves that is comparable to the residual bottom displacement.

The fourth consequence can hardly be reflected in testimonies by witnesses—its registration requires special instrumentation; therefore, in dealing with the first three of the above consequences, and having become acquainted with the descriptions of events, we consider it appropriate to single out the four following groups of features that point to possible intensification of vertical exchange in the ocean, resulting from an underwater earthquake:

1. Unusual agitation, rough behaviour of the sea, 'boiling' water and so on
2. The appearance of fish (dead, deep-water, unusual) or of sea 'monsters', a decrease (increase) in the fish catch, etc.
3. Weather phenomena (change of wind, fog, rain, hail etc.)
4. Change in colour of the sea, the water becomes turbid

We shall present some excerpts from the descriptions of those events, that are characterized by at least one of the indicated features.

The Mediterranean sea

1. *79, August 24(23), 7h. Tyrrhenian Sea, Bay of Naples. 40.48° N, 14°27' E.*

The next day after the powerful eruption of the volcano Vesuvius a strong earthquake occurred with its source, possibly, in the Bay of Naples. In a letter to the historian Tacitus, the writer Pliny the Elder, an eyewitness of the events, mentioned that 'the sea was violent and hostile'.

2. *551, spring or July 7. Aegean Sea. Eastern Greece, Strait of Euboea (between the Island of Euboea and the continent). 38.4° N, 22.3° E.*

A destructive earthquake that embraced both coasts of the Gulf of Corinth. A tsunami. After the water went away, fish, including rare species, were left stranded.

3. *1202 (possibly, 1222) May 22, morning (at sunrise). Levantian Sea, Middle East.*

A destructive earthquake in Palestine with its source, possibly, near Nablus, where no single wall remained intact. Near the coast of Syria, after the earthquake, the water in the sea opened up slightly in various places and was divided into hill-like masses ... a great number of fish were thrown out onto the shore.

7.1 Manifestations of Seaquakes 261

4. *1456, December 4, between 20 and 21 h. Tyrrhenian Sea, Italy, Bay of Naples.* 41°18′ N, 14°42′ E.

A catastrophic earthquake, mostly embracing the mountainous regions of the Apennine Range, continued for approximately 6 min. In the Bay of Naples, the sea was so rough that people on the shore 'felt as if they were being attacked by thousands of devils', while people on ships and galleys felt doomed.

5. *1494, July 1, 16 h. Hellenic Arc, Island of Crete.* 35.5° N, 25.5° E.

A strong earthquake shook the island. In the harbour, great waves tore ships off their anchors. ... The water changed colour many times.

6. *1564, July 20(27), 23 h. Ligurian Sea, Italy, France.* 44° N, 7°17′ E.

An extremely strong earthquake with its source apparently in the Maritime Alps, possibly, in the valley of river La Vésubie. It embraced Nice, Provence, Villefranche, San Remo, Porto Maurizio and so on.

The sea near Nice and Villefranche receded (became lower by 'one spear') and its bottom was uncovered; many fish were left there, including, also, some unknown species (deep-water fish?).

7. *1693, January 11, 21 h. Calabrian Arc. Island of Sicily, the eastern coast.* 37°10′ N, 15°01′ E.

At 04 h on January 9, Sicily was enveloped by a strong earthquake preceded by a boom. ... Sailors in one of the boats near the Island of Malta reported that the sea suddenly became rough (stormy) without any reason. The seismic process reached its culmination on January 11, when, according to chronicles, a catastrophic earthquake originated and caused, together with the fires that arose, the death of 60,000 inhabitants. ... No fish were caught during 15 days after the earthquake.

8. *1694, September 8, 17 h 45 min. Adriatic Sea, eastern Italy.* 40°48′ N, 15°35′ E.

A catastrophic earthquake occurred. ... The sea carried a stink of silt, mud, and it lasted for over half an hour.

9. *1707, May. Aegean Sea, Island of Thira, volcano Santorini.*

Formation of a new islet. The first underground shocks on Santorini were quite frequent on May 18, 21 and 24. A new islet began to appear at dawn of the 23rd between the Islands of Palaia and Micra Kameni. Fishermen landed on it several days later. But the rocks on the islet were in motion, everything shook under their feet. ... Numerous elevations appeared above the sea waves, only to vanish and then reappear. The colour of the sea around this islet, which was called 'white', changed time and again.

10. *1742, january 16, 19, 20–27 (maybe February 9). Ligurian Sea, Tuscany district.* 01.19(20) 43°5′ N, 10°2′ E, I = VII; 01.27, 43°32′ N, 10°15′ E.

Livorno was subjected to intensive seismic influence. The state of the sea was continuously changing: it sometimes rose and then immediately became lower, then turned stormy, then quietened down. Fishermen, who were at sea between the banks Meloria and Gorgona on January 19, saw the sea run very high within a small area and white foam rise to a great height with a terrible roar. The 'high–running' part of the sea rushed toward the old fortress (Livorno), so that for several moments the fortress could not even be seen.

The captain of one of the ships said he was surprised to see, several miles from Cape Corso, a great number of currents running rapidly with a frightening force in different directions (i.e. he observed maelstrom).

After the January 27 powerful underground shock. ... the weather was quiet and fine, but at the time of the shock the sea was very stormy... and after the earthquake 'most high' waves rose. According to other data, a hurricane south-western wind accompanied by rain started to blow just before the earthquake.

On January 27 the sea boiled with such fury and rage, that its noise was like the roar of an enormous cannon.

11. *1766, May 22, 5 h 30 min. Zone of the Sea of Marmara, Strait of Bosphorus.* $40°8'$ N, $29°0'$ E, I = IX.

An underground boom passed through Istanbul from south to north, after which strong underground shocks immediately followed in the same direction; they lasted for 2 min without interruption. The town was destroyed. The sea was unusually turbulent.

12. *1783, February 5, 12 h ±30 min. Calabrian Arc.* $38°25'$ N, $15°50'$ E.

The catastrophic Calabrian earthquake, which started the long period of seismic activity in the south-west of Italy that lasted for several years.

Unusual phenomena in the sea have been described, which could be considered short-term precursors of the earthquake. First, the deep-water small fish cicirella, which usually does not leave the sea bottom and buries itsels in the silt, began to appear in abundance near the sea surface not far from Messina and in other places in the first days of February. Second, on February 5 the sea was calm in the immediate vicinity of the coast in the region of Monteleone di Calabria, but it was stormy and it 'boiled up' in windless weather far off the coast, according to fishermen's accounts.

According to words of the officer, who was commander of the Messina Citadel and stayed inside it during the events, the sea 'swelled' in an absolutely unusual manner, within a quarter of a mile from the fortress, during the earthquake and three subsequent days, and it boiled with an incredibly gloomy and terrible roar, inspiring indescribable fear, while at the same time the sea on the other side of the light-house was calm.

13. *1846, August. Ligurian Sea, Italy, France.* $43°30'$ N, $10°30'$ E.

A hollow underground rumble started to be heard from the sea near a light-house in Livorno since July 25, and it grew louder on August 12 and 13. The rumble was accompanied by an unusual choppiness of the sea water, which could not be explained by the character and force of the winds, that blew at the time. Thus, on August 11, at 12 h, in the vicinity of the Fancale di Livorno the sea level significantly and quite instantaneously rose, and, where the bottom was silty and covered with seaweed, the water become turbid. ... The underground rumble and change in sea level increased by 10 h 30 min on August 14, and the main earthquake broke out at 12 h 52 min.

A strong choppiness occurred throughout lake Massaciuccoli, where the water was covered with ripples and became turbid.

14. *1846, September–November. Ligurian Sea, Italy, Tuscany district, Livorno.*

September 12. A rumble and movements of the sea water resumed under a clear sky.

7.1 Manifestations of Seaquakes

September 19. An underground rumble was heard, the sea became stormy owing to unusual heaves and lowerings of the water and to eddy flows.

October 4, 22 h 15 min. A strong underground rumble was heard, movements of the sea water continued. Similar events took place till the end of November.

15. *1863, March 22, 22 h 15 min. Aegean Sea, Island of Rhodes.* 36°30' N, 28° E.

A destructive earthquake occurred. The Greek ship 'Panagia' experienced such a strong seaquake near the Island of Kasos, that two masts were broken off.

The earthquake was followed by many aftershocks, noticed on the island of Rodos, in Izmir and other places. It coincided with a storm that raged on March 22 in the East Mediterranean, therefore, reports on the earthquake were mixed up with information on the storm and calamities at sea, the information being of not quite clear origin. Thus, the French newspaper Le Moniteur wrote, that the earthquake gave rise to a terrible storm at sea, which resulted in many accidents.

16. 1866, January 19 or 22, 12 h 30 min. Aegean Sea, Island of Chios. 38°15' N, 26°15' E.

A strong earthquake. Accompanied by numerous shocks. On the same day, intensive boiling of the sea water was noticed approximately in the middle of the strait separating the Island of Chios and Asia Minor. A cloud vapour seemed to rise from the waves.

17. *1867, March 7, 8. Aegean Sea, Island of Lesbos.* 39°2' N, 26°4' E, $M = 7.0$.

A destructive earthquake occurred on the Island of Lesbos and in its vicinity at 18 h 15 min–18 h 30 min. The earthquake lasted for about 40 s in Mitilini, the main city of the island. A dense cloud of dust rose, owing to the shaking.

An eyewitness wrote that the sea in the harbour of Mitilini rose and was covered with foam, like in the case of an underground explosion; according to another eyewitness: 'the sea swelled, as if its bottom rose, and started to boil'.

18. *1878, April 19 (15). Zone of the Sea of Marmara.* 40°8' N, 29°0' E, $M = 6,7$.

An earthquake occurred in Bursa and Izmit and spread up to the European shore of the Bosphorus. The Sea of Marmara was in a state of unusual agitation, and the water seemed to boil.

19. *1887, February (March) 23, 6 h 20 min. Ligurian Sea, Italy, France.* 43°42' N, 08°03' E.

A strong earthquake covered an area of 570,000 km² Deep-water fish, and fish rarely encountered in the winter season, were found thrown out onto the beaches of Nice, San Remo, Savona.

20. *1888–1892. Tyrrhenian Sea, Lipari Islands.*

The volcano Vulcano erupted on the island with the same name. The underwater cable laid approximately 3 miles to the east of the island at a depth of 700–1,000 m and connecting the Lipari Islands with Milazzo on Sicily, became unserviceable five times owing to severances (November 19 and 22, 1888; March 30 and September 11, 1889; December 14, 1892). As a rule, the water at the severance points 'boiled' strongly; silt was present at the water surface, ash appeared at the sea bottom (the points, where the cable was severed, were considered to indicate the sites of local underwater volcanic eruptions).

21. *1894, November 16, 18 h 52 min. Calabrian Arc, Calabria, the shore of the Tyrrhenian Sea. A destructive earthquake.* 38°15′ N, 15°52′ E.

The sea was absolutely calm; then the water started to bubble violently, and all of a sudden it became calm again. At Pizzo, near Villa S.Giovanni, the sea was calm during the earthquake; a boat seemed to sink, but then it came up again. The water in the sea was bubbling, its movement was like the movement of water caused by the paddle-wheels of a steamboat. Fishermen from Cannitello also affirmed that the sea was unusually teeming with fish after the earthquake and for some of the following days. An eyewitness from Villa S. Giovanni said that the sea in the Strait of Messina boiled up very strongly along the tidal line several hours before the earthquake, and the movement of the waves was irregular and unusual.

22. *1908, December 28, 5 h 20 min. Calabrian Arc.*

Catastrophic Messina earthquake and tsunami. The source was under the bottom of the Strait of Messina; its magnitude was 7.

The tsunami stirred up bottom sediments; bubbles of gas rose up to the surface of the strait from the bottom; sea animals and fish, unknown to fishermen, including inhabitants of great depths—down to 1,600 m, were thrown out onto the beach

A fisherman from the village of Contemplazione near Messina told that he was fishing in the sea, when the earthquake occurred: 'Suddenly, the sea began to boil and to rise with sharp waves.'

23. 1953, August 9–12. Ionian Sea, Ionian Islands.

Three destructive earthquakes occurred with increasing strength on the Islands of Kefallonia, Ithaca, Zakynthos. During the third earthquake, people on the deck of the transport 'Alfeios' in the Bay of Argostoli were thrown up into the air.

French tourists on a small beach of the south-western coast of Kefallonia Island observed a rough 'boiling' sea with sharp furious wind waves.

24. *1961, May 23. Aegean Sea, Asia Minor, Izmir.* 36.7° N, 28.5° E.

The colour of the water in the Gulf of Izmir changed after the earthquake, and it was filled with algae.

25. *1989, October 29. Sea of Alboran, Algeria. An earthquake occurred with its epicentre 80 km to the West of the city of Algiers, $M = 6.1$.*

The earthquake occurred in Central Algeria. From interviews with inhabitants of coastal villages it became clear, that some people, who had been in the port during the earthquake, noticed unusual vibrations of their vessels. Fishermen felt a strong seaquake on their vessels, which were shifted by anomalous sea waves.

The Pacific Ocean

JAPANESE ISLAND ARC

1. *1854, December 23, around 9 h.* 34.1° N, 137.8° E.

An extremely strong earthquake in the Tokai region. Area of most significant destructions: 260 × 120 km.

'I was in my cabin at about 9 o'clock, when I felt a slight shudder. A quarter of an hour after this earthquake the water in the vicinity of the city seemed to start to boil—the flow of the river, that suddenly started to strengthen, produced breakers and splashes on sand-bars. At the same time the water from the sea started to rise strongly and, having assumed a dirty aspect, to gurgle' (from the report by Putyatin, the captain of the frigate 'Diana').

7.1 Manifestations of Seaquakes

In the vicinity of the town of Koga, an eyewitness, walking along a path in the mountains, saw that at a distance of 2.2–2.3 km from the shore the water became turbid, while farther away the sea was clean and blue.

2. *1854, December 24, around 17 h. 33.2° N, 135.6° E.*

The catastrophic Nankaido earthquake, accompanied by a destructive tsunami. This was one of the strongest earthquakes at the south of Japan.

Prefecture Wakayama, Tanabe Bay. On December 24 the weather was clear. At about 17 h a strong earthquake took place. From the sea a boom, similar to cannon shots, was heard, and then a high wave appeared, that flooded the town and its surroundings. Everything seemed to be obscured by fog.

3. *1929, May 22, 01 h 35 min. 31.8° N, 131.8° E.*

At the south of Miyazaki prefecture a shock caused noticeable destruction in Miyazaki and its surroundings. At the same time as the shock a strong boom was heard, and, according to questionable data, a typhoon passed.

4. *1939, May 1, 14 h 58 min. 39.95° N, 139.8° E.*

A very strong earthquake on the Oga peninsula (Akita prefecture). Houses were destroyed over most of the peninsula. A tsunami was observed.

According to a fisherman, the fish catch was good before the earthquake in Yusiri, but afterwards it became very bad.

People, collecting seaweed, said that on Cape Oppa (close to Wakimoto) some sounds were heard one day before the earthquake at about 15 h, and by the evening the waves had thrown out onto the beach shell-fish, octupuses, trepangs, crabs, seaweed and sea cabbage.

In Wakimoto, the evening before the earthquake, an octupus, trepangs and sea cabbage were thrown out onto the shore.

5. *1940, August 2, 00 h 08 min. 44.1° N, 139.5° E, $M = 7.0$.*

Quite a strong earthquake was felt throughout the entire western part of Hokkaido Island and at a number of sites of the Soviet seaside. A tsunami was observed, which caused significant damage.

The tsunami arrived at Haboro. The sea water was very turbid and 300 m from the shore it acquired a dirty colour.

NAMP'O, MARIANA, CAROLINE, MARSHALL ISLANDS

6. *1850 (without month and date). 20°36′ N, 134°45′ E.*

The American military ship 'Mary', en route from the Hawaiian islands to Hongkong and being at the point with the above coordinates, apparently happened to witness an underwater eruption. A moderate eastern wind was blowing and the sea was calm. All of a sudden, the wind was reduced, and agitation of the sea started: the temperature of the air increased, and someone felt the smell of sulphur. Several gusts of wind arrived from different directions, but as soon as the sails were raised, the wind subsided.

THE PHILIPPINE ARCHIPELAGO

7. *1653, between June 26 and July 23.*

According to one source, not confirmed by others, the vessel 'San-Francisco-Xavier' en route to Manila, having passed through the San-Bernardino Strait and cast anchor at Mindoro island, experienced an earthquake that caused storm waves at sea.

8. *1824, October 26.*

Earthquake in Manila and surrounding provinces. Shoals of dead fish floated down the fish. The earthquake was accompanied by a hurricane.

9. *1852, September 16, 18 h 30 min.*

A destructive earthquake in the central part of Luzon Island.

The earthquake lasted about 3 min in Manila. The sea rose significantly, and the weak wind varied direction, the water was phosphorescent.

10. *1863, June 3, 19 h 20 min.*

Catastrophic earthquake in Manila and neighbouring provinces.

At 19 h 30 min a wave from the south-east approached the vessels in the Manila Bay. It hit the vessels and rolled over the decks, completely flooding them. The frigates shook and trembled as if they were being battered against the sea bottom. The water in the vicinity bubbled and was covered with foam. It was also said that the steamboat 'Esperanza' vanished together with the entire crew.

Sources note that after the earthquake the coast was continuously visited by storms, which nearly paralyzed the navigation.

11. *1880, July 18, 12 h 40 min.*

A destructive earthquake with its source near the Pacific coast of the central part of Luzon Island.

The water in rivers became dirty and assumed a state of continuous, agitation, that lasted, at least, for 24 h.

12. *1901, September (or October) 10, 8 h 30 min.*

A strong earthquake in the eastern part of Tayabas province, accompanied by an underground boom.

Yawning fissures formed along the coast, and the water became very turbid; dead fish was also found. The water becoming turbid could be due to a tsunami; no straightforward data exist concerning the generation of waves.

13. *1924, April 15, 00 h 20 min. 6.5° N, 126.5° E.*

A destructive earthquake with its source south-east of Mindanao Island.

A strong seaquake was experienced on board a steamboat anchored in Karaga Bay. All of a sudden the sea became very rough.

THE INDONESIAN ARCHIPELAGO

14. 416.

The Javanese 'Book of kings' ('Rustaka Rajah') representing a chronicle of the island, contains the following description of an eruption of the mountain Kapi. In the year 416, from the depths of Pulosari mountain there arrived a thunderous boom, which was answered by an identical boom coming from the inside of Kapi mountain. An enormous blinding fire, reaching the skies, shot up from Kapi mountain. The entire Universe shuddered. Peals of thunder was heard, a storm broke out, it started to rain.

15. *1722, October, no date, 8 h.*

Jakarta (Batavia), a strong earthquake. Water in the roadstead was being thrown up, 'like in red-hot salt-works'.

16. *1757, August 24, around 2 h.*

A strong wave-like shaking lasting 5 min occurred in Jakarta (Batavia). At 2 h 05 min, during the strongest shock, a wind from the North-East started blowing.

7.1 Manifestations of Seaquakes

17. *1860, August, no date.*

Light earthquakes and plenty of rain, accompanied by strong winds and floods, took place on Minahasa Peninsula.

18. *1897, March 15, 6 h 30 min.*

A strong earthquake on Kayuadi Island, accompanied by a thunder-like roar coming from the water.

An eyewitness of the events said that half an hour after the shaking began someone shouted that the water was rising. ... The third wave rolled up the shore... The water was of an unusual green colour.

19. *1904, September 7.*

Shaking of the land was felt at Cilacap. At noon, an eyewitness on the southern coast of Java noticed, that the ocean water turned white (became similar to milk). By 23 h the phenomenon stopped, but resumed in 2 h.

20. *1913, March 14, 16 h 45 min.* 4.2° N, 126.5° E.

An extremely strong earthquake with its source in the vicinity of Sangihe Island. After the earthquake pouring rains took place.

A local captain from the Talaud islands claimed that at the same time as he heard a boom he saw how in the sea, several kilometers west of the island, a large elevation formed.

Sailors in boats near Surigao felt the earthquake; waves rose on the sea surface.

21. *1922, February 22.*

An earthquake of strength 4, preceded by a boom, occurred in Amahae at 19 h 45 min. According to reliable fishermen's accounts, by midnight two more underground shocks took place, while the sea was very agitated all this time. No instruments registered the earthquake.

ISLANDS NEW GUINEA, NEW BRITAIN, NEW IRELAND

22. *1857 (1856), April 17, immediately after sunset.*

A powerful earthquake occurred on Umboy Island (Ruk), located in the strait separating New Guinea and New Britain. Shocks were repeated.

To the west of island Tolokiva (Lotten), a volcano seemed to rise from the water and to smoke for a long time. After the shaking 'the sea started to move'.

23. *1937, May 28/29.*

The volcanoes Raluan and Tavurvur in the Blanche-Bay erupted. The first strong earthquake broke out on May 28 at about 13 h 30 min. A new shock was felt in Rabaul on May 29 at about 5 h 30 min, and it was followed by a series of shocks of varying intensities, that lasted right up to the volcano's eruption. At 13 h 10 min, the water at the western coast of Raluan island became turbid and started to move.

SOLOMON ISLANDS, SANTA CRUZ, NEW HEBRIDES

24. *1863, August 17.*

A weak underground shock occurred on August 15 at 19 h 30 min in Noumea (Port-de-France) on New Caledonia. A more noticeable shock took place on August 17 at 20 h. It caused significant derangement of the instruments of the astronomical observatory. Subsequently, news was received that the shocks could be related to eruption of the volcano on the Tanna island.

The schooner 'Ariel', anchored at the nearby Eromanga Island, brought the news, that the shock of August 17 was felt on this island with a great strength and was accompanied by unusual movement of the sea. A large wave made the schooner 'dance' on the anchors, subjecting it to extreme danger.

25. *1892, August, first week (no date). Around 00 h 20 min.*

A very strong shock occurred in Liuganville (Island Espiritu-Santo). Approximately at the same time an eyewitness, who was sleeping on the upper deck of a small schooner on the Rekena traverse, all of a sudden woke up and saw an enormous foamy billow, that had appeared on the hitherto calm ocean surface and was approaching the schooner. The weather was quiet, with a gentle south-east breeze.

26. *1893, July 31, 11 h 30 min.*

A strong very rapid shock occurred at Port-Vila (Efate island). The weather was bad, with rain and strong gusts of wind. There was a heavy swell at sea.

27. *1926, September 17, around 4 h 20 min.*

At Ovi (Guadalcanal island) a boom similar to the peals of distant thunder, that arrived from the west and strengthened while approaching, was followed by oscillations and rotations of the Earth's surface. Then, they became weaker, but did not stop, upon which the main shaking started, it proceeded mainly up and down and lasted a minute. Shakings continued all day and, then, for a whole week, but with lesser intensity.

The sea water was ink-black, and all the sea inhabitants floated bellies-upward under different degrees of shock, and some even dead, so the stink was horrible during ebb-tides.

28. *1958, October 7, 21 h.*

The inhabitants of Epi island noticed pronounced choppiness south-east of the island, like in the case of the eruption of an underground volcano. From an airplane flying over this region of the sea spots and streams were revealed of unusual colours (oily-yellow and pale green, 500, 2,000 m in diameter).

FIJI, SAMOA, TONGA, KERMADEC ISLANDS

29. *1865, November 18, 5 h 40 min.*

On November 18, at 4 h 20 min the English vessel 'John Wesley' happened to be stranded on the soil of the small islet (of a coral reef) Tau, but after several underground shocks it was completely free in the water. The ocean raged with such a strength, that the waves rolled over the deck.

A strong earthquake occurred at 5 h 40 min, and the waves became even more dangerous.

On the Tonga islands, the earthquake was accompanied by 'sudden and terrible movement of the sea', which flooded the land and washed away everything in its way.

30. *1866, September 12.*

The eruption of an underwater volcano, accompanied by numerous earthquakes, took place in the vicinity of the island of Samoa.

On September 12, after lunch, agitation of the sea started at a distance of about 3 km from Oloseg island and 8 km from Tau island, at the point with coordinates 14°13′ S, 169°34′ W, and it continued all that day and part of the next day. Then, an underwater eruption followed. The sea was in a state of strong agitation and for 20 km it shone with phosphorescent light. Much dead fish floated on the ocean surface was thrown out onto the beach, including monsters, never seen before, from 2 to 3.5 m long.

31. *1953, September 14, 12 h 17 min.*

An earthquake and tsunami took place on the Fiji islands. The earthquake was accompanied by a large number of weak aftershocks.

Eyewitnesses observed that at the entrance to the harbour of Suva the sea level started to drop immediately after the seismic shocks. In about 10 s a large brown 'bubble' appeared in between the buoys, indicating the fairway, and the ring-like wave, caused by this perturbation, rolled onto the reefs. The height of the wave amounted to 2 m.

NEW ZEALAND, AUSTRALIA, SOUTHERN PART OF THE PACIFIC OCEAN

32. *1855, February 14.*

Two strong earthquakes occurred in Wellington, the first lasted more than a minute. A 'terrific' tide gushed onto the shore. Numerous small bubbles of sulphurous gas came up from the bottom; there was plenty of dead fish floating.

33. *1866, August 15–21.*

A 'striking' movement of water was observed in the harbour of Sidney.

CHILE, PERU, SOUTH ECQUADOR

34. *1604, November 23 or 24, 13 h 30 min.*

Peru and the northern part of Chile, an earthquake and tsunami. Shaking embraced the coast along 1,650 km.

The ocean was so agitated, that evaporation rose and covered the entire coast.

35. *1633, May 14, at sunset.*

The inhabitants of fortress Carelmapu on Chiloe island were woken up by a thunderous noise. At the same time a strong earthquake destroyed the port.

The sky became covered by dense clouds; it was dark. For many hours there was a hailstorm, thunder rumbled, and lightning flashed.

36. *1647, May 7 (date unreliable).*

Catastrophic earthquake in Santiago and adjacent regions of Chile.

In all ports of this coast fishermen observed strange and unusual movements of the ocean.

37. *1687, October 20.*

Very strong earthquakes in Lima.

A vessel, that was 600 km (!) from the shore at the latitude 12°30′ S,[1] experienced such a terrible seaquake, that it nearly perished; several sailors were thrown out of their hammocks. The usually green water seemed to have whitened. When it was scooped up, it was seen to be mixed with sand.

According to the testimony of Forster, the captain of the vessel 'Davis', the seaquake was felt even at 1,800 km from the shore.

38. *1828, March 20 or 30, 7 h 30 min.*

The strongest earthquake in Lima since 1746.

The vessels anchored in Callao felt a strong earthquake. A boom-like distant thunder was heard. The surrounding sea suddenly started to boil up, bubbles of hydrogen sulphide rose from the ocean bottom, much dead fish rose to the surface.

---

[1] The depth of the ocean at this point exceeds 4 km (authors' comment).

The ocean surface that was absolutely mirror-like before the earthquake became agitated, and the water became dirty.

At Trujillo the earthquake was followed by pouring rain that damaged building structures. The rain in Lambayeque and Chiclayo was even stronger, and continued without interruption for 4 days. In the Sechura desert, where no single drop of water ever falls, a river appeared.

39. *1835, February 20, approximately at 11 h 30 min.*

A destructive earthquake and tsunami took place in the central part of Chile with its source in the vicinity of Concepcion. Maybe it was preceded by foreshocks.

From the very beginning of the earthquake the water over the entire surface of the Bay of Concepcion (Talcahuano) was boiling up. Bubbles of air or gas were rising rapidly. The water became dark and smelled of the very unpleasant smell of hydrogen sulphide.

After the earthquake and tsunami at the coasts of the central part of Chile, it is possible that the catch of fish and of whales decreased.

40. *1840, January 28, 3 h.*

In Lima a strong earthquake occurred, that was also felt in Chorrillos. It was accompanied by a hurricane wind blowing from the south-west, and by a heavy shower. In Chorrillos the ocean was so agitated, and the temperature of the air fell so low, that on the shore bathing stopped and people looked for a shelter.

41. *1868, August 13, 16 h 45 min.*

A destructive earthquake and catastrophic tsunami occurred with the source near the coastal cities of south Peru. In literature, the tsunami is known as the Arica tsunami.

During the earthquake the feeling was such as if the vessel were shaken by some giant. Above the city there rose a dense cloud of dust, which soon spread out to the ships (from the memoirs of Billings, who observed the events from the deck of the American ship 'Watery').

When it became dark, at about 20 h 30 min, an enormous 'wall' of phosphorescenting and foaming water, mixed with sand, approached from the ocean with a thunderous sound.

On the steamboat 'Taranaki', that approached Littleton on the 15th, it was noticed at 55 km from the shore that the water became turbid; logs and fragments of constructions floated.

Lake Titicaca was agitated as never before.

42. *1869, August 24, about 13 h 15 min.*

The area, situated to the north of Itica along about 500 km, experienced a strong earthquake that lasted a minute.

At the point $19°17'$ S, $70°21'$ W 5.5 km from the coast and 90 km from Arica at a depth of 135 m the steamboat 'Le-Paita' experienced a strong seaquake that lasted about 50 s. During the seaquake it was difficult to stay on one's legs; someone fell off board; heavy objects fixed to the deck were jumping up and down a decimeter. The ocean around the vessel, as far as one could see, seemed to boil; jets 40–60 cm high were thrown up with a noise similar to the noise made by a strong rain at sea. At the same time a low underground grumble was heard.

43. *1871, August 21, 20 h 32 min.*

A strong earthquake lasting 15 s. occurred at Callao. Approximately at the same time the vessel 'Colon' felt a strong seaquake to the west of Cape Chala, and waves immediately appeared on the surface.

At Callao, the ocean, that was unusually calm till then, suddenly became extremely agitated; a strong southern wind started to blow. At Serro-Asul the ocean remained very agitated for 2 days.

44. *1871, December 28, immediately after midnight.*

An earthquake at Puerto-Monte. The ocean was very agitated.

45. *1877, June 15(?).*

At Pisagua, not far from the shore, an enormous water column rose in the ocean, and strong vortexes formed. It was announced that at the same time in Caleta Pabellon de Pica and Chanaral long shakings were felt, accompanied by a boom.

46. *1878, February 14.*

At 4 h the steamboat 'Chile' experienced in the Bay of Concepcion (Talcahuano) a 'terrific' shock. The vessel happened to be in between three such large waves, accompanying this shock that it nearly perished.

47. *1880, August 15, 8 h 48 min.*

A strong earthquake occurred with its source to the north of Santiago.

There were strong shakings in Cocimbo. Large columns of water rose in the ocean, owing to which the anchor chain of one of the vessels was torn away.

48. *1906, August 16, 20 h 40 min.*

A most strong earthquake took place in the central part of Chile with its source near Valparaiso.

From small coastal villages near Constitucion, such as Buchupureo, Putu, Lico and others, the news arrived that during the earthquake the ocean assumed a state of 'boiling' or 'gurgling', so the process of regular wave formation was violated.

49. *1930, December 29, 3 h 26 min. 24 s.*

In Copiano strong vibrations were felt. The earthquake was felt in Cocimbo and La Serena. It was preceded by a shock at 23 h 55 min on the 28th and was accompanied by subsequent shocks noted up to the 30th.

In the region of Cocimbo and up to 550 km to the north during 24 h after the first earthquake unusually large tides were observed together with strong agitation of the ocean. Giant waves were coloured deep green, the water had an unpleasant smell and was saturated with plankton.

NORTH ECUADOR, COLOMBIA, PANAMA

50. *1877, October 11, 9 h.*

The steamboat 'Paita' was torn away from its anchors in the port of Esmeraldas by a sudden storm and had a narrow escape. At the same time a similar phenomenon was observed in the port of Buenaventura, where it was accompanied by an earthquake.

51. *1906 January, 31, 10 h 30 min.*

A catastrophic earthquake and strong tsunami occurred with its source at the coast of Ecuador and Colombia.

The soil of Cape Manglares cracked all over; water fountained strongly.

The tsunami was recorded well by the mareograph on Naos island, situated 5 km to the south of Panama. With the arrival of the tsunami the water surrounding the island became turbid.

CENTRAL AMERICA (FROM COSTA RICA TO MEXICO)

52. *1787, April 3, 9,5–10 h.*

Apparently, a most strong aftershock of the 28 April 1787 earthquake with the source near San Marcos

In some coastal points, situated 18 km from Tehuantepec, at the same time as the earthquake, an unusual movement of the water was observed, which was accompanied by a menacing roar. Fish and shells of enormous sizes, not seen before, were thrown onto the shore. The same phenomenon took place on the coast at Pochutla and Jucila.

53. *1883, March 12.*

According to news from Cuautla (Mexico, state of Jalisco), the ocean left its usual bed along a significant part of the coast and receded from the shore by a significant distance, so that various elevations and depressions of the ocean bed were uncovered and dried. How long the ebb tide continued is not known, but after some time the ocean returned with quite a loud noise and quite rapidly to its previous state.

No information on any earthquake was received. The next day a hurricane with pouring rain broke out in this locality.

USA, CANADA, SOUTH-EAST OF ALASKA

54. *1812, December 21, 10 h 30 min.*

A strong earthquake occurred at the south of California, accompanied by an 'enormous' tsunami in the strait at Santa Barbara.

The sea at Santa Barbara was calm.

55. *1851, November 13, 19 h.*

A shock was felt in San Francisco; people on board a vessel experienced 'unusual movement of the water'.

56. *1899, September 10, 12 h 15 min.*

Catastrophic earthquake with source at the top of Yakutat Bay. Tectonic movements caused numerous falls and slides, tsunamis.

The waves left plenty of dead fish on the shore.

Vortexes formed in the Bay, in which trees, flotsam and all kind of debris spinned around so fast, that it was difficult to keep an eye on individual objects. The water foamed and was covered all over with white caps.

On September 12, one of the inhabitants going by steamboat from Yakutat to Juneau noticed a belt of turbid water and a mass of floating trees in the ocean between Yakutat and the Fairweather mountain range.

57. *1946, June 23, 10 h 13 min.*

A most strong earthquake occurred in the west of Canada and the north-west of the USA with its source in the central part of the eastern coast of Vancouver Island.

On Quadra Island, the forest-guard noticed that the water, which before was transparent, had become turbid; fish bit badly for 2 weeks.

HAWAIIAN ISLANDS

58. *1868, April 2, 15 h 40 min.*

A destructive earthquake and tsunami occurred with source to the south or southeast of island Hawaii, which was manifested as lasting (2–3 min) long-period oscillations.

A vessel, that happened to be in the source area had its jib-boom broken by a blow from the water.

In Punalu'u immediately after the earthquake, or even at the same time as the earthquake, the ocean became very agitated, 'as if an enormous amount of red-hot lava had been poured into it at a certain distance from the shore'. The water boiled and frantically tossed about in all directions.

59. *1903, October, 5, noon.*

On the English vessel 'Ormsry', approaching the western coast of Hawaii Island, it was noticed, that the ocean 'started to boil', as if powerful sources were in action under the water surface. The temperature increased noticeably. The vessel received a blow, as if from the tidal wave travelling from the coast, and it was turned stern first. On October 6 the Mauna Loa volcano threw up a column of smoke; maybe lava also poured out.

As it was correctly noted in reports compiled in Honolulu, here, most likely, it was not a tsunami that took place, but a seaquake or convection flows.

60. *1903, November 24.*

In Punalu'u the ocean started moving. On the water surface, that had been calm, there suddenly appeared waves lasting 10 min; their origin was inexplicable. At the same moment a black column of unusual dimensions rose above the Mauna Loa Volcano.

Several not very striking, but typical, phenomena are described in the Pacific Tsunami Catalogue [Soloviev et al. (1986)], which embraces the period from 1969 up to 1982 and contains information on 85 events. In Appendix 3 to this catalogue there are Addenda to the 'Catalogue of tsunamis on the West coast of the Pacific Ocean' [Soloviev, Go (1974)].

61. *1977, August 19.*

A catastrophic earthquake on the south of Sumbawa Island embraced a wide region—from the southern coast of Indonesia to the north-western coast of Australia.

It was unusual that during the period between the origination of the earthquake and the tsunami arrival the inhabitants of the villages on Sumbawa and Lombok Islands heard three sounds similar to explosions, with intervals between them from several seconds up to 1 min and more. Nearly in each village it was said that the water became black, some people mentioned an increase of temperature and an unpleasant smell.

62. *1920, February 2, 21 h 12 min.*

A very strong earthquake took place on Gasmata Island (the south of New Britain). Its strength amounted to about 8–9, and it lasted for about 1 min 10 s.

In Gasmata, after the main shock the water left the harbour in a rapid stream through the main entrance toward the sea, but did not return, as a result of which the tides were 50–60 cm lower than the previously. Much dead fish was left on the reefs, and its smell was brought by the wind from the sea. Although the weather was good, there was a heavy swell on the sea all day.

63. *1923, November 4, 10 h 20 min.*
Several strong earthquakes occurred on the northern coast of New Britain.
... The water surface was in a very agitated state.

64. *1930, December 24, 7 h 30 min.*
A strong earthquake shook Mal Island and the surrounding islands. Immediately after the earthquake there were two tidal waves.

From the side of the south-eastern end of the island a loud sputtering was heard, that was similar to the noise of escaping vapour; two large dense clouds of smoke or vapour hung over the sea.

In the middle part of Lau Island all foliage was destroyed, as if the trees had experienced a strong drought, and they became covered with a thin coating of yellow-white sandy loam.

A similar picture, but even to a larger extent, was observed on Pihun island, the inhabitants of which left the island and went to Amot Island. People said that they could not return to their island owing to the terrible stink of dead fish.

65. *1959, August 18, 08 h 05 min.*
A seismic sea wave was excited, most likely, somewhere in the region of the west coast of Ranongga Island.

An eyewitness said that in the morning on the east coast of Velia Island there had been a 'tidal whirl'. From Simbo Island it was also said that in the morning there was a strong and chaotic agitation on the sea.

7.1.2 Analysis of Historical Testimonies and the Physical Mechanisms of Vertical Exchange

The phenomena, described in Sect. 7.1.1, testify in favour of the possibility of significant enhancement of the vertical exchange in the ocean in the case of an underwater earthquake. We shall present possible mechanisms (scenarios) leading to enhancement of the vertical exchange in the water column in the case of seismic movements of the bottom. In reality, the mechanisms may combine arbitrarily.

1st scenario. Direct generation of turbulence by a tsunami wave (when it passes over shallow water, in case of wave breaking and run up on the shore). The mechanism is efficient only in the case of shallow water, since at large depths the velocity of motion of water particles in the wave is small. It can be assumed that, with the exception of cases of smooth inundation of the shore or of waves of small amplitude, the mechanism considered is always manifested. The existence of such a mechanism gives rise to no doubt that sea sand has been repeatedly seen to be carried up onto the shore. Modern studies of paleotsunamis are based on the analysis of intact ('conserved' among layers of soil, peat and volcanic ash) sediments of sea sand brought up the coast by waves [Pinegina et al.(1997); Pinegina, Bourgeois (2001)].

7.1 Manifestations of Seaquakes

2nd scenario. When a fault outcrops on the surface (of the bottom), or in the case of horizontal movements of the bottom, turbulence generation is possible in the near-bottom region owing to the shear instability. It may be assumed that intense turbulent mixing will involve the near-bottom region of water of insignificant thickness of the order of the height of the bottom inhomogeneity or of the vertical displacement within the fault.

3rd scenario. Seismic movements lead to formation on the water surface of standing wave structures (parametric resonance). In a number of cases, such waves are characterized as extremely violent storm waves. The excitation of turbulence occurs when the waves collapse. Turbulent mixing only embraces the upper layer (several tens of meters). In shallow water the entire water column will evidently be involved in intense mixing. For standing waves to arise during an earthquake it is necessary for the bottom to oscillate with infrasonic frequencies ($\sim 0.1-1$ Hz). An alternative, here, can be represented by elastic oscillations of a water column several kilometers thick, which arise in the case of any (not necessarily oscillatory) movements of the bottom and proceed with the same infrasonic frequencies.

4th scenario. The development of turbulence in the case of cavitation effects.

5th scenario. Intensive movements of the bottom form non-linear currents (including acoustic wind [Ostrovsky, Papilova (1974)]). Vertical transfer is realized directly by these currents and by the turbulence resulting from the instability of these currents. This phenomenon takes place at any depths and may involve the entire thickness of the water column.

For *scenarios 1 and 2* the ocean depth is a critical parameter in the sense that surface effects of mixing will not be present where the depth is sufficiently great. Thus, for example, an SST anomaly will be related, here, to small depths along the coastline, and, consequently, the area of the anomaly will be small.

The *3rd scenario* is already capable of providing surface effects over significant areas, comparable to the area of the pleistoseist zone of the earthquake. But, owing to mixing only involving the near-surface layer, no significant change in the surface temperature is to be expected to accompany the total stratification destruction.

Cavitation effects in the case of underwater earthquakes are little studied till now. Therefore, we shall be prudent and only make the conclusion that the *4th scenario* may, probably, provide insignificant SST deviations over areas inferior to the area of the pleistoseist zone.

The most widespread SST anomalies with significant temperature deviation from the initial values can be provided for by the *5th scenario*. It is possible that the fifth scenario includes the case, when on board a vessel 600 km from the coast the sea water was found to be mixed with sand. Another striking event, which, most likely, demonstrates the transfer of cold depth waters up to the ocean surface, is to be considered the case, when 'temperature of the air fell so low, that on the shore bathing stopped and people looked for a shelter'.

Often, in the case of underwater earthquakes the water is seen to churn and foam, or 'to boil'. The origin of such a phenomenon is due to non-linear currents, cavitation effects or to convection flows (for example, in the case of underwater eruptions). Of course, in the latter case, an enhancement will be observed of the temperature of the water and the air.

Weather phenomena, accompanying earthquakes, merit separate consideration. In a number of cases one cannot exclude a simple coincidence of the earthquake and the observed weather phenomenon. But it would be quite difficult to place in this category the weather anomalies, described in the aforementioned case **PO-40** and in cases **PO-38** (it rained, 'In the Sechura desert, where no single drop of water ever falls, a river appeared'), **PO-35** ('For many hours there was a hailstorm, thunder rumbled, and lightning flashed'). In case **PO-10** news was received of 'storms, which nearly paralyzed the navigation' after the catastrophic earthquake in Manila. Less intense weather anomalies were observed in cases **PO-2, PO-34** (fog) **PO-6** (change of wind).

The anomalous weather phenomena, that accompanied the well-known Erzincan earthquake of December 26, 1939 of magnitude ~ 8.0, are described in [Ranguelov, Bearnaerts (1999)]. The earthquake caused a tsunami in the Black Sea. Numerous newspaper publications of the time communicated extremely low temperatures, strong snow-falls, ice-cold winds and great storms. In southern and south-eastern Turkey the earthquake was followed by pouring rain, that resulted in a large inundation. We note, that in the case considered anomalously low temperatures began to be noticed starting from December 9–10, 1939, but the peak of weather anomalies in the region was observed precisely after the earthquake. In the work considered the hypothesis is put forward of the possible influence of sea water mixing on weather conditions.

We shall further present an extensive quotation from Chap. XVI of the well-known book 'The Voyage of the Beagle', [Charles Darwin (1839)], in which the relationship between earthquakes and the weather is discussed.

'The connection between earthquakes and the weather has been often disputed: it appears to me to be a point of great interest, which is little understood. Humboldt has remarked in one part of the Personal Narrative, that it would be difficult for any person who had long resided in New Andalusia, or in Lower Peru, to deny that there exists some connection between these phenomena: in another part, however he seems to think the connection fanciful. At Guayaquil it is said that a heavy shower in the dry season is invariably followed by an earthquake. In Northern Chile, from the extreme infrequency of rain, or even of weather foreboding rain, the probability of accidental coincidences becomes very small; yet the inhabitants are here most firmly convinced of some connection between the state of the atmosphere and of the trembling of the ground: I was much struck by this when mentioning to some people at Copiapo that there had been a sharp shock at Coquimbo: they immediately cried out, 'How fortunate! there will be plenty of pasture there this year.' To their minds an earthquake foretold rain as surely as rain foretold abundant pasture. Certainly it did so happen that on the very day of the earthquake, that shower of rain fell, which I have described as in ten days' time producing a thin sprinkling of grass. At other times rain has followed earthquakes at a period of the year when it is a far greater prodigy than the earthquake itself: this happened after the shock of November, 1822, and again in 1829, at Valparaiso; also after that of September, 1833, at Tacna. A person must be somewhat habituated to the climate of these countries to perceive the extreme improbability of rain falling at such

7.1 Manifestations of Seaquakes

seasons, except as a consequence of some law quite unconnected with the ordinary course of the weather. In the cases of great volcanic eruptions, as that of Coseguina, where torrents of rain fell at a time of the year most unusual for it, and 'almost unprecedented in Central America,' it is not difficult to understand that the volumes of vapour and clouds of ashes might have disturbed the atmospheric equilibrium. Humboldt extends this view to the case of earthquakes unaccompanied by eruptions; but I can hardly conceive it possible, that the small quantity of aeriform fluids which then escape from the fissured ground, can produce such remarkable effects. There appears much probability in the view first proposed by Mr. P. Scrope, that when the barometer is low, and when rain might naturally be expected to fall, the diminished pressure of the atmosphere over a wide extent of country, might well determine the precise day on which the earth, already stretched to the utmost by the subterranean forces, should yield, crack, and consequently tremble. It is, however, doubtful how far this idea will explain the circumstances of torrents of rain falling in the dry season during several days, after an earthquake unaccompanied by an eruption; such cases seem to be speak some more intimate connection between the atmospheric and subterranean regions'.

Charles Darwin did not provide any reason for phenomena of this kind, in his time. From our point of view an earthquake influences the atmosphere indirectly—via the ocean. A cold SST anomaly causes cooling and thickening of the air, wind arises from the region of the anomaly (an analogue of a breeze). The cooling of damp air leads to the formation of fog or rain. Of course, it is possible that this mechanism for earthquakes to influence the weather is not the only one.

In most cases a strong underwater earthquake results in the mass death of fish or its possible migration. Here, the appearance at the coast or on the ocean surface of deep-water or unknown fish is often noted. Probably, the instinct of deep-water fish makes it rise to the surface or migrate to shallow-water areas, since it is precisely there, as it is shown in Sect. 3.1.5, that the risk of being subjected to the influence of intense acoustic waves, excited by bottom slides, is minimal. Note that the death of fish may be related not only to acoustic waves, but also to poisoning by hydrogen sulphide (in a number of cases the presence of hydrogen sulphide is noted directly **PO-32, PO-38, PO-39**).

From our point of view, timely purposeful migration of fish from quite a broad (hundreds of kilometers) pleistoseist zone of a catastrophic earthquake is quite improbable. The version of fish migration to the surface or to shallow-water areas seems more plausible. In this case, only a relatively small distance has to be covered after the first signs of an earthquake in preparation. Thus, the absence of a catch of fish after an earthquake may be explained either by its mass death, or by its migration from its traditional habitat. It is interesting that an opposite case is known **MS-21**, when 'after an earthquake and during subsequent days the sea was unusually abundant with fish'. It is possible, that this abundance was related to the explosive development of phytoplankton; one cannot exclude, also, the migration of fish from another region. The case **PO-49** must be mentioned, when eyewitnesses directly pointed to saturation of the sea water with plankton after an earthquake. We recall that we were interested in the behaviour of fish as a possible indicator of the content of phytoplankton, an enhancement of the concentration of which could be initiated by an earthquake owing to enhancement of the vertical exchange in the ocean.

All the identified events are related to known zones of seismic activity. In most cases eyewitnesses of the events noted the first group of events (unusual agitation, excited behaviour of the sea, 'boiling' water and so on). No peculiarities are observed in the geographic distribution of event groups (1–4).

At present, in connection with the development of satellite oceanography, a large number of 'in situ' measurements from vessels and buoy stations, and the accessibility of data via the Internet, a real possibility has arisen of searching for and identification of cases of transformation of the stratification structure of the ocean resulting from underwater earthquakes. The following three ways exist of searching for such cases, which do not require the introduction of special changes into the existing techniques:

1. Temperature of the ocean surface is quite reliably determined from satellites by analysing IR images within the range of the transparency window of the atmosphere.
2. Concentration of phytoplankton is also determined from satellites by the colour of the ocean, i.e. by the concentration of chlorophyll 'a' [Behrenfeld, Falkowski (1997)].
3. Data from buoy stations on the transformation of the vertical temperature profile together with meteodata would provide the most complete and conclusive information on an event, if such a station happened to be in the seaquake zone; moreover, such a station can register the passage of an internal wave caused by evolution of the area of disrupted stratification.

7.1.3 Instrumental Observations of Variations of the Ocean's Temperature Field After an Earthquake

The phenomena, taking place in the ocean above the pleistoseist zone of an underwater earthquake, remained little studied for a long time. Such a situation was due to the absence of any whatsoever instrumental observations. The scarce testimonies of seaquake witnesses could certainly not serve as a source of objective quantitative information. The difficulty in observing effects in the epicentral zone of an underwater earthquake is related to these effects, like the earthquake itself, being transient and practically unpredictable.

In the past decade, satellite oceanography has reached a new level, available databases of geophysical data have appeared together with effective means for analysing them. The new possibilities, that have opened up in this connection, have permitted to reveal a whole series of events, when underwater earthquakes happened to cause the transformation of hydrophysical fields in the ocean [Nosov (1996), (1998a, b); Filonov (1997); Levin et al. (1998), (2001), (2006); Luchin et al. (2000); Degterev (2001); Zaichenko et al. (2002); Ouzounov, Freund (2004); Yurur (2006); Agarwal et al. (2007)].

7.1 Manifestations of Seaquakes

In this section we shall not present the entire list of events, but dwell upon the first case, that was revealed already 10 years ago, and on the results of one of the most recent works in this direction.

The formation of cold SST anomalies, resulting from a series of strong underwater earthquakes, was first observed in 1996 near Bougainville Island (Solomon Islands). The phenomenon was revealed by a joint analysis of the maps of SST anomalies (FNMOC, United States Fleet Numerical Meteorology and Oceanography Centre, http://www.fnmoc.navy.mil/) and of the operative issue of the NEIC seismological bulletin (National Earthquake Information Centre, http://earthquake.usgs.gov/).

In the region of Bougainville (within a radius of 300 km from the epicentre of the earthquake of maximum magnitude 6.52S, 155.00E) during the period from 20.04.1996 to 31.05.1996 there were 145 earthquakes, that were mainly shallow. Preceded by several foreshocks, the main earthquake of magnitude 7.5 occurred on April 29 at 14 h 40 min Greenwich time. The location of the epicentres is shown in Fig. 7.1.

In Fig. 7.2, the blue line, marked by squares, presents the SST dependence on time at the point with coordinates 7S, 156E (the centre of temperature anomalies). The time dependence is reconstructed from maps of SST anomalies (FNMOC) at our disposal. Analysis of the global ocean surface temperature distribution is performed twice a day (at 00 and 12 GMT), but, regretfully, not all the data turned out to be available. The presence of a marker (square) signifies the existence of data. As an example, six fragments of anomaly maps are shown in the figure. The time and amplitude of seismic events are indicated by red triangles.

The dark-blue spot, the appearance of which in the vicinity of Bougainville Island is especially well seen on the third and fourth fragments (May 6 and 12), represents a cold SST anomaly. The maximum deviation of temperature amounts to $\sim 3°C$ in the first case and to $\sim 2°C$ in the second. The characteristic horizontal size of the anomaly is 300–500 km. Note, that manifestation of the anomaly after the main shock (second fragment) is just noticeable, and only after a whole series of seismic events, from April 25 to May 5, the anomaly becomes extremely pronounced.

Fig. 7.1 Earthquake epicentres in the vicinity of Bougainville Island for the period April 20 to May 31, 1996. The centre of the red circle of radius 300 km is at the point 6.528 S, 155.00 E (epicentre of earthquake of maximum amplitude). The black circles indicate the location of TAO Array stations (see also Plate 16 in the Colour Plate Section on page 324)

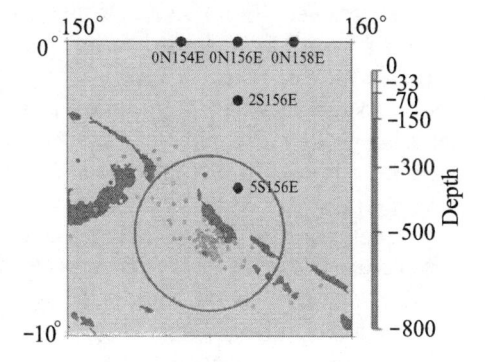

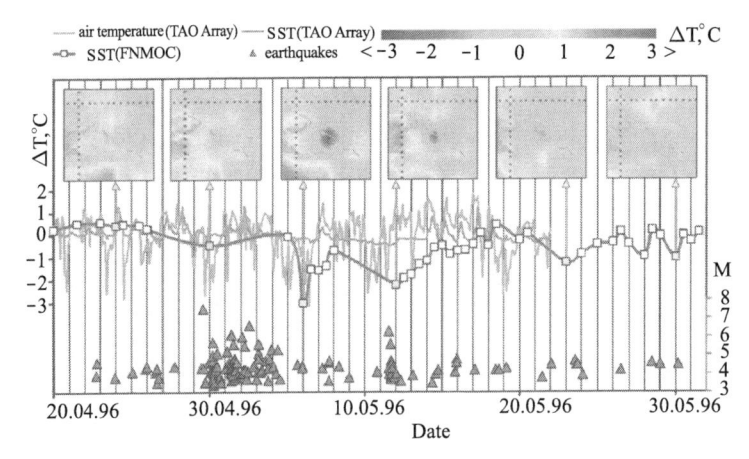

Fig. 7.2 Time dependence of sea surface temperature (SST) and of temperature of the near-the-water atmospheric layer; time and magnitude of seismic events (red triangles) that occurred within a radius of 300 km from the epicentre of earthquakes of maximum magnitude. The blue line corresponds to the SST at the point 7 S, 156 E, which was restored from maps of SST anomalies (FNMOC), the green and purple curves are data from the TAO Array buoy station with coordinates 5 S, 156 E. Fragments of maps of SST anomalies and the coloured temperature scale in degree Celsius. The dates indicated along the timescale correspond to 00 h 00 min Greenwich time (see also Plate 17 in the Colour Plate Section on page 324)

Recurrence of the anomaly on May 12 took place immediately after the earthquake of May 11. The anomalies persisted for several days.

In the region of Bougainville Island there are several autonomous buoy stations TAO Array (Tropical Atmosphere Ocean Array, NOAA, http://www.pmel.noaa.gov/tao/data_deliv/). The location of the five stations nearest to the island is shown in Fig. 7.1. During the time period of interest to us from April 20 to May 31, 1996, only the following three stations of the ones shown in the figure were functioning: 5S156E, 2S156E, 0N156E (a station is indicated by its coordinates). Note that not a single station TAO Array was in the immediate vicinity of the temperature anomaly. The station 5S156E, nearest to the site of the event, was only at the north-eastern periphery of the anomaly. The stations 2S156E and 0N156E are situated too far away, and recorded no significant effects.

Figure 7.2 shows the time dependence of the temperature of air at a height of 3 m (the green curve) and of the water at a depth of 1 m (purple curve), obtained from the data of station 5S156E. Note, that starting from May 23 this station was out of operation. On the curve, describing time changes in temperature of the water surface layer, a process of a 24-h periodicity, representing the daily variation in temperature, is quite noticeable. Station 5S156E registered a clear violation of this process, which took place precisely on the day of the earthquake of maximum magnitude (April 29). On May 2 the daily course, namely, the daily warming, is restored within one cycle; however, a long period subsequently sets in (up to May 13), when no daily heating at all is observed, and the temperature variation turns out to be 'smoothed out'. Moreover, during this period a general tendency toward a decrease in temperature

7.1 Manifestations of Seaquakes

(approximately by 0.5°C) is noted. Violation of the daily course of temperature can be explained by intensification of the vertical exchange. Satellite images (GMS-5) show that during the period of interest the sky was never completely covered with clouds, which could have restricted the arrival of solar radiation.

Earlier, in Sect. 7.1.2, the assumed peculiarities of atmospheric circulation due to the formation of cold SST anomalies were already described. Over the region of reduced temperature, cooling and thickening of the air should take place, and, consequently, a 'cold wind' should start blowing from the region of the anomaly. From Fig. 7.2 it is seen that in the second half of April 29 (i.e. immediately after the earthquake) the temperature of the air fell sharply. A significant drop in the air temperature is also noticeable in the period from May 4 to 6, when the most pronounced SST anomaly was observed.

In a recent work [Levin et al. (2006)], a description is given of several cases of the transformation of vertical temperature profiles in the ocean, which took place after strong underwater earthquakes. To reveal events of this type the authors of the work indicated created an integrated database (IDB), that includes vertical profiles of oceanographical observations in the World Ocean and seismic catalogues.

Several sources were made use of in filling up the IDB with data. The database of the World Ocean, version of year 2001 (DB WO-01) [Conkright et al. (2002)], served as the main source, precisely it made up the main volume of the informational base. The database of the World Ocean of 2001 presents a further development of a series of products concerning oceanographic data in the World Ocean, published earlier—the 1994 Atlas of the World Ocean and the 1998 database of the World Ocean. The database contains oceanographic observations at standard and measured horizons in the form of vertical profiles. It includes practically all observations available for the time period of observation, which involved 7 037 213 stations. Since the data in the DB WO-01 are limited to the year 2001, in order to supplement the IDB with the most recent data it turned out to be necessary to take advantage of the Global Temperature-Salinity Profile Program—GTSPP.

Samples of data of the National Earthquake Information Centre (NEIC) of the USA were used as initial data. The first set of data comprises historical earthquakes of magnitudes above $M = 6.5$ during the period from 1900 to 2004, and the second set includes all significant historical earthquakes starting from 1973 up to the current day.

In analysis of events, information was sought on measurements made near the epicentre of an underwater earthquake before and after the seismic event. Besides, climatic characteristics were calculated: the average temperature profiles over many years and the respective root-mean-square deviations.

Figure 7.3 presents a case of the vertical thermal structure of the ocean changing after the underwater earthquake of magnitude $M = 7.7$, which took place on December 4, 1972. The earthquake epicentre was to the south of Honshu Island at the point with coordinates 33.0° N and 140.7° E. The sites, at which measurements of the temperature profiles were made before and after the event, were located at distances of approximately 40–50 km from the epicentre. The dates of temperature profile measurements made before the event were November 19–20 and after the

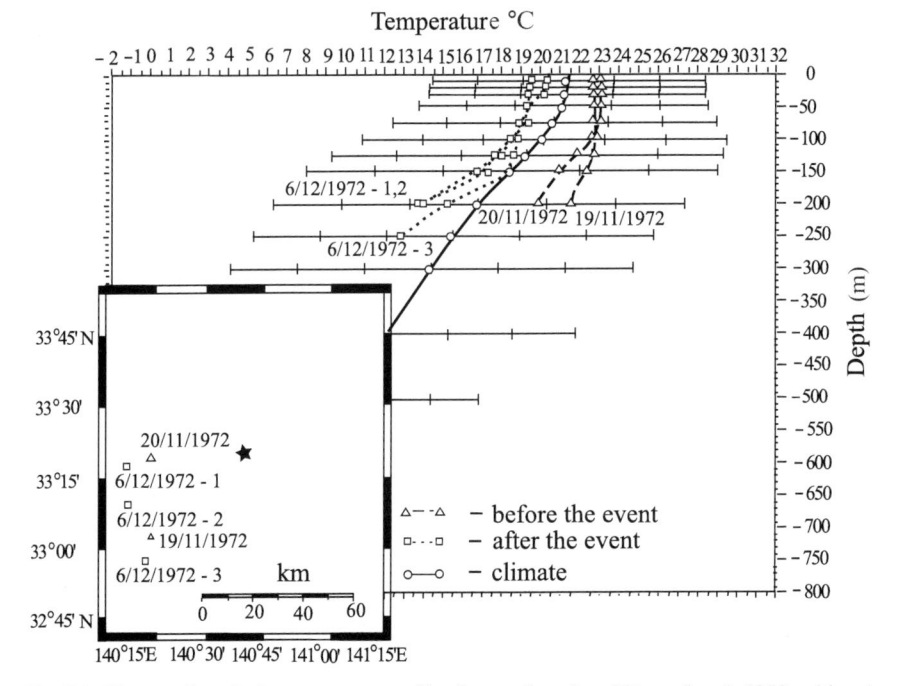

Fig. 7.3 Change of vertical temperature profile after earthquake of December 4, 1972, with epicentre to the south of Honshu Island (Japan)

event—December 6, 1972. From Fig. 7.3 the temperature in the upper 20-m layer is seen to be reduced, on the average, by 4–6°C.

Without presenting illustrations, we shall briefly describe two more events. The second case was observed during the November 5, 1938 earthquake of magnitude $M = 7.7$ at the eastern coast of Honshu Island (Japan). In the upper 50-m layer, a reduction of temperature by 2°C was noted, while on the 250 m horizon it amounted to approximately 4°C.

The third case was noted during the earthquake of magnitude $M = 7.5$ that occurred at the point with coordinates 8.5° S and 123.5° E on the Indonesian Island of Flores on March 22, 1944. The sites of temperature profile measurements before and after the event were situated next to each other within a radius of 5–10 km and at a distance of 40–50 km from the epicentre. The temperature at the surface decreased from 29°C down to 26.5°C.

We note that the prospects of making use of instrumental observations of seaquakes may be related not only to improvement of the quality and volume of information (for instance, enhancement of the spatial resolution of satellite images, an increase of the number of buoy stations, etc.). An analysis of oceanographic measurements, made in the past, but performed at a qualitatively new level, may still yield new interesting results.

7.2 Estimation of the Possibility of Stable Stratification Disruption in the Ocean Due to an Underwater Earthquake

In this section we investigate whether significant transformation of the stratification structure of the ocean can, in principle, result from an underwater earthquake. As the stratification disruption mechanism we shall consider vertical turbulent exchange. The concrete mechanism of energy transfer from the moving ocean bottom to the turbulent motion remains outside the framework of our estimations. We shall be interested in the parameters of the hypothetical source of turbulence, which are necessary for noticeable transformation of the stratification structure, and in the correspondence of these parameters to the possibilities of the seismic energy source.

The characteristic horizontal size of the region, within which transformation of the stratification structure takes place in the case of an underwater earthquake, significantly exceeds the ocean depth, therefore, it is justifiable to consider a one-dimensionable problem along the $0z$ axis, directed vertically downward. We shall set the origin of the reference frame on the water surface. The base set of equations includes the equation of balance for the turbulence energy b ($[b] = \text{m}^2\text{s}^{-2}$) for a stratified liquid and the equation of turbulent heat transfer [Nosov, Skachko (1999)]:

$$\frac{\partial b}{\partial t} = -\frac{g}{\rho_0} K_\rho \frac{\partial \rho}{\partial z} + \frac{\partial}{\partial z} K_b \frac{\partial b}{\partial z} - \frac{b^{3/2}}{L} + \beta(z,t), \tag{7.1}$$

$$\frac{\partial T}{\partial t} = \frac{\partial}{\partial z} K_T \frac{\partial T}{\partial z}, \tag{7.2}$$

where g is the acceleration of gravity, ρ_0 and ρ represent the respective average and current densities of the liquid, L is the turbulent motion scale, K_ρ, K_b and K_T are the turbulent exchange coefficients of mass, turbulence energy and heat. The first term in the right-hand part of equation (7.1) describes the energy spent on work against the forces of buoyancy, the second term describes turbulence energy transfer, the third—dissipation of turbulence energy. Generation of turbulence energy is described by function $\beta(z,t)$, which we have chosen in the following simple form:

$$\beta(z,t) = \beta_0 \Big(\theta(t) - \theta(t-\tau) \Big),$$

where $\theta(t)$ is the Heaviside step function. Generation (pumping) of turbulence energy is characterized by power β_0 ($[\beta_0] = \text{m}^2\text{s}^{-3}$) and action duration τ. Note, that attempts at any further determination of the detailed structure of function $\beta(z,t)$, when resolving the estimation problem, is not expedient, since they will only result in an increase of the number of free parameters. Consider all the turbulent exchange coefficients to be equal to each other, and assume dimensionality arguments to make it possible to express them in terms of the turbulence energy b and the turbulence scale L:

$$K_\rho = K_b = K_T \approx L b^{1/2}. \tag{7.3}$$

Having chosen the equation of state of water in the linearized form [Monin, Ozmidov (1981)] (assuming salinity not to vary with depth):

$$\rho = \rho_0 \left(1 - \alpha (T - T_0)\right),$$

where $\alpha = 2 \cdot 10^{-4} (°C)^{-1}$ and $\rho_0 = 1{,}000 \, kg \cdot m^{-3}$, we close the initial set of equations (7.1), (7.2).

As boundary conditions we adopt the conditions, that there be no flows of turbulence energy and of heat on the surface and on the bottom:

$$\frac{\partial b}{\partial z} = 0, \qquad z = 0, H; \qquad (7.4)$$

$$\frac{\partial T}{\partial z} = 0, \qquad z = 0, H. \qquad (7.5)$$

The initial conditions assume the energy of turbulence pulsations to equal zero at time moment $t = 0$ and the existence of an initial temperature distribution determining stable stratification (Fig. 7.4):

$$b(z, 0) = 0, \qquad (7.6)$$

$$T(z, 0) = T_0(z). \qquad (7.7)$$

The set of equations (7.1) and (7.2) with boundary conditions (7.4) and (7.5) and initial conditions (7.6) and (7.7) was solved numerically by the method of finite differences with respect to functions $b(z,t)$ and $T(z,t)$.

In all calculations, the temperature difference $T_2 - T_1 = 10°C$. The quantity h_1, characterizing the depth at which the thermocline was to be situated, was set to 10, 45 and 100 m. The thermocline thickness $h_2 - h_1 = 10$ m. The duration of the process at the earthquake source usually lies within the range from 1 to 100 s. To cover the range indicated the action of the turbulence energy source was considered to last for duration times equal to 1, 10 and 100 s.

Typical values for turbulence energy generation in the ocean, β_0, usually lie within the limits of 10^{-5}–10^{-9} m^{2}s^{-3} [Monin, Ozmidov (1981)]. For the purposes

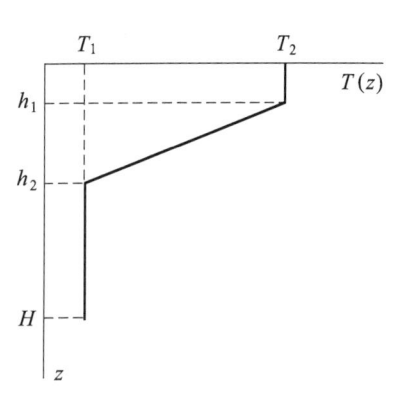

Fig. 7.4 Initial temperature profile

7.2 The Possibility of Stable Stratification Disruption

of the present study the upper limit of β_0 values was chosen from the condition $\beta_0 \tau = 1\ \text{m}^2 \cdot \text{s}^{-2}$. As it will be shown below, this approximately corresponds to 1–10% of the energy of a strong earthquake, if it is uniformly distributed over the water column over the pleistoseismic zone. The maximum scale of turbulent motion, L, having a natural upper limit equal to the depth H, did not exceed 100 m in calculations. The choice of lower limits for the values of parameters β_0 and L was based on the absence of noticeable variations in the initial temperature profile for times up to 1,000 s.

The main purpose of investigating the system (7.1)–(7.7) consists in determining the values of parameters β_0, L and τ, for which noticeable transformation of the initial temperature profile occurs. Such a transformation is conveniently traced either by variation in the surface temperature, δT_s, or by variation in the centre-of-mass position of the water column, δz.

$$\delta T_s = T(0,t) - T(0,0),$$
$$\delta z = (z(t) - z(0)),$$

where

$$z(t) = \int_0^H z \rho(z,t)\,dz \left(\int_0^H \rho(z,t)\,dz \right)^{-1}.$$

Variation of the centre-of-mass position is a more universal characteristic, since it permits to trace changes in the density (temperature) profile in such cases, when these changes do not influence the surface. Moreover, knowledge of the quantity δz makes it possible to calculate the potential energy of local stratification disruption (per unit area).

$$P = \rho_0 H g \delta z. \tag{7.8}$$

Local disruption of the vertical density distribution should, clearly, become the source of internal waves, the energy the upper limit of which can be estimated applying formula (7.8). Probably, the registration of internal waves of precisely such nature was described in [Filonov (1997)].

Figure 7.5 presents an example of the evolution of temperature and turbulence energy profiles. Profile $b(z,0)$ coincides with axis $0z$ and is not indicated in the figure. With time, the temperature profile smooths out, and the surface temperature gradually decreases. The turbulence energy profile exhibits a local minimum in the region of the thermocline, the existence of which is related to enhanced energy losses, spent on work against buoyancy forces. As the temperature profile smooths out, the minimum on profile $b(z)$ becomes less pronounced.

Figure 7.5 illustrates only one of the two main scenarios for development of the process in the system investigated. Thus, the first scenario (high-power source of turbulence) is characterized by a decrease in the maximum temperature gradient and by significant variations in the surface temperature. In the case of a source of relatively low power (Fig. 7.6) the maximum temperature gradient monotonously increases with time, while the surface temperature does not change.

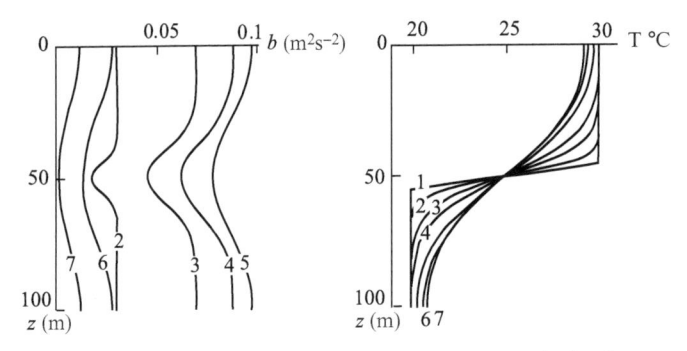

Fig. 7.5 Vertical profiles of temperature and of turbulence energy, calculated at successive moments of time for $L = 10\,\text{m}$, $\beta_0 = 0.003\,\text{m}^2\,\text{s}^{-3}$ and $\tau = 100\,\text{s}$. Curves 1–7 correspond to $t = 0, 10, 30, 50, 100, 150$ and $200\,\text{s}$

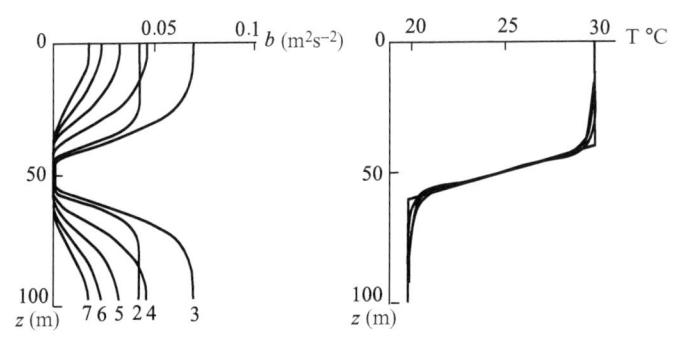

Fig. 7.6 Vertical profiles of turbulence energy and of temperature, calculated at successive time moments for $L = 10\,\text{m}$, $\beta_0 = 0.0001\,\text{m}^2\,\text{s}^{-3}$ and $\tau = 100\,\text{s}$. Curves 1–7 correspond to $t = 0, 50, 100, 150, 200, 250$ and $300\,\text{s}$

In both cases, after pumping is switched off, the energy of turbulence pulsations and, consequently, the exchange coefficient in the region of the jump in temperature rapidly decrease, practically down to zero. However, above and below the region of the jump in density turbulence pulsations still persist for some time (up to 1,000 s), thus influencing the temperature profile.

A phenomenon, similar to the second scenario takes place, when a thin vertical structure of hydrodynamic fields forms in the ocean. In the non-linear system considered one observes the most simple, and at the same time typical, case of self-organization: the energy flux brought to the system leads to its stucturization in space (enhancement of the gradient).

Figure 7.7 presents typical dependences, describing variation in time of the maximum temperature gradient $f_1(t)$, the energy of turbulence pulsations averaged over depth, the surface temperature $f_3(t)$ and the centre-of-mass position $f_4(t)$:

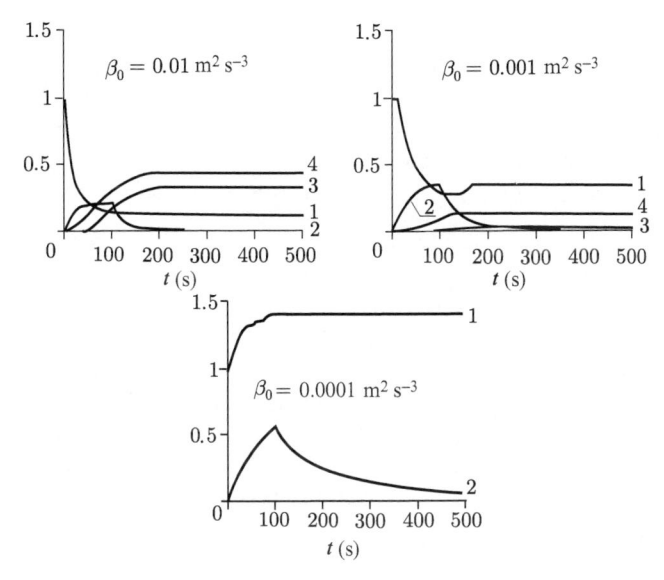

7.2 The Possibility of Stable Stratification Disruption

Fig. 7.7 Typical time dependences, describing variation of the maximum temperature gradient (1), of the turbulence energy averaged over depth (2), of the surface temperature (3), of the centre-of-mass position (4). Calculations are performed for $L = 10$ m, $t = 100$ s and different β_0 values (shown in the figure)

$$f_1(t) = \max_{0<z<H} \text{grad}\, T(z,t) \left(\max_{0<z<H} \text{grad}\, T(z,0) \right)^{-1},$$

$$f_2(t) = (H \beta_0 \tau)^{-1} \int_0^H b(z,t)\, dz$$

$$f_3(t) = \left(T(0,t) - T(0,0) \right) \left(H^{-1} \int_0^H T(z,0)\, dz - T(0,0) \right)^{-1},$$

$$f_4(t) = (z_{cm}(t) - z_{cm}(0))\,(0.5 H - z_{cm}(0))^{-1},$$

where

$$z_{cm}(t) = \int_0^H z \rho(z,t)\, dz \left(\int_0^H \rho(z,t)\, dz \right)^{-1}.$$

The maximum gradient is normalized to its value at the initial moment of time. The turbulence energy averaged over the depth is normalized to the total energy arriving in time τ per unit mass. The changes in surface temperature and centre-of-mass position are normalized to their maximum possible values, which are achieved, when the water is mixed until its state is completely uniform.

Both the position of the centre of mass and the change in surface temperature characterize the degree of mixing. Owing to the centre-of-mass position starting to change, unlike the surface temperature, immediately after generation of turbulence energy 'switches on', it is more convenient, in a number of cases, to make use of function $f_4(t)$, instead of $f_3(t)$.

In Fig. 7.7 the two aforementioned development scenarios for the process in the system investigated are clearly seen:

1. When $\beta_0 = 10^{-3} \, \text{m}^2 \, \text{s}^{-3}$ and $\beta_0 = 10^{-2} \, \text{m}^2 \, \text{s}^{-3}$ the maximum gradient decreases and noticeable changes of surface temperature and of the centre-of-mass position occur, which signifies destruction of the thermocline and arrival of depth waters at the surface.
2. When $\beta_0 = 10^{-4} \, \text{m}^2 \, \text{s}^{-3}$, the maximal temperature gradient increases monotonously with time, no noticeable changes of the surface temperature and of the centre-of-mass position occur, i.e. the thermocline is not destroyed and it serves an obstacle in the way of depth waters toward the surface.
3. When $\beta_0 = 10^{-3} \, \text{m}^2 \, \text{s}^{-3}$, the maximum gradient changes nonmonotonously with time: after the action of the source terminates a little enhancement of it takes place; this is due to the energy of turbulence pulsations and, consequently, the exchange coefficient in the region of the temperature jump rapidly dropping practically down to zero. At this time the turbulent exchange in the region above and below the jump still persists for some time, which precisely leads to enhancement of the maximum gradient.

The duration of the transient process of 'switching on' ('switching off') pumping, which can be traced by the behaviour of function $f_2(t)$, depends both on the turbulence scale L and on the value of parameter β_0. As it is seen from Fig. 7.7, noticeable changes in the surface temperature, the centre-of-mass position and the maximum gradient can occur not only during pumping of turbulence energy, but also when it is switched off.

In Fig. 7.8 diagrams are presented that in the $\beta_0 \tau, L$ plane show the change in

1. Maximum temperature gradient $\delta \left(\text{Max} \left[\text{grad} \, T \right] \right)$ by a quantity exceeding 10% of the initial gradient value
2. Surface temperature δT_s
3. the Centre-of-mass position δz.

The data presented in the figures correspond to the final changes in the temperature profile and are not related to any time moment. The calculation is performed for three different durations of the source action, $\tau = 1, 10$ and 100 s, for $h_1 = 10$ m and $H = 100$ m. Similar diagrams, but for a thermocline at the greater depth $h_1 = 45$ m, are presented in Fig. 7.9.

Analysis of the figures permits to make the following conclusions. Enhancement of the duration of the turbulence source action, τ, for the same total amount of energy entering the system, $\beta_0 \tau$, leads to an insignificant spread of the range of noticeable variations in all three parameters studied $(\delta \left(\text{Max} \left(\text{grad} \, T \right) \right), \delta T_s, \delta z)$. It is interesting to note that this range spreads exclusively toward smaller turbulent

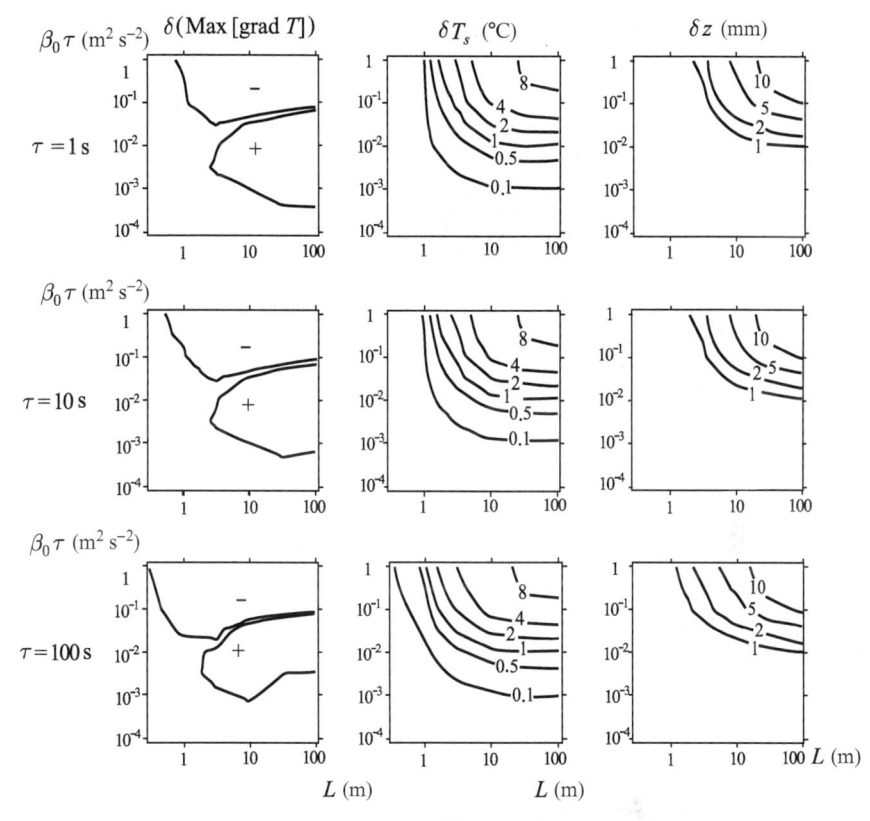

7.2 The Possibility of Stable Stratification Disruption

Fig. 7.8 Range of parameters $\beta_0\tau$ and L, for which turbulence energy pumping during a time τ (the values are indicated in the figure) results in (1) an increase (+), a decrease (−) of the maximum temperature gradient by a quantity, exceeding 10% of the initial value; (2) a change of the surface temperature; (3) a change of the centre-of-mass position. The figures at the isolines indicate the change in the surface temperature in degrees Celsius and in the centre-of-mass position in millimeters. Calculations are carried out for $h_1 = 10$ m and $H = 100$ m

motion scales, but not towards smaller energies. Such an effect is readily explained. The point is that the lifetime of a vortex increases with its scale L. Therefore, in the case of a large-scale turbulence a source action of duration τ within the range from 1 to 100 s can be considered instantaneous; only the total amount of energy $\beta_0\tau$ is essential. But a small-scale turbulence, generated by a source of short action, dissipates before it has time to contribute noticeable changes in the temperature profile.

The influence of the thermocline depth on the change in the centre-of-mass position is insignificant: differences are observed only in the case of large energies $\beta_0\tau$ and scales L, i.e. when the water column is mixed until it is practically in a uniform state. The characteristic value of quantity δz amounts to several millimeters, which, however, represents a significant value. By calculations in accordance with formula

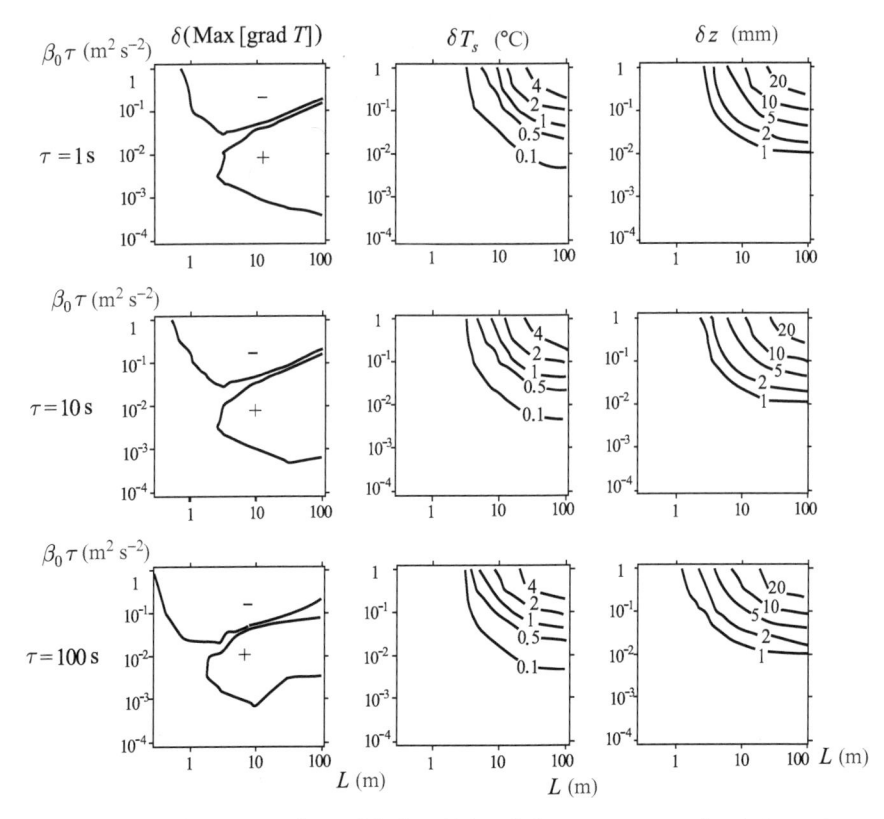

Fig. 7.9 Range of parameters $\beta_0 \tau$ and L, for which turbulence energy pumping during a time τ (the values are indicated in the figure) results in (1) an increase (+), a decrease (−) of the maximum temperature gradient by a quantity, exceeding 10% of the initial value; (2) a change of the surface temperature and (3) a change of the centre-of-mass position. The figures at the isolines indicate the change in the surface temperature in degrees Celsius and in the centre-of-mass position in millimeters. Calculations are carried out for $h_1 = 45$ m and $H = 100$ m

(7.8), it is not difficult to demonstrate that a change in the centre-of-mass position by 1 mm approximately corresponds to an energy of $1\,\text{kJ}\,\text{m}^{-2}$.

As to changes in the surface temperature, they are primarily determined by the depth of the thermocline: the closer the cold layers are to the surface, the more noticeable are the changes in the T_s value, given all other conditions being equal. This assertion is demonstrated in Fig. 7.10. Here, we no longer present diagrams for the centre-of-mass position, since calculations were performed for various depths H. In each case the depth H was chosen to be the minimum possible one, but so that variations in the temperature profile did not reach the bottom.

Now, we shall estimate the power (energy), which can be provided by a seismic source. Estimations will be based on the known correlative relationship between an earthquake energy E [J] and its magnitude M [Puzyrev (1997)]:

$$\lg E = 1.8M + 4. \tag{7.9}$$

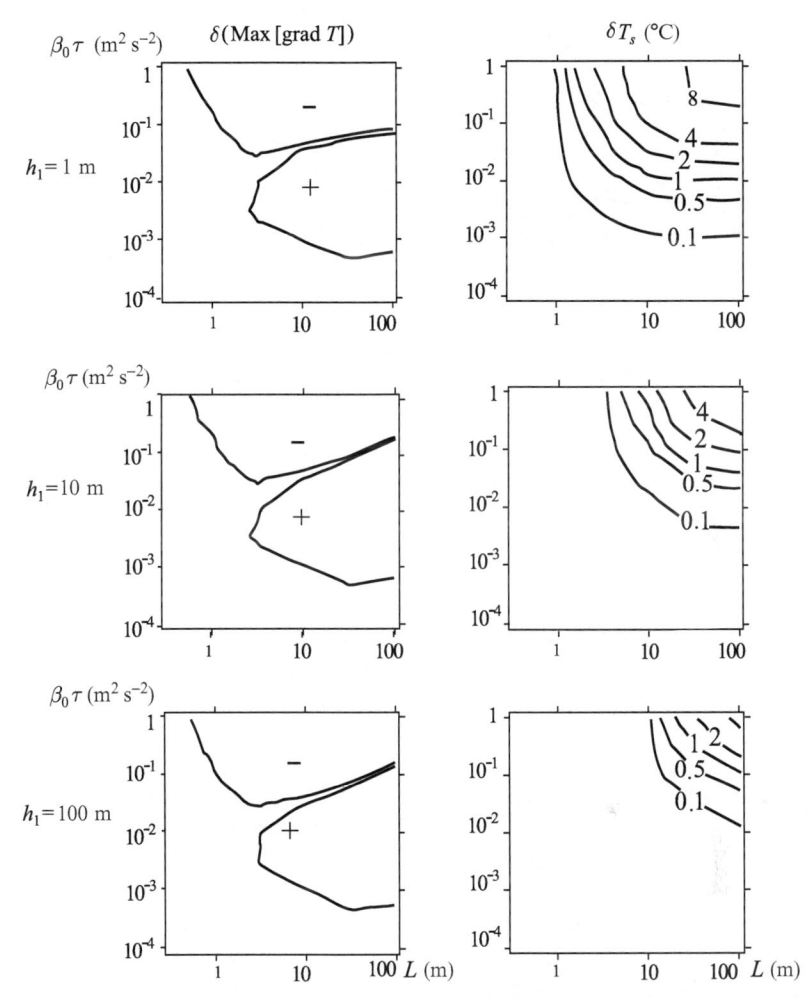

7.2 The Possibility of Stable Stratification Disruption

Fig. 7.10 Range of parameters $\beta_0 \tau$ and L, for which turbulence energy pumping during a time τ (the values are indicated in the figure) results in (1) an increase (+) a decrease (−) of the maximum temperature gradient by a quantity, exceeding 10% of the initial value; (2) a change of the surface temperature. The figures at the isolines indicate the change in the surface temperature in degrees Celsius. Calculations are carried out for various depths of the thermocline, $h_1 = 10$, 45 and 100 m for $\tau = 10$ s

Strong underwater earthquakes are known to be accompanied by residual deformations of the bottom, which serve as the source of tsunami waves. It would be natural to assume the activation process of vertical exchange to take place over the region of the bottom, subjected to the strongest deformations. The area of this region is unusually identified with the area of the tsunami source, the average radius of which, R (km), is related to the magnitude M by the empirical dependence:

$$\lg R = 0.5 M - 2.1. \tag{7.10}$$

Distributing the earthquake energy E uniformly over the area of the tsunami source and the ocean depth H, we estimate the upper limit of the energy of the source of turbulence, $W^{\lim} = \beta_0 \tau$ [m^{2}s^{-2}]:

$$W^{\lim} = \frac{E}{\pi R^2 H \rho_0}. \qquad (7.11)$$

Substituting the empirical dependences (7.9) and (7.10) into expression (7.11) and taking into account that $\rho_0 = 1{,}000$ kg·m^{-3} and $\lg \pi \approx 0.5$, we obtain the following estimation formula:

$$W^{\lim} \approx H^{-1} 10^{0.8M-1.3}. \qquad (7.12)$$

In Fig. 7.11 the dependence of the quantity $W^{\lim}$ upon the earthquake magnitude is presented in a logarithmic scale. From the figure it is seen that the quantity $W^{\lim}$ can amount to 100 and even 1,000 m^2 s^{-2}. On the basis of data presented in Figs. 7.8–7.10, the conclusion can be made that strong earthquakes have a sufficient reserve of energy for essential transformation of the ocean stratification structure. Tenths of a percent of the energy of an earthquake is sufficient for formation on the ocean surface of a temperature anomaly with a characteristic horizontal dimension, measured by hundreds of kilometers and with a temperature deviation of the order of 1°C. Note, that a comparable amount of energy (less than 1% of the earthquake energy) is spent on the formation of tsunami waves.

The formation of a temperature anomaly of the ocean surface is most probable in the case of a shallow thermocline and for seismic events, characterized by a persistent process at the source or by a large number of aftershocks. The most striking manifestation of the effect is to be expected in the case of realization of the turbulence generation mechanism with a scale exceeding 10 m. Local variations of the vertical temperature distribution should serve as a source of internal waves even in those cases, when temperature variations are insignificant.

The estimates obtained show that noticeable transformation of the stratification structure in the ocean, including formation of a temperature anomaly on its surface,

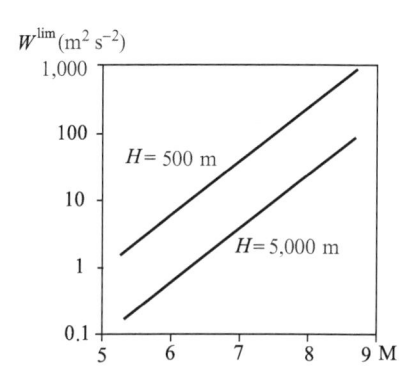

Fig. 7.11 Upper energy limit for a hypothetical turbulence source $W^{\lim}$ versus the earthquake magnitude

is a process, permitted from the point of view of energy. Moreover, the process considered requires a negligible part of the earthquake energy.

7.3 Parametric Generation of Surface Waves in the Case of an Underwater Earthquake

One of the most widespread effects, described by eyewitnesses of seaquakes, consists in formation on the ocean surface of standing waves of large amplitude. To reveal the causes resulting in generation of such waves and to calculate their length it is quite sufficient to apply linear theory. To determine the amplitude of the waves, their shape, their type of space symmetry it is necessary to take non-linearity into account.

Assume that during an earthquake the ocean bottom undergoes periodic movements, in accordance with the following law

$$\eta(t,x,y) = \eta_0 \cos(\omega t). \tag{7.13}$$

Since the horizontal scale of the pleistoseismic zone significantly exceeds the ocean depth, we assume the amplitude of oscillations, η_0, not to vary along the horizontal plane. If the ocean depth satisfies the condition $H < c\pi/2\omega$, where c is the velocity of sound in water (see Sect. 3.2.1), then the water column behaves like an incompressible medium, undergoing induced oscillations, that repeat movements of the bottom. In a deep ocean elastic oscillations of the water column at normal frequencies $v_k = 0.25c(1+2k)H^{-1}$, where $k = 0, 1, 2, \ldots$, can arise in the case of any vertical movements of the bottom (not necessarily periodic movements) (see Sect. 3.1.3). In the second case we shall consider the movements of a certain upper layer of the ocean, of thickness $h < H$, to proceed according to the law (7.13). We shall choose the thickness of this layer so as to be able to describe its behaviour as the motion of an incompressible liquid $h < c\pi/2\omega$.

Thus, as initial conditions we have a layer of incompressible liquid with a free surface, in the field of gravity. The layer undergoes vertical oscillations according to the law (7.13). We shall show that such a system is not stable, and inside it there develop standing surface gravitational waves. It must be underlined that in the case dealt with the layer of liquid is not limited horizontally, therefore, the nature of the standing waves considered here has little in common with the traditional standing waves, which arise in a restricted region.

We now pass to a noninertial reference system, the origin of which oscillates in accordance with the law (7.13). In this case the gravitational field is supplemented with the periodical in time, but uniform in space, component

$$a(t) = \eta_0 \omega^2 \cos(\omega t).$$

The main equation of linear potential wave theory, (2.29), will remain without changes. But in the boundary condition on the free surface there will now appear a time-dependent coefficient,

$$(g+a(t))\frac{\partial F}{\partial z} = -\frac{\partial^2 F}{\partial t^2}, \quad z = 0. \tag{7.14}$$

At the lower boundary in the new noninertial reference system it is necessary to set the vertical velocity component equal to zero,

$$\frac{\partial F}{\partial z} = 0, \quad z = -H \text{ (or } z \to -\infty\text{)}. \tag{7.15}$$

The general solution of the problems (2.29), (7.14) and (7.15) is given by the following formula (analogue of expression (2.34)):

$$F(x,y,z,t) = \int_{-\infty}^{+\infty} dm \int_{-\infty}^{+\infty} dn \exp(imx - iny)$$
$$\times \Big(A(t,m,n)\cosh(kz) + B(t,m,n)\sinh(kz)\Big). \tag{7.16}$$

Substituting formula (7.16) into the boundary conditions (7.14) and (7.15), we obtain that the potential is determined by the expression

$$F(x,y,z,t) = \int_{-\infty}^{+\infty} dm \int_{-\infty}^{+\infty} dn \exp(imx - iny)$$
$$\times A(t,m,n)\Big(\cosh(kz) + \tanh(kH)\sinh(kz)\Big), \tag{7.17}$$

and the coefficient $A(t,m,n)$ can be found from the solution of the known Mathieu equation

$$\frac{\partial^2 A}{\partial t^2} + gk\tanh(kH)\left(1 + \frac{\eta_0 \omega^2}{g}\cos(\omega t)\right)A = 0. \tag{7.18}$$

The properties of the solution of this equation are such, that, when the following equality is satisfied:

$$gk\tanh(kH) = \left(\frac{n\omega}{2}\right)^2, \quad n = 1,2,3\ldots, \tag{7.19}$$

in the system there arises a parametric resonance, in the case of which the solution increases exponentially with time. The most rapid growth takes place, when $n = 1$. The growth of oscillations is possible not only when relationship (7.19) is satisfied exactly, but also within certain finite intervals of the pumping frequency values, termed instability zones. The widths of these zones increase with the coefficient $\eta_0 \omega^2/g$ in equation (7.18).

7.3 Parametric Generation of Surface Waves

Thus, if the layer of liquid oscillates as a unique whole vertically with a frequency ω, then standing waves arise on its surface, that are characterized by the frequency $\omega/2$. Note that for parametric resonance to be realized the water column does not have to oscillate with a certain definite frequency. The oscillation frequency can be arbitrary, but the length of the waves formed depends on it.

If quite high oscillation frequencies (~ 10 Hz are considered, then effects of surface tension must be taken into account and the general formula must be applied:

$$\left(gk + \frac{\alpha}{\rho}k^3\right)\tanh(kH) = \left(\frac{n\omega}{2}\right)^2, \qquad (7.20)$$

where α is the surface tension coefficient.

The relationship (7.20) is presented in Fig. 7.12 in the 'wavelength–frequency' plane. The calculation was performed for $n = 1$ and an ocean depth $H = 4,000$ m. The dotted line shows the dependence in the case of small depths ($H = 10$ m), when the surface waves start 'to feel the bottom'. Deviation of the dependence from linearity at small wavelengths is explained by the action of surface tension forces.

To characteristic seaquake frequencies of 0.1–1 Hz there correspond wavelengths from tens up to hundreds of meters, which is in good agreement with the testimonies of eyewitnesses of these events. The characteristic dimensions of the space structures (~ 1 cm), observed in laboratory experiments at high frequencies (10–50 Hz), are also in good accordance with theory.

In accordance with the Floke theorem, the general solution of the Mathieu equation can be represented in the form

$$A(t) = C_1 \exp\{\alpha t\}\varphi(t) + C_2 \exp\{-\alpha t\}\varphi(-t),$$

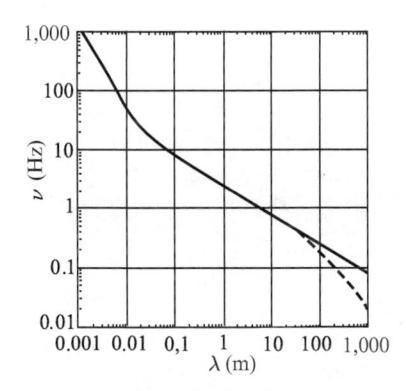

Fig. 7.12 Relationship between the frequency of bottom oscillations (or oscillations of the water column) and the wavelength of standing waves, formed on the water surface in the case of parametric resonance. The calculation is carried out for $n = 1$ and for ocean depths 4,000 m (solid line) and 10 m (dotted line)

where $C_{1,2}$ are coefficients determined by the initial conditions, $\varphi(t)$ is a periodic function. The growth increment of the oscillation amplitude in the case of parametric resonance ($n = 1$) is determined by the following formula [Rabinovich, Trubetskov (1984)]:

$$\alpha = \frac{\eta_0 \omega^3}{8g}.$$

The development of the parametric resonance may be hindered by various effects. But one such effect is always active—the viscosity of water. Gravitational surface waves in a viscous liquid are characterized by an amplitude damping decrement expressed by the formula [Landau, Lifshits (1987)]

$$\gamma = 2\nu k^2,$$

where ν is the kinematical viscosity of water. Assuming the length of waves, formed in the case of parametric resonance, to be significantly smaller than the ocean depth ($kH \gg 1$), we obtain from formula (7.19) the relationship between the frequency and wave number, $\omega^2 = 4gk$. The decrement of viscous damping can now be readily expressed via the frequency,

$$\gamma = \nu \frac{\omega^4}{8g^2}.$$

Equating the increment α and the decrement γ, we obtain the threshold condition for development of the parametric resonance,

$$\eta_{0c} = \nu \frac{\omega}{g}.$$

For development of parametric waves it is necessary for the amplitude of bottom oscillations to exceed the critical value η_{0c}. At frequencies peculiar to seismic oscillations of the bottom ((0.1–1 Hz), the critical amplitude turns out to be extremely small, $\eta_{0c} \sim 1\,\mu\text{m}$. Therefore, one can conclude that the conditions for the parametric resonance to develop are fulfilled in the case of any earthquake that is just felt. A strong earthquake usually lasts for several minutes. Will the amplitude of parametric waves reach a noticeable value in this time depends on the increment. The characteristic growth time of the amplitude is defined as the quantity inverse to the increment,

$$\tau \equiv \frac{1}{\alpha} = \frac{8g}{\eta_0 \omega^3}.$$

The quantity τ is seen to be determined by the amplitude and frequency of oscillations, but its dependence upon the frequency is stronger. Thus, for example, in the case of an oscillation frequency of 1 Hz and amplitude of 0.1 m, the development time of parametric waves will only amount to 3 s. Therefore, during an earthquake the amplitude has a high probability of reaching a significant value. Note that to a frequency of 1 Hz there corresponds a wavelength of the order of 10 m (see Fig. 7.12). Standing waves of precisely such lengths have been repeatedly observed by eyewitnesses of seaquakes.

A reduction of the oscillation frequency down to 0.1 Hz leads to an increase in the quantity τ by three orders of magnitude (up to 3,000 s for $\eta_0 = 0.1$ m), which makes it doubtless larger, than the earthquake duration. In the case of small frequencies, parametric resonance will have time to develop only if the oscillation amplitude is very large, $\eta_0 \sim 10$ m. Such amplitudes can, apparently, be achieved, when the frequency of seismic movements of the bottom happens to be close to one of the normal frequencies of elastic oscillations of the water column.

Another mechanism restricting the exponential growth of amplitudes may be the collapse of waves exceeding the critical steepness. In this case, the wave energy starts to be effectively transferred to the turbulent motion, which contributes to intensification of vertical exchange in the water column [Ermakova et al. (2007)].

7.4 Experimental Study of Wave Structures and of Stable Stratification Transformation in a Liquid in the Case of Bottom Oscillations

Processes arising in the ocean in the case of underwater earthquakes have been studied experimentally with the aid of a simple physical model [Levin, Trubnikov (1986); Alexandrov et al. (1986); Levin (1996); Nosov et al. (1996); Nosov, Ivanov (1997); Nosov, Skachko (2000, 2004)].

The experimental set-up represented a round basin of diameter 0.45 and height 0.25 m, fixed on a massive foundation. Its layout is shown in Fig. 7.13. In the central part of the basin there was a circular opening with a piston of diameter 0.15 m. The piston underwent vertical oscillations with controlled amplitude and frequency. The basin was filled with fresh water. In some of the experiments a stable temperature stratification was created in the basin. To avoid intense cooling of the water, due to evaporation, the basin was tightly covered with a specially adjusted lid of plexiglas.

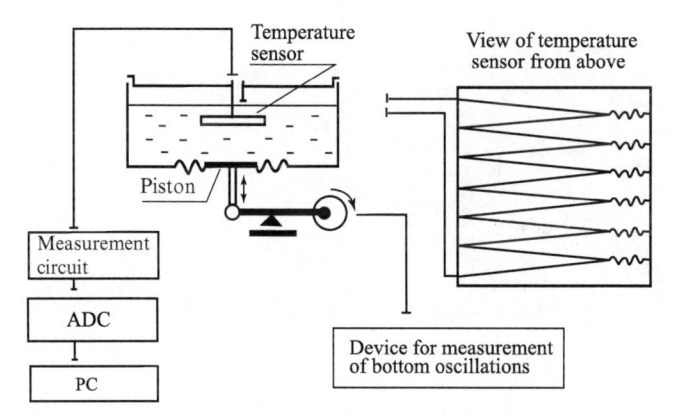

Fig. 7.13 Layout of experimental set-up. Distributed temperature sensor

A distributed resistance thermometer, made of copper wire 0.06 mm in diameter, was used as the temperature sensor. During measurements the frame was set in the horizontal plane. The dimension of the active area of the sensor amounted to 10×10 cm. The time constant of the temperature sensor was ~ 0.1 s. The sensor continuously scanned the area above the piston along the vertical, its velocity was 1.8 mm s^{-1}. Registration of the readings of the resistance thermometer was performed with the aid of an online personal computer.

To reveal dynamic modes that may exist in the liquid, in the case of bottom oscillations, 'scanning' was performed over the amplitude and frequency within the ranges of 0.22–0.97 mm and 0–70 Hz, respectively. Three main dynamic modes were revealed:

1. Linear wave formation
2. Dissipative structures—systems of standing waves exhibiting characteristic hexagonal (Fig. 7.14a) or orthogonal (Fig. 7.14b) symmetry (Faraday ripples)
3. Irregular (chaotic) motion with drops torn away and intense 'fountaining' of liquid (Fig. 7.14c)

When speaking of 'linear wave formation' we intend that waves peculiar to the first mode can be described as the linear response of an ideal liquid to oscillations of the bottom.

Determination was to be performed of two most clearly traceable bifurcation boundaries: 'linear wave formation–dissipative structures' (boundary 1–2) and 'separate drops–intense fountaining' (boundary 2–3). The results of experiments are presented in a summary form in Fig. 7.15, from which the bifurcation portrait of the system, drawn in 'amplitude–frequency' coordinates, reveals three regions, corresponding to regular linear wave formation (1), to the existence of dissipative structures (2) and to chaotic motion (3). The boundaries of the regions indicated correspond to bifurcation combinations of oscillation frequencies and amplitudes, at which, when achieved, a change of dynamic mode is observed. Note that the respective amplitude–frequency bifurcation combinations were achieved both by smooth

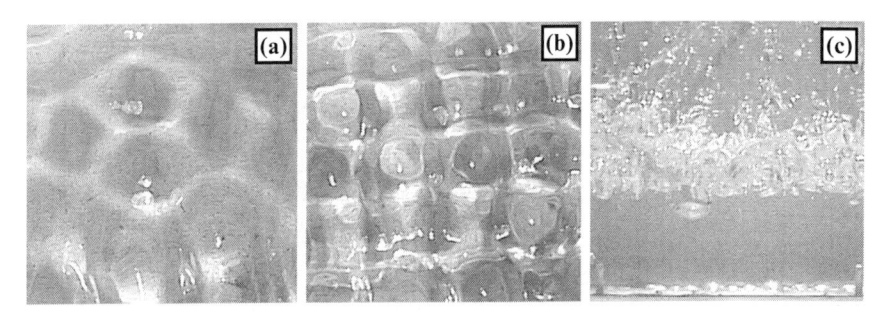

Fig. 7.14 Structures of standing waves of hexagonal (**a**) and orthogonal (**b**) symmetry on the surface of the liquid in the case of bottom oscillations. Chaotic motion of the surface (view through the transparent lateral wall of the basin) (**c**). The photographs are obtained with the aid of the modified version of the set-up

7.4 Experimental Study of Wave Structures

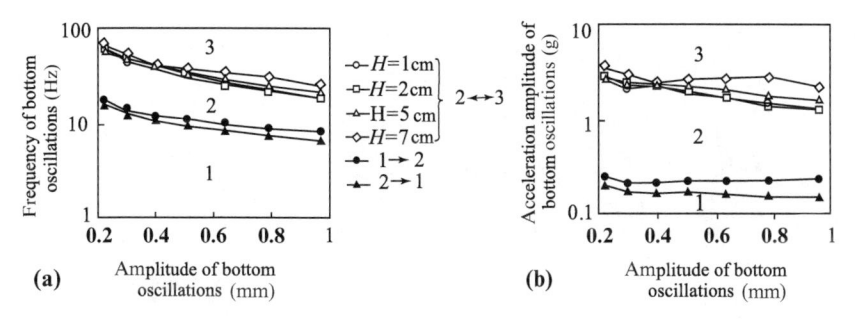

Fig. 7.15 Bifurcation portrait of a column of liquid of thickness H on an oscillating bottom in co-ordinates 'amplitude–frequency' (**a**) and 'amplitude–acceleration' (**b**). 1—linear wave formation, 2—dissipative structures, 3—chaotic motion

enhancement of the frequency at a constant amplitude and by smooth enhancement of the amplitude at a fixed frequency. No difference between the two ways of arriving at the bifurcation point was revealed.

It is remarkable, that the height of the column of liquid in a vessel influences its dynamic behaviour weakly: when this quantity is altered by nearly an order of magnitude (from 1 up to 7 cm), the shift of the three described regions in the parametric plane turns out to be insignificant in the case of transition across boundary 2–3 and absolutely indistinguishable in the case of transition across boundary 1–2.

The creation of bifurcation boundaries in the 'amplitude–acceleration' plane permits to assume that precisely the acceleration amplitude of bottom oscillations serves as the main control parameter in the system investigated, which is especially apparent in the transition from linear wave formation to structures (the two lower curves are practically parallel to the x-axis). This actually signifies that the boundaries of the regions, where different dynamic modes in the 'amplitude–frequency' plane exist, have the form of inverse quadratic dependences.

The stably reproducible hysteresis character of bifurcation boundary 1–2 must also be noted. When the oscillation frequency (amplitude) is increased gradually, structures develop in the system at an acceleration amplitude (0.26 ± 0.004) g. When the frequency (amplitude) is reduced, a 'collapse' of the wave structures is observed at a noticeably smaller value of this quantity—(0.19 ± 0.004) g.

Note, also, that transition from stable dynamic structures on water to a state of chaotic surface motion takes place at different threshold frequencies for waves with square and hexagonal cells. Figure 7.16 shows photographs of the state of a vibrating water surface with square wave cells in the case of cyclic oscillation frequencies $100\,\text{s}^{-1}$ (a), $160\,\text{s}^{-1}$ (b) and $180\,\text{s}^{-1}$ (c); here, also, presented are variations in the mode of hexagonal cells at frequencies $180\,\text{s}^{-1}$ (d), $200\,\text{s}^{-1}$ (e) and $220\,\text{s}^{-1}$ (f). Figure 7.17 presents a comparison of measurement results for the parameters of hexagonal cells (triangles) and of square cells (circles) with theoretical dispersion laws for gravitational waves (1), gravitational–capillary waves (2), and parametric waves (3). These results were first obtained and described in [Levin, Trubnikov (1986); Alexandrov et al. (1986)] and [Levin (1996)].

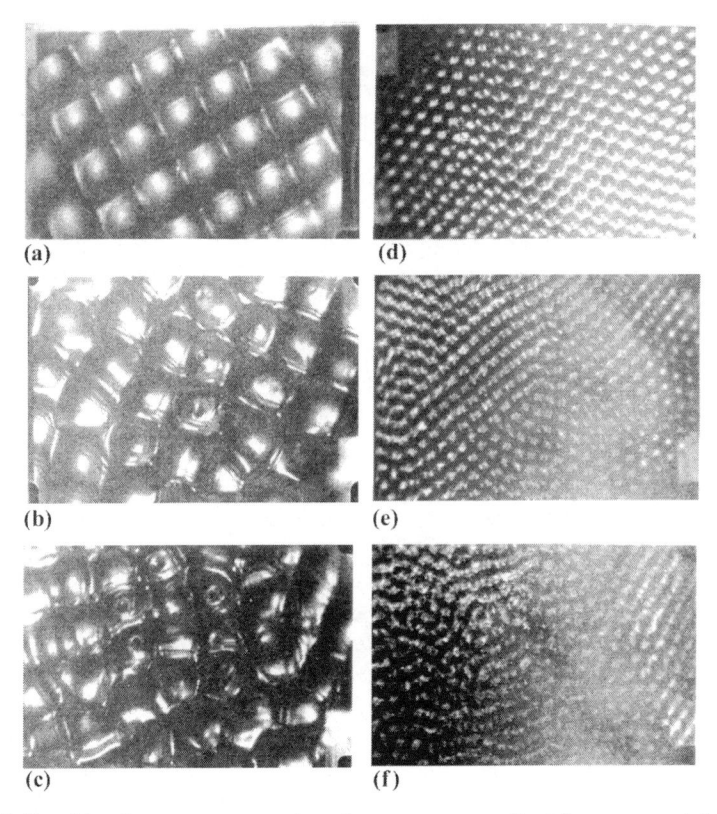

300 7 Seaquakes: Analysis of Phenomena and Modelling

Fig. 7.16 Transition from structures to chaos for square wave cells at frequency variations from $100\,\text{s}^{-1}$ (**a**) via $160\,\text{s}^{-1}$ (**b**) to $180\,\text{s}^{-1}$ (**c**) and for hexagonal cells at frequency variations from $180\,\text{s}^{-1}$ (**d**) via $200\,\text{s}^{-1}$ (**e**) and further up to $220\,\text{s}^{-1}$ (**f**)

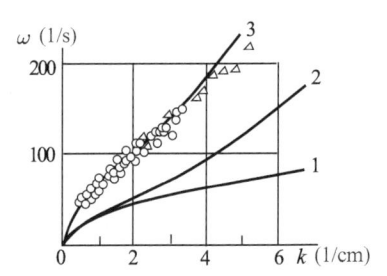

Fig. 7.17 Comparison of theoretical dispersion laws for gravitational waves (1), gravitational–capillary waves (2), and parametric waves (3) with measurement results for parameters of hexagonal wave cells (triangles) and square cells (circles). The respective axes represent wave numbers (k) and frequencies (ω)

The most important feature of the dynamic behaviour of the system considered is apparently the existence of a sole external parameter—the amplitude of the bottom acceleration, which determines the character of this behaviour within a broad range

7.4 Experimental Study of Wave Structures

of heights of the column of liquid. This fact opens up the possibility of using the obtained results in studying seaquakes, arising in the case of underwater earthquakes or other local perturbations of the ocean bottom. The results of the experiment are relevant to shallow areas of the ocean or to its upper layer. The mechanisms of turbulence generation in the ocean column in the case of underwater earthquakes are, most probably, of another nature.

The vertical exchange intensity in a liquid column in the case of a section of the bottom undergoing oscillations was estimated by the transformation of vertical temperature profiles. Processing of the temperature profiles was performed in accordance with the formula

$$K(z) = \frac{1}{\frac{\partial T}{\partial z}} \left(\int_0^z \frac{\partial T}{\partial t} dz + K_0 \frac{\partial T}{\partial z} \bigg|_{z=0} \right) \tag{7.21}$$

The quantity K_0 was chosen to be the molecular coefficient of temperature conductivity of water, equal to $0.0014\,\text{cm}^2\text{s}^{-1}$.

In Fig. 7.18 the examples are presented of successively measured temperature profiles and of the respectively calculated profiles of the vertical turbulent exchange coefficient. The two presented cases correspond to weak (close to molecular) and intense turbulent exchange at bottom oscillation frequencies of 3.4 and 39.5 Hz, respectively.

For a quantitative analysis it is necessary to have a scalar quantitative characteristic, reflecting the vertical exchange intensities. Significant (more than an order of magnitude) variations of the quantity $K(z)$ over the ocean depth make simple averaging of this coefficient over the depth unacceptable. It has been noted that most profiles of the turbulent exchange coefficient expressed in semilogarithmic coordinates $(z, \ln K)$ are described not badly by a linear dependence. In this connection, their exponential approximation was constructed in the form

$$K(z) = C_0 \exp\{\alpha z\} \tag{7.22}$$

The coefficients of approximation (7.22) were determined by the method of least squares for a group of profiles obtained during a single experiment for fixed frequencies and amplitudes of bottom oscillations. Then, the constructed regression dependence was applied for calculating the turbulent exchange coefficient close to the water surface (at point $z = 7$ cm). The error in the estimate of $K(7)$ was determined in the standard way.

Figure 7.19 presents, in a double logarithmic scale, values of the turbulent exchange coefficient $K(7)$ versus the acceleration amplitude of bottom oscillations, corresponding to different oscillation frequencies and amplitudes. The $K(7)$ values are normalized to the molecular coefficient of temperature conductivity in water, K_0, and the acceleration amplitude is normalized to the free-fall acceleration. From the figure it is seen that right up to acceleration amplitudes of about 0.2 g (i.e. before the rise moment of structures), independently of the piston's oscillation amplitude,

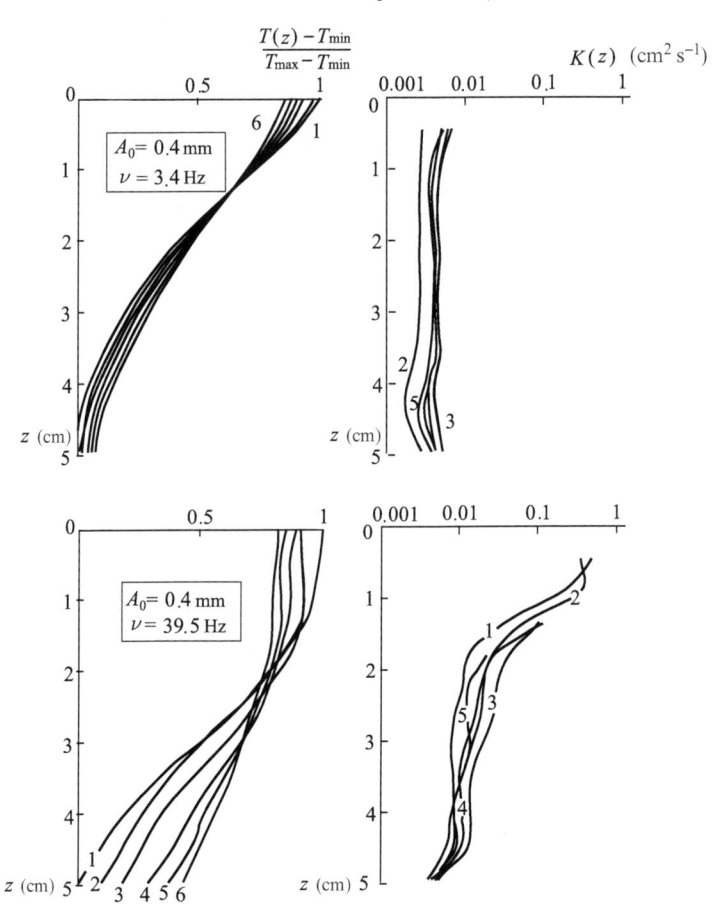

Fig. 7.18 Examples of successively registered (with an interval of 56 s) vertical temperature profiles (curves 1–6) and of respectively calculated profiles of the vertical turbulent exchange coefficient (curves 1–5)

the quantity $K(7)$ assumes values close to the molecular value. Then, an increase of this parameter, in certain cases quite significant, is observed, which is related to the collapse of waves: at frequencies of the order of 10 Hz the amplitude of waves in the structures could amount to 3–5 cm, i.e. a value comparable to the depth of the liquid. Finally, further enhancement of the acceleration amplitude, in the region of $a \sim g$, is accompanied by a sharp increase of $K(7)$ values.

Here it must by noted that large values of the turbulent exchange coefficient are achieved at large amplitudes of bottom oscillations. Of interest, also, is the fact that $K(7)$ values can exceed the molecular temperature conductivity coefficient by more than three orders of magnitude.

The results obtained with a model system can, naturally, not be applied straightforwardly to a real ocean. At any rate, they permit to reveal the general character of processes that can take place in natural conditions. From this point of view,

7.4 Experimental Study of Wave Structures

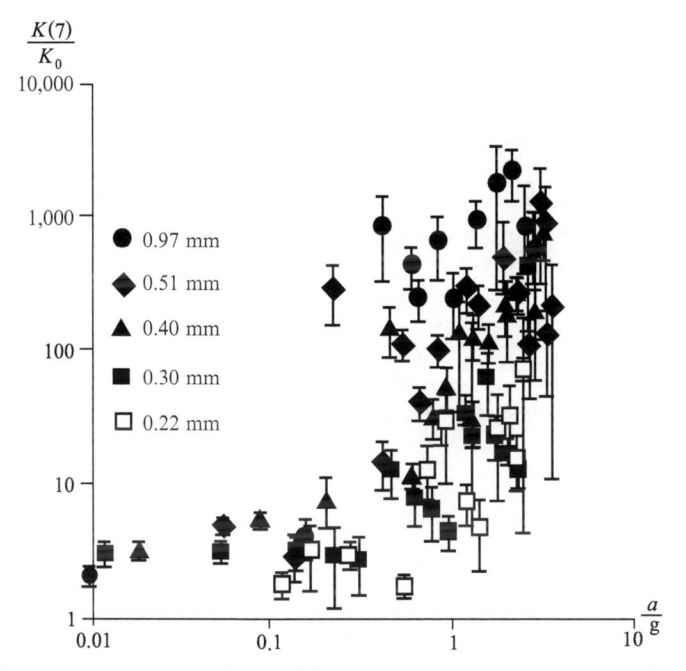

Fig. 7.19 Turbulent exchange coefficient $K(7)$ close to the surface versus acceleration amplitude of bottom oscillations at various oscillation amplitudes (indicated in figure)

the data considered point to the possibility of significant transformation of the stratification structure in the case of an underwater earthquake. Besides, a most important feature of the dynamic behaviour of the system considered must be recognized to be the existence of a sole external parameter—the bottom acceleration amplitude, which determines the character of this behaviour within a broad range of heights of the column of liquid. This fact opens the possibility of applying the obtained results in studies of the seaquake mechanism.

A logical development of the construction of the experimental set-up consisted in it being equipped with a basin with transparent lateral walls, permitting to observe the processes taking place in the column of liquid subjected to the influence of the oscillating bottom. The modified experimental set-up (Fig. 7.20) represented a rectangular basin with internal dimensions $0.22 \times 0.22 \times 0.22$ m. The transparent lateral walls of the basin were made of plexiglas 30 mm thick, which excluded the possibility of their vibrating. At the centre of the basin there was a circular opening, in which there was a piston 0.07 m in diameter, that underwent vertical harmonic oscillations of given amplitude and frequency. Experiments were carried out within the following ranges of oscillation amplitudes and frequencies: 5–35 Hz and 0.6–2.65 mm, respectively.

The basin was continuously filled with stratified liquid (a water solution of NaCl) exhibiting a vertical density distribution close to linear. The density gradient in different experiments varied between 30 and 500 kg m^{-4}. The water depth in the basin

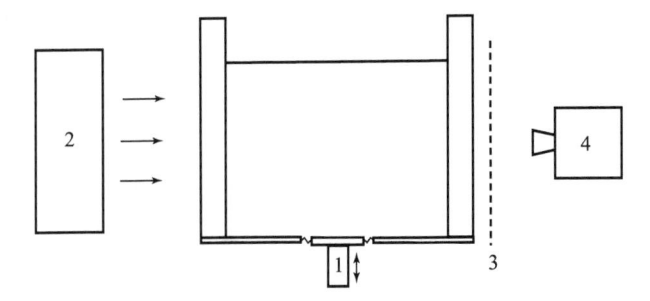

Fig. 7.20 Layout of experimental set-up. 1—piston undergoing vertical oscillations, 2—source of light (quasiparallel rays), 3—semitransparent screen, 4—digital photocamera

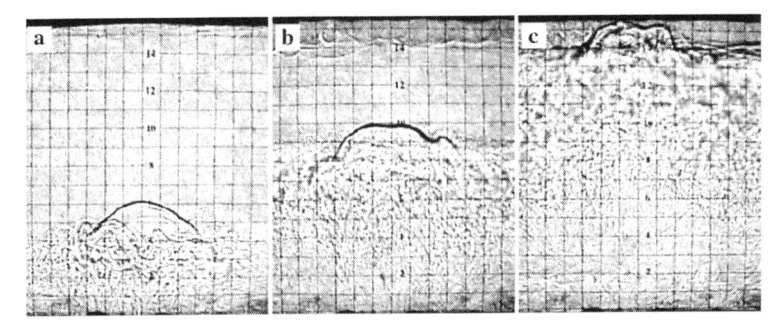

Fig. 7.21 Shadow pictures of the destruction of continuous stratification in the case of harmonic oscillations of a section of the bottom. The step of the net is 1 cm. Pictures (**a**), (**b**) and (**c**) are arranged according to the increase of the piston oscillation velocity

in all cases amounted to 0.16 m. Visualization of dynamic processes in the basin was performed by the shadow method. Registration of shadow pictures was done with the aid of a digital photocamera.

Figure 7.21 presents typical shadow pictures, reflecting the dynamics of processes in a stratified liquid over an oscillating section of the bottom (the piston). Above the piston a 'dome' forms, the height of which is determined by the density gradient and the oscillation velocity of the piston. At fixed oscillation frequency and amplitude of the piston the height of the dome changes slowly.

The reason the dome arises consists in the extremely pronounced turbulized flow, concentrated above the central part of the piston inside a cylindrical region 2–3 cm in diameter. The flow is directed vertically upward. In absence of stratification the maximum flow velocity (a visual estimate) amounted to $10\,\text{cm}\,\text{s}^{-1}$. Near the walls of the basin relatively slow lowering of the liquid proceeded.

On the water surface structures of standing waves formed, which also led to destruction of the stratification in the surface layer of thickness about 2 cm, which is well seen in the shadow picture. In the case of high oscillation velocities of the piston (Fig. 7.21c) the upper mixed layer and the turbulized flow closed: 'the depth waters reached the surface'.

It has already been noted above that the turbulence exchange coefficient decreases as the distance from the surface increases (towards the depth) according to a law close to exponential. The present experiment explains this fact. The point is that the source of turbulence consists in the intensive motion of water in standing surface waves (Faraday ripples). Obviously, as the distance from the surface increases, the turbulence decreases. If there existed no upward going flow, then vertical transfer would be accomplished only by turbulent diffusion. In calculations of the turbulent exchange coefficient $K(z)$ the possibility of adequate transfer was not taken into account. The large spread of experimental data in Fig. 7.19 is explained by the contribution of the flow to transfer processes being significant in some cases.

Thus, experiments have revealed two mechanisms conducive to the enhancement of vertical exchange. The first is related to the development of turbulence due to intensive motion of the liquid in standing waves; it only concerns the surface layer. The second mechanism consists in a rising non-linear flow, forming above the oscillating bottom. Such a flow can realize the transfer directly from depth layers of the ocean towards the surface.

References

Agarwal V. K., Mathur A., Sharma R., Agarwal N., Parekh A. (2007): A study of air-sea interaction following the tsunami of 26 December 2004 in the eastern Indian Ocean. Int. J. Remote Sens. **28**(13–14) 3113–3119

Alexandrov V. E., Basov B. I., Levin B. W., Soloviev S. L. (1986):On the formation of parametric dissipative structures during seaquakes. DAN SSSR (in Russian), **289**(5) 1071–1074

Behrenfeld M. J., Falkowski P. G. (1997): Photosynthetic rates derived from satellite-based chlorophyll concentration. Limnol. Oceanogr. **42**(1) 1–20

Conkright M. E., Antonov J. I., Baranova O., et al. (2002): In: Sydney Levitus (ed) World Ocean Database 2001, vol. 1. Introduction. NOAA Atlas NESDIS. US Government Printing Office, Washington, DC, **42**

Charles Darwin (1839): Darwin Charles: The Voyage of the Beagle. Wordsworth Editions, UK

Degterev A. Kh. (2001): Earthquake effect on hydrogen sulfide contamination of the Black Sea. Russian Meteorology and Hydrology (12) 38–42

Ermakova O. S., Ermakov S. A., Troitskaya Yu. I. (2007): Izvestiya – Atmos. Ocean Phys. **43**(1) 86–94

Ezersky A. B., Korotin P. I., Rabinovich M. I. (1985): Chaotic automodulation of two-dimensional structures on the surface of a liquid in the case of parameteric excitation. Lett. JETP (in Russian), **41**(4) 129–131

Filonov A. E. (1997): Researchers study tsunami generated by Mexican Earthquake. EOS. **78**(3) 21–25

Higuera M., Vega J. M., Knobloch E. (2000): Coupled amplitude-streaming flow equations for nearly inviscid faraday waves in small aspect ratio containers. J. Non-linear Sci. **12** 505–551

Landau L. D., Lifshitz E. M. (1987): Fluid Mechanics, V.6 of Course of Theoretical Physics, 2nd English edition. Revised. Pergamon Press, Oxford-New York-Beijing-Frankfurt-San Paulo-Sydney-Tokyo-Toronto

Levin B. W. (1996): Non-linear oscillating structures in the earthquake and seaquake dynamics. Chaos **6**(3) 405–413

Levin B. W., Nosov M. A., Skachko S. N. (2001): SST and Chlorophyll Concentration Anomalies due to Submarine Earthquakes: Observations, Consequences and Generation Mechanism.

In: Proceedings of Joint IOC–IUGG International Workshop Tsunami Risk Assessment Beyond 2000: Theory, Practice and Plans, Moscow, pp. 105–109
Levin, B., Kaistrenko V., Kharlamov A., Chepareva M., Kryshny V. (1993): Physical processes in the ocean as indicators for direct tsunami registration from satellite, In: Proceedings of the IUGG/IOC International Tsunami Symposium Wakayama, Japan, pp. 309–319
Levin B. W. (1996): Tsunamis and seaquakes in the ocean (in Russian). Priroda (5) 48–61
Levin B. V., Likhachiova O. N., Uraevskii E. P. (2006): Variability in the thermal structure of ocean waters in periods of strong seismic activity. Izvestiya – Atmos. Ocean Phys. **42**(5) 653–657
Levin B. W., Soloviev S. L. (1985): Variations of the field of mass velocities in the pleistoseist zone of an underwater earthquake (in Russian). DAN SSSR, **285**(4) 849–852
Levin B. W., Trubnikov B. A. (1986): 'Phase transitions' in the lattice of parametric waves on the surface of an oscillating liquid (in Russian). Lett. JETP **44**(7) 311–315
Levin B. V., Nosov M. A., Pavlov V. P., Rykunov L. N. (1998): Cooling of the ocean surface as a result of seaquakes. Doklady Earth Sci. **358**(1) 132–135
Luchin V. A., Levin B. W., Nosov M. A., et al. (2000): Variations of water temperature at sea surface, caused by tectonic movements of the bottom (in Russian). Ln: DVNIGMI Jubilee issue. Dal'nauka, Vladivostok, pp. 172–182
Monin A. S., Ozmidov R. V. (1981): Oceanic turbulence (in Russian). Hydrometeoizdat, Leningrad
Nosov M. A. (1996): On the Influence of Submarine Earthquakes on the Stratification Structure of the Ocean (in Russian). In: Theses of reports to All-Russian scientific conference 'Interaction in the lithosphere–hydrosphere–atmosphere system', Moscow, pp. 70–71
Nosov M. A. (1998a): Effect of submarine earthquake on a stratified ocean. Moscow Univ. Phys. Bull. **53**(4) 23–27 (1998)
Nosov M. A. (1998b): Ocean surface temperature anomalies from underwater earthquakes. Volcanol. Seismol. **19**(3) 371–375
Nosov M. A., Ivanov P. S. (1997): Location of different flow regime domains in oscillating fluid. Volcanol. and Seismol. **19** 123–128
Nosov M. A., Ivanov P. S. (1994): Dynamic modes in a hydrodynamic system with an oscillating lower boundary (in Russian). In: Theses of reports to Third International Symposium 'Physical engineering problems of new equipment'. Centre of applied physics of the N. E. Bauman Moscow State Technical University, Moscow, p. 151
Nosov M. A., Skachko S. N. (1999): Transformation of the stratification structure of the ocean as a result of the submarine earthquake. Moscow Univ. Phys. Bull. (5) 51–55
Nosov M. A., Skachko S. N. (2000): Mechanism of transformation of the stratification structure of the ocean by the seismic bottom movements. Moscow Univ. Phys. Bull. (4) 66–68
Nosov M. A., Skachko S. N. (2004): Non-linear steady flows generated by bottom domain oscillations. Vestnik Moskovskogo Universita. Ser. 3 Fizika Astronomiya (5) 57–60
Nosov M. A., Ivanov P. S., Shelkovnikov N. K. (1996): Modeling of thermal water stratification behavior in a system with a mobile bottom. Volcanol. Seismol. **17** 689–692
Ouzounov D., Freund F. (2004): Mid-infrared emission prior to strong earthquakes analyzed by remote sensing data. Adv Space Res **33** 268–273
Ostrovsky L. A., Papilova I. A. (1974): On Non-linear acoustic wind (in Russian). Acous. J. (1) 79–86
Pinegina T. K., Bourgeois J. (2001): Historical and paleo-tsunami deposits on Kamchatka, Russia: long-term chronologies and long-distance correlations. Nat. Hazards Earth Sys. Sci. **1**(4) 177–185
Pinegina T. K., Melekestsev I.V., Braitseva O. A., et al. (1997): Traces of prehistoric tsunamis on the eastern coast of Kamchatka (in Russian). Priroda (4) 102–106
Puzyrev N. N. (1997): Methods and objects of seismic studies (in Russian). Publishing House of RAS Siberian Branch, NITs OIGGM, Novosibirsk
Rabinovich A. B., Trubetskov D. I. (1984): Introduction to the theory of oscillations and waves (in Russian). Nauka, Moscow

Ranguelov B., Bearnaerts A. (1999): The Erzincan 1939 earthquake—a sample of the multidisaster event. In: Book of Abstracts, 2nd Balkan Geoph. Congr. and Exhibition. Istanbul, pp. 62–63, 5–9 July

Richter C. F. (1963): Elementary seismology (in Russian). Foreign literature Publishing House Moscow

Soloviev S. L., Go C. N. (1974): Catalogue of tsunamis on the western coast of the Pacific Ocean (173–1968) (in Russian). Nauka, Moscow

Soloviev S. L., Go C. N. (1975): Catalogue of tsunamis on the eastern coast of the Pacific Ocean (1513–1968) (in Russian). Nauka, Moscow

Soloviev S. L., Go C. N., Kim Kh. S. (1986): Catalogue of tsunamis in the Pacific Ocean, 1969–1982 (in Russian). Izd. MGK, USSR AS, Moscow

Soloviev S. L., Go C. N., Kim Kh. S., et al. (1997): Tsunamis in the Mediterranean Sea, 2000 BC–1991 AD (in Russian), Nauchnyi mir, Moscow

Vega J. M., Knobloch E., Martel C. (2001): Nearly inviscid Faraday waves in annular containers of moderately large aspect ratio. Physica D **154** 313–336

Yurur M. T. (2006): The positive temperature anomaly as detected by Landsat TM data in the eastern Marmara Sea (Turkey): possible link with the 1999 Izmit earthquake. Int. J. of Remote Sens. **27**(6) 1205–1218,

Zaichenko M. Yu., Levin B. V., Pavlov V. P., Yakubenko V. G. (2002): Cooling effect of the Black Sea active layer recorded after the earthquake. Izvestiya – Atmos. Ocean Phys. **38**(6) 695–699

Colour Plate Section

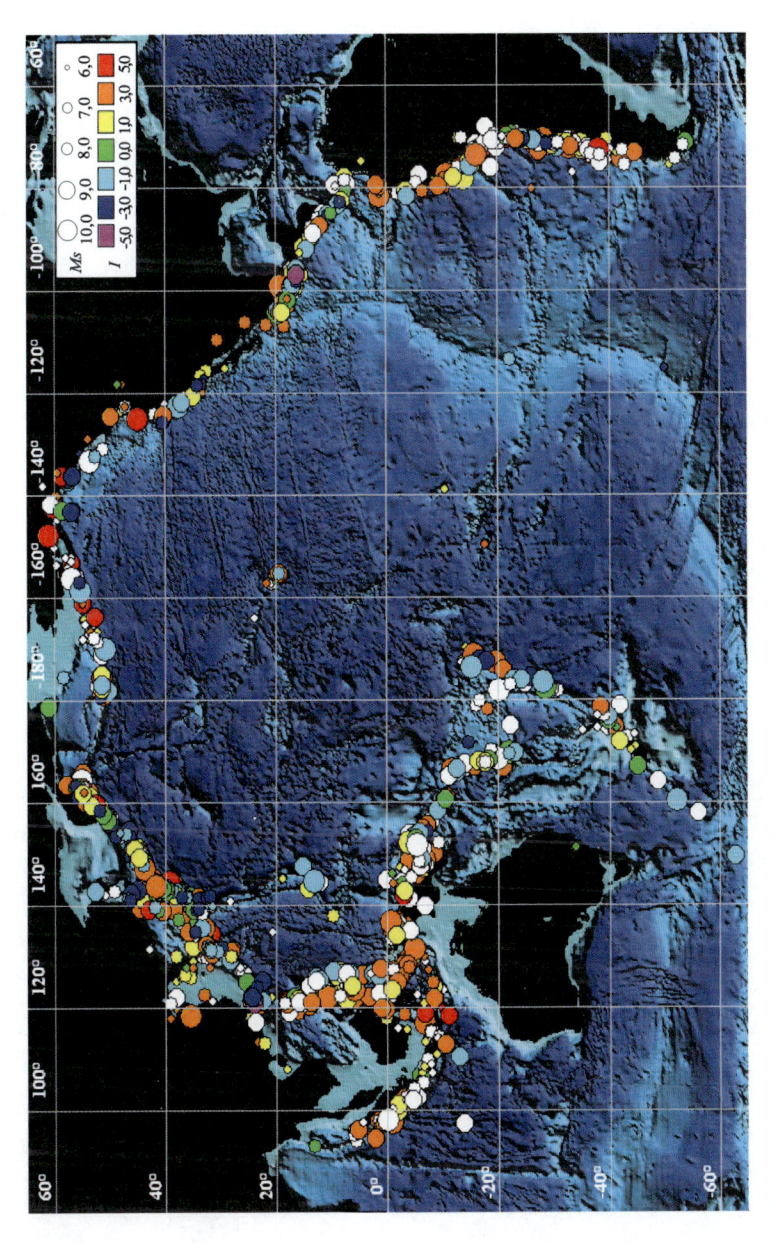

Plate 1 Distribution of tsunami sources in the Pacific region within the period from 47 BC up to 2004. The sizes of the circles correspond to earthquake magnitudes and their colours to the tsunami intensities (see also Fig. 1.4 on page 18)

310 Colour Plate Section

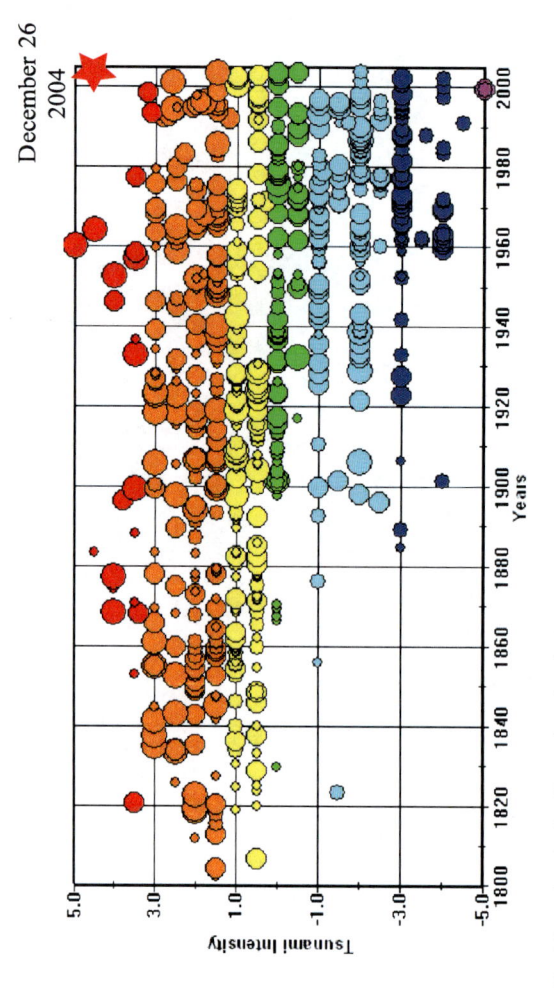

Plate 2 Tsunamis of the Pacific region in the 'intensity–time' plane (see also Fig. 1.6 on page 19)

Colour Plate Section

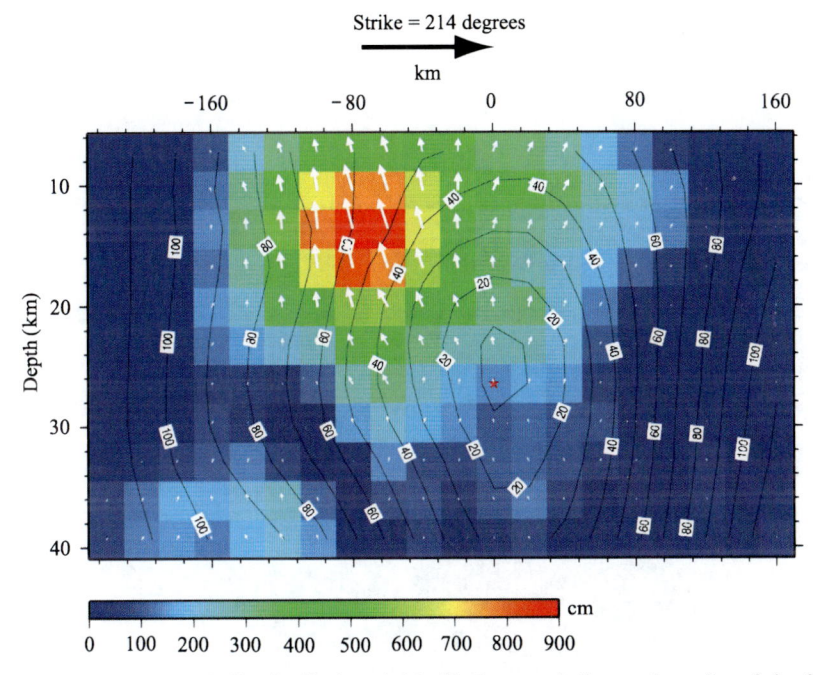

Plate 3 Cross section of slip distribution. A big black arrow indicates the strike of the fault plane. The colour shows the amplitude of dislocations and white arrows represent the motion of the hanging wall relative to the footwall. Contours show the rupture initiation time in seconds and the red star indicates the hypocentre location. The figure is taken from http://earthquake.usgs.gov/eqcenter/eqinthenews/2006/usvcam/finite_fault.php (see also Fig. 2.6 on page 49)

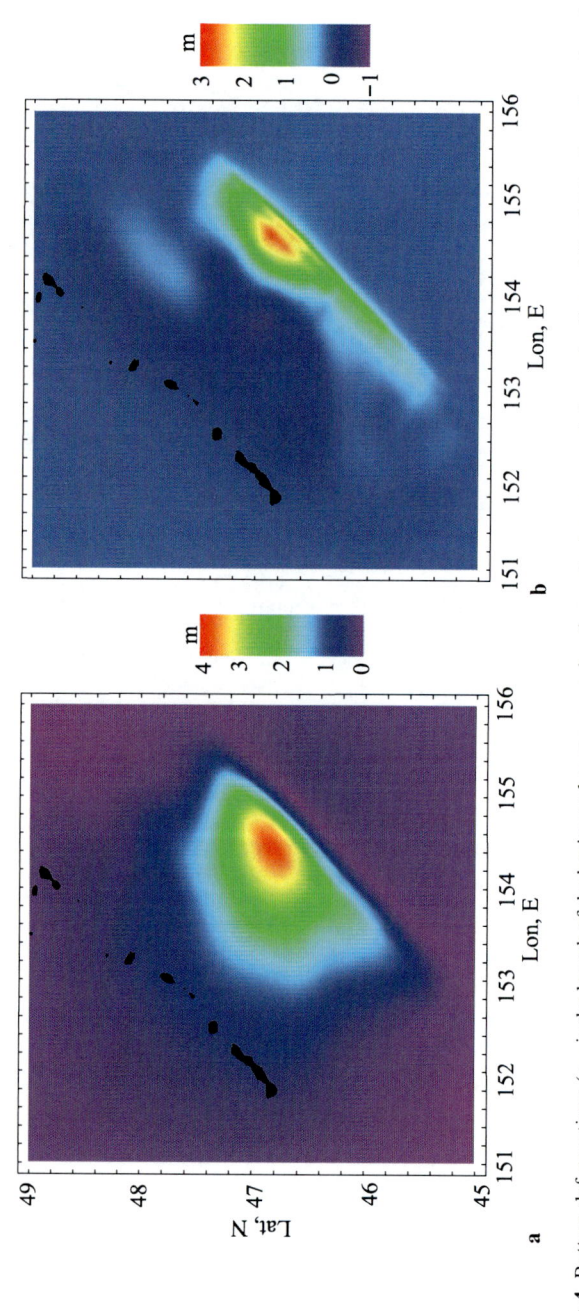

Plate 4 Bottom deformations (a—is the length of the horizontal component, b—the vertical component) due to the Central Kuril Islands earthquake of November 15, 2006. Calculations are performed applying Yoshimitsu Okada's formulae in accordance with USGS NEIC slip distribution data (see also Fig. 2.7 on page 50)

Colour Plate Section 313

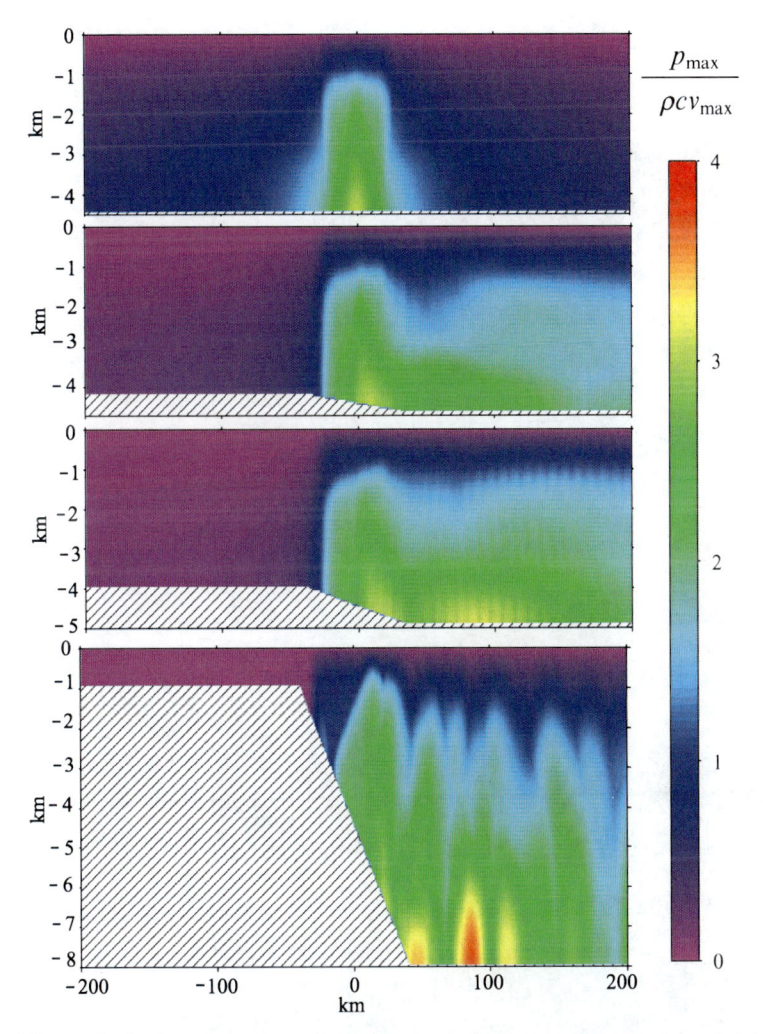

Plate 5 Space distribution of maximum dynamic pressure. The calculation is performed at $\tau = 10$ s for various profiles of the ocean bottom (see also Fig. 3.12 on page 121)

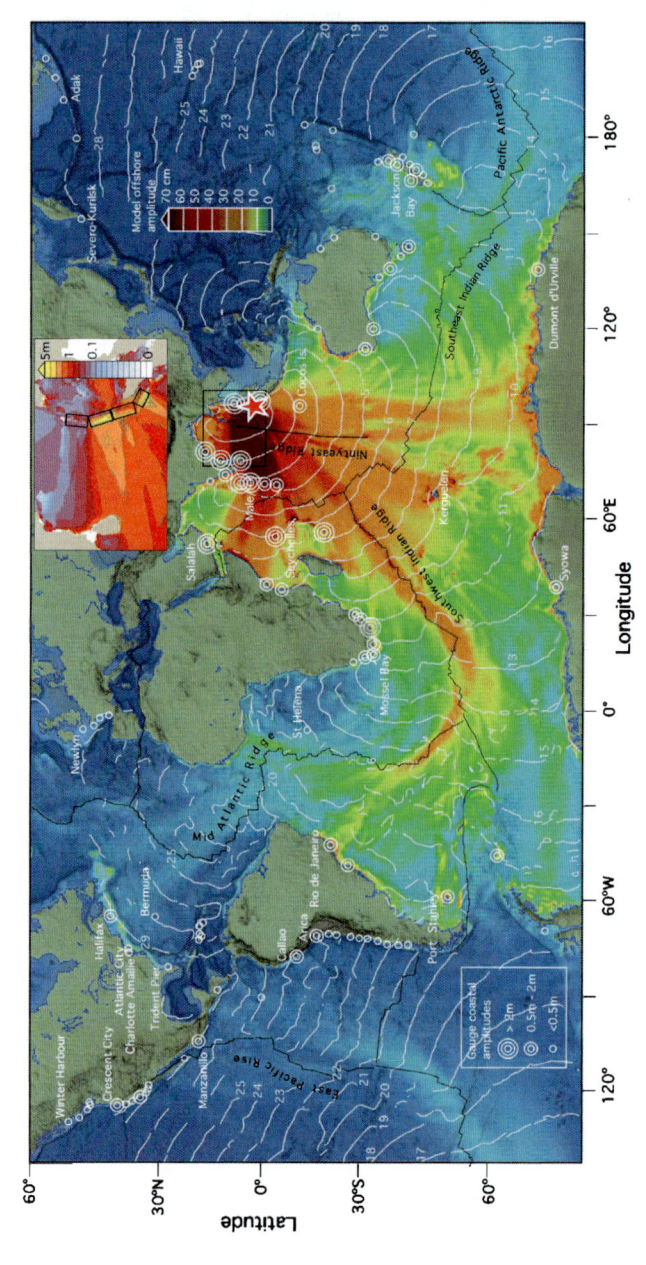

Plate 6 Global chart showing energy propagation of the 2004 Sumatra tsunami calculated from MOST. Filled colors show maximum computed tsunami heights during 44 hours of wave propagation simulation. Contours show computed arrival time of tsunami waves. Circles denote the locations and amplitudes of tsunami waves in three range categories for selected tide-gauge stations. Inset shows fault geometry of the model source and close-up of the computed wave heights in the Bay of Bengal. Distribution of the slip among four subfaults (from south to north: 21 m, 13 m, 17 m, 2 m) provides best fit for satellite altimetry data and correlates well with seismic and geodetic data inversions (Reprinted from [Titov et al. (2005)] by permission of the publisher) (see also Fig. 5.10 on page 221)

Colour Plate Section 315

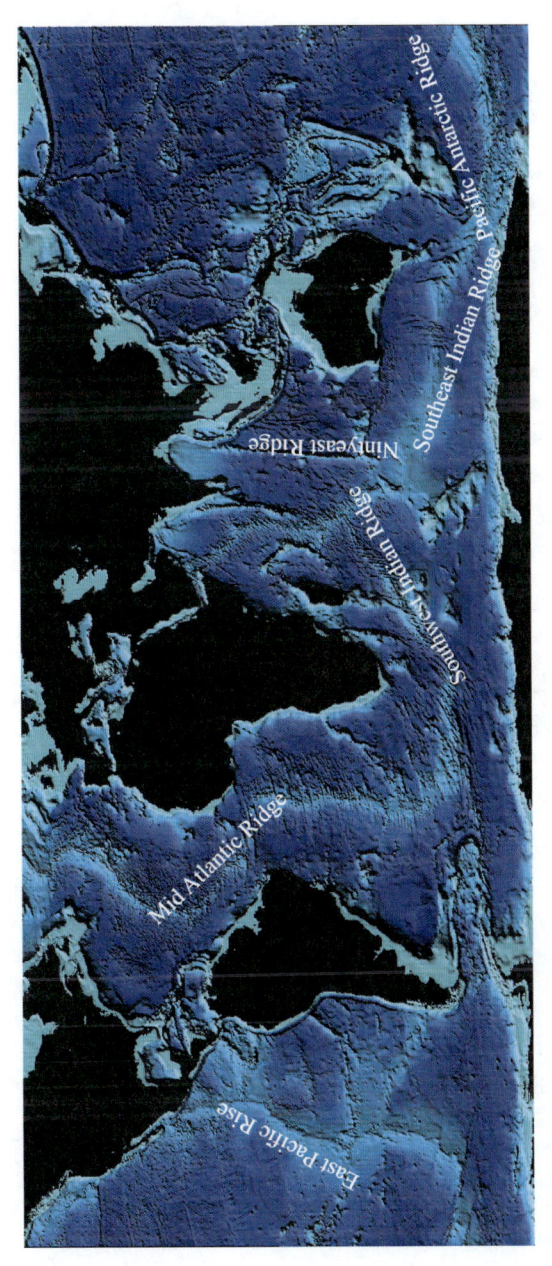

Plate 7 Bottom topography of World Ocean. Mid–oceanic ridges that essentially influenced propagation of the Indonesian tsunami of December 26, 2004 (see also Fig. 5.11 on page 222)

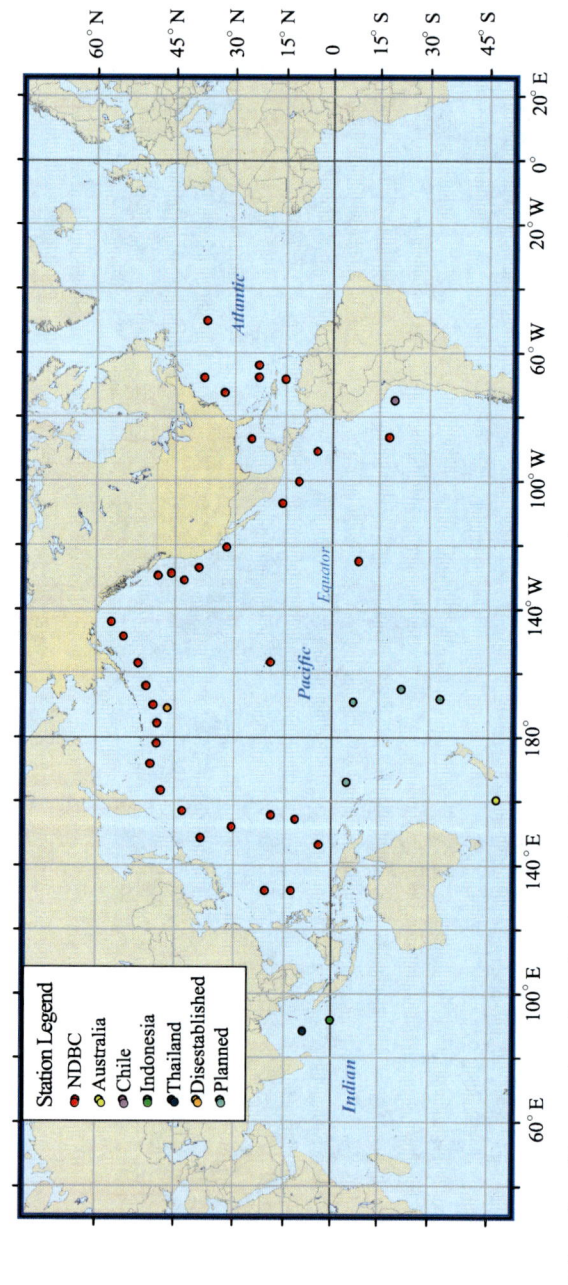

Plate 8 Location of buoy stations, for registering tsunamis in the open ocean (as of November 2007). The figure is taken from http://www.ndbc.noaa.gov/dart.shtml (see also Fig. 6.4 on page 238)

Colour Plate Section

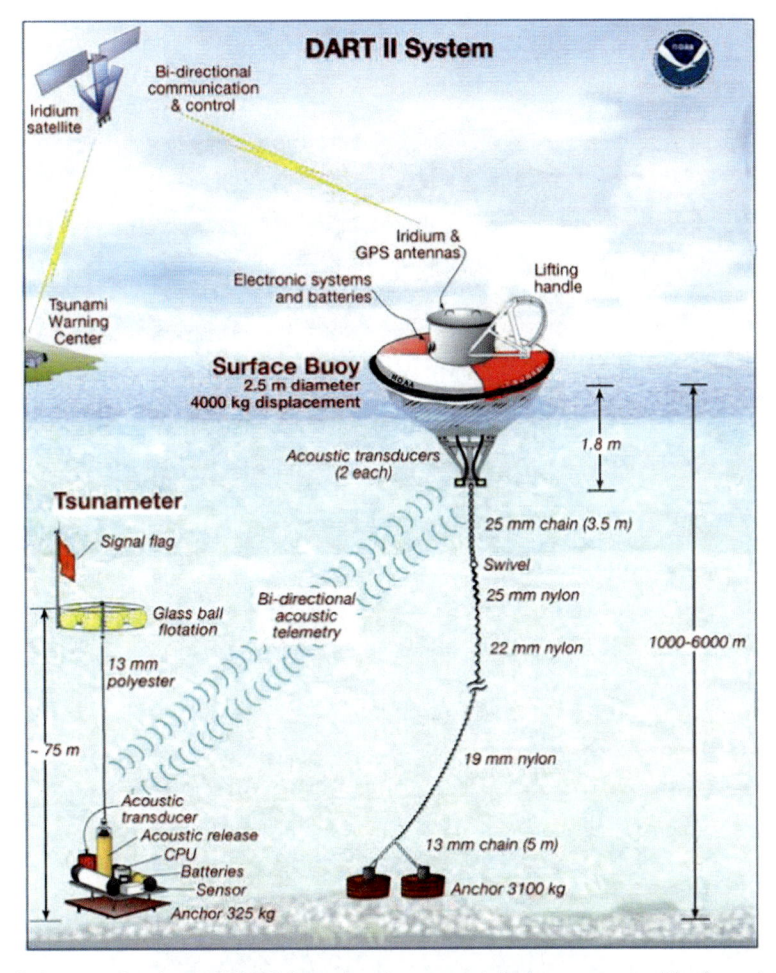

Plate 9 Layout of station (DART II) for registering tsunamis in the open ocean. The figure is taken from http://www.ndbc.noaa.gov/dart.shtml (see also Fig. 6.5 on page 239)

Plate 10 City Banda Aceh (North of island Sumatra) destroyed by the tsunami of December 26, 2004. The photograph was taken at a distance of 3 km from the shore, in the very beginning of the zone, where material and fragments of buildings displaced from the shore were deposited (Photo by T. K. Pinegina) (see also Fig. 6.7 on page 241)

Colour Plate Section

Plate 11 Segment of shore 15 km south-west of the northern end of island Sumatra. The tsunami height amounted to 15–20 m, here, while the heights of individual run-ups were up to 35 m above the sea level. In this region the tsunami exerted a strong erosive and abrasive influence on the shore and the nearby hills. In the background of the photograph one can see an abrasion niche, formed after the tsunami. The distance from the shoreline is 300 m (Photo by T. K. Pinegina) (see also Fig. 6.8 on page 242)

Plate 12 Photograph of wall of bore with strata of tephra and tsunamigenic horizonts in the soil. The bore is located at the 20th terrace 500 m from the shoreline of today (peninsula of the Kamchatka cape, Kamchatka) (see also Fig. 6.10 on page 245)

Colour Plate Section

Plate 13 Example of coseismic deformation on the island Simeulue, village of Busung, due to repetition of earthquake on March 28, 2005 at the coast of Indonesia (Photo by V. Kaistrenko) (see also Fig. 6.11 on page 246)

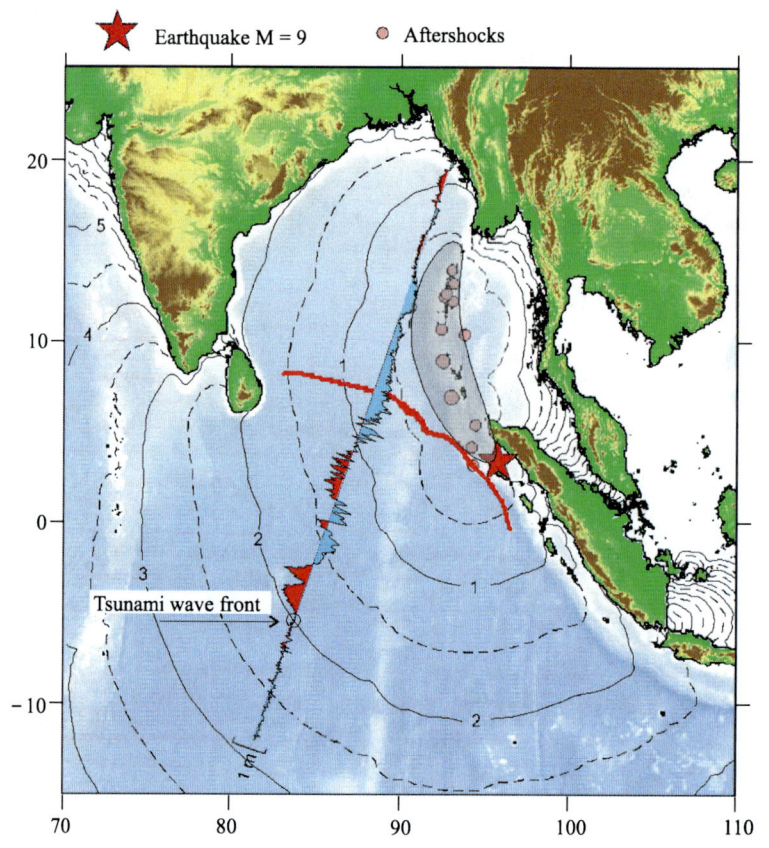

Plate 14 Map of north-eastern part of the Indian Ocean with isochrones, showing the calculated front position of the tsunami of December 24, 2004 (with an interval of 0.5 an hour). 1—epicentre of the main earthquake, 2—epicentres of main aftershocks. Route of JASON-1 satellite (circuit 109-129). The profile of the ocean level determined from altimetry data is shown along the route (Courtesy of E.A. Kulikov) (see also Fig. 6.14 on page 252)

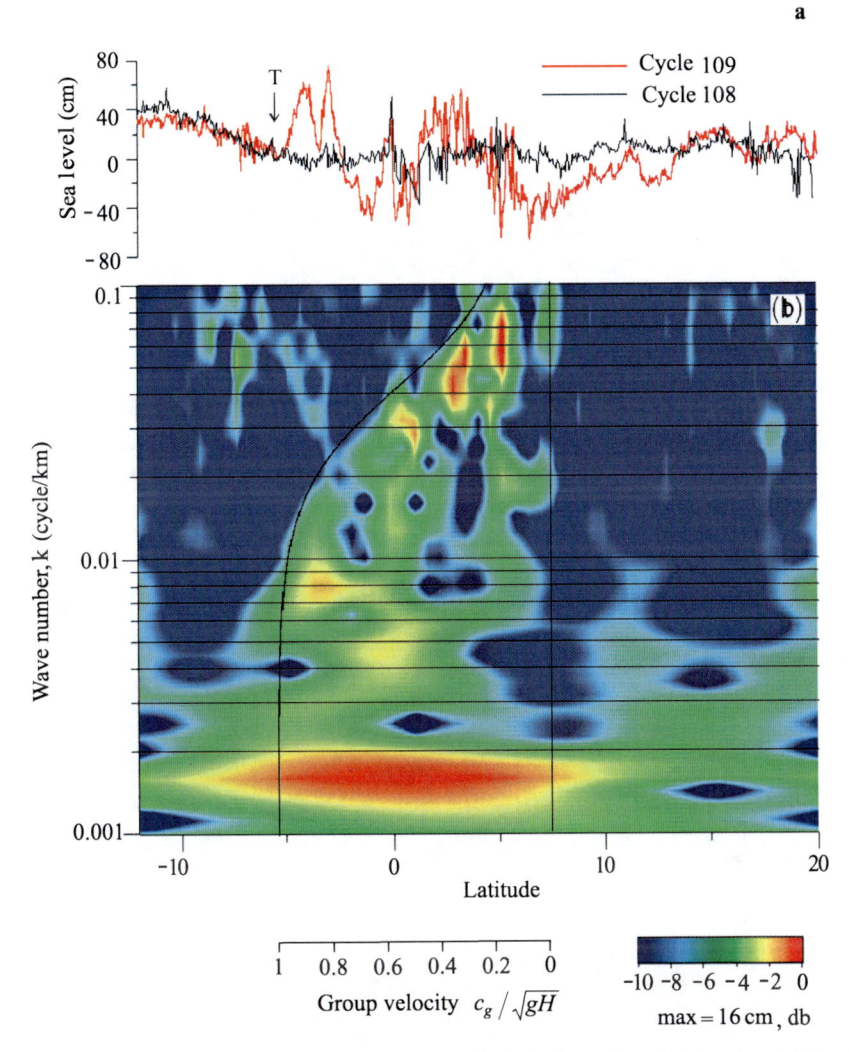

Plate 15 (a) Altimetry sea-level taken along track 129 of the Jason-1 satellite for Cycle 109 and along the same track 10 days earlier for Cycle 108; and (b) wavelet analysis of the sea level profile in (a). The theoretical curve, calculated in accordance with the linear dispersion law for surface gravitational waves, shows the calculated 'onset' moments of the respective spectral components. Letter "T" indicates the tsunami wavefront (Courtesy of E.A. Kulikov) (see also Fig. 6.15 on page 253)

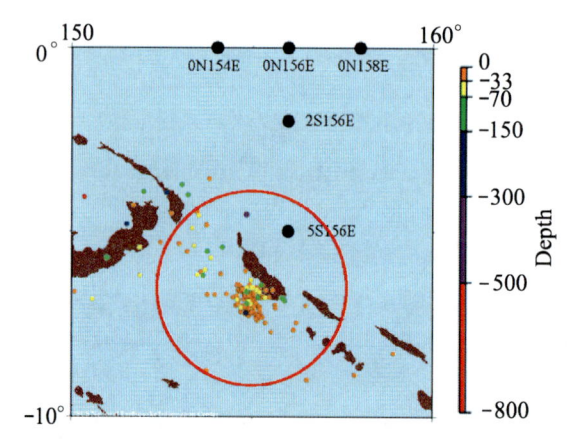

Plate 16 Earthquake epicenters in the vicinity of Bougainville Island for the period of 20.04.1996–31.05.1996. The center of the red circle of radius 300 km is at the point 6.528 S, 155.00 E (epicenter of earthquake of maximum amplitude). The black circles indicate the location of TAO Array stations (see also Fig. 7.1 on page 279)

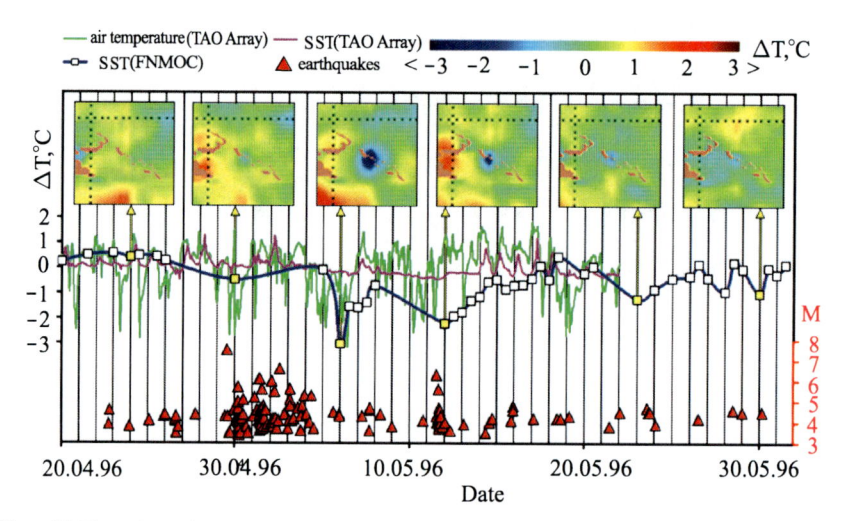

Plate 17 Time dependence of sea surface temperature (SST) and of temperature of the near-the-water atmospheric layer; time and magnitude of seismic events (red triangles) that occurred within a radius of 300 km from the epicentre of earthquakes of maximum magnitude. The blue line corresponds to the SST at the point 7 S, 156 E, which was restored from maps of SST anomalies (FNMOC), the green and purple curves are data from the TAO Array buoy station with coordinates 5 S, 156 E. Fragments of maps of SST anomalies and the coloured temperature scale in degree Celsius. The dates indicated along the timescale correspond to 00 h 00 min Greenwich time (see also Fig. 7.2 on page 280)

Index

A
Abrasion, 241, 242
Acoustic-gravity waves, 103, 107
Amplitude dispersion, 201
Analytical solution, 44, 49, 77, 116
Anemobaric waves, 171
Asteroid, 183–185, 188
Atmospheric pressure, 171–183

B
Bathymetry, 208, 218
Biogenes, 260
Bottom displacement, 41, 42, 70, 76, 77, 82, 84, 87, 90, 92, 94, 100, 109–111, 116, 127, 131, 132, 144–146, 149
Bottom friction, 210, 214, 215, 224, 226
Bottom oscillations, 295–305
Bottom pressure, 124–126, 129–131, 233, 237, 238, 240, 252
Boundary conditions, 217, 218, 225
Breaking waves, 226, 228
Burger's vector, 44, 45, 48

C
Caldera collapse, 167, 168
Cartesian reference system, 44, 57, 103, 157, 197, 214
Cavitation, 24
Coriolis force, 41, 214, 224
Cosmogenic tsunami, 183–192

D
Damage, 5, 8, 10–12, 14, 15, 21
DART. *See* Deep-ocean Assessment and Reporting of Tsunami
Deep-ocean Assessment and Reporting of Tsunami, 236, 238–240

Dip angle, 45, 48
Dip-slip fault, 45
Directional diagrams, 87–89, 93, 94
Dispersion, 39, 44, 66, 70, 73, 93, 170, 171, 188, 189, 191, 249, 251–253
Dissipation, 163, 188, 197, 202, 208, 209, 211, 212, 283
Dissipative structures, 298, 299
Duration of earthquake, 34, 38, 39
Dynamic modes, 298, 299

E
Earthquake, 32–39, 43, 44, 48–50, 71, 153–156, 159, 160, 166, 167, 172, 185
Earthquake energy, 290, 292, 293
Elastic oscillations, 101–103, 112, 117, 120, 122–132, 144, 145, 148
Environment pollution, 5
Epidemics, 5
Equivalent source, 170
Erosion, 5, 14, 155, 164, 241, 242
Eruption, 153, 155, 165–171
Euler's equations, 133
Experimental set-up, 297, 303, 304
Explosive, 165–167, 170

F
Faraday ripples, 258, 298, 305
Fault, 32, 34, 36, 38, 39, 44, 45, 48, 49, 71, 92, 94
Fires, 5, 14, 15
Fourier transformation, 52, 104
Freak waves, 25
Frequency spectra, 129, 144
Froude number, 165

325

G
Gravitational surface wave, 154
Green's law, 203

H
Historical testimony, 259, 274–278
Historical tsunami database, 5, 17
Human casualties, 4, 9
Hydroacoustic signals, 22–24

I
Impact of waves, 5
Increment, 296
Initial elevation, 34, 40, 42, 43, 94, 215–217
Instability, 275, 294
Internal waves, 41–43, 179, 260, 278, 285, 292
Inundation, 243, 245

J
JAMSTEC. *See* Japan Agency for Marine-Earth Science and Technology
Japan Agency for Marine-Earth Science and Technology, 124, 130, 236, 240

K
Kinetic energy, 35, 42, 74, 185–187

L
Laboratory experiments, 228
Lame constants, 47
Landslide, 153–165, 167
Laplace transformation, 77, 106
Linear potential theory, 62, 69, 72, 73, 136
Local tsunami, 5, 23, 24
Long waves, 2–4, 165, 168, 169, 171–173, 175, 177, 179, 180, 189, 191
Long-wave theory, 167, 175, 202, 215, 225

M
Magnitude, 32–34, 36, 38, 39, 50
Manning coefficient, 214
Mareograph, 233–237
Mathieu equation, 294, 295
Meteorite, 153, 176, 183, 185, 186, 188, 190, 192
Meteotsunami, 171–183
Mixing, 275, 276, 288
Mud flow, 154, 156

N
Non-linear currents, 275
Non-linear effects, 100, 132, 133, 142, 143, 147
Non-linear long-wave theory, 225

Non-linear tsunami source, 134, 136, 144, 146
Normal modes, 122
Numerical simulation, 102, 116, 125, 159, 161, 187, 188
Numerical simulations, 206, 213, 218–220, 223, 224, 228
Numerical tsunami models, 213, 219

P
Paleotsunami, 234, 241–246
Parametric resonance, 275, 294–297
Parametrization, 187
Period of tsunami, 35, 37, 39, 65
Phase dispersion, 198, 199, 202, 213, 226
Potential energy, 35, 40, 42, 43, 64, 74, 87, 94
Proudman resonance, 177, 180, 182
Pyroclastic flow, 165, 167

Q
Quarter-wave resonator, 108

R
Rake(slip) angle, 48
Reflection, 202–204, 208, 217, 225
Refraction, 203, 207, 208
Regional tsunami, 5
Residual deformation, 34, 36, 37, 42, 44, 70, 73, 77, 86, 91, 94
Resonance, 163, 165, 172, 173, 177, 179, 180, 182, 183
Ridge, 206–208, 221–223
River tsunami, 154
Run-up, 4, 6–10, 15–17, 19
Run-up height, 220, 222, 225, 227, 228
Run-up heights, 243, 244

S
Satellite altimetry, 234, 247–253
Satellite communications, 236, 238, 240
Scattering, 212, 213
Sea colour, 260, 261
Sea level, 233–240, 242–244, 247
Seaquake, 2, 6, 20–22, 257–259, 263, 264, 266, 269–271, 273, 278, 282, 293, 295, 296, 301, 303
Sedimentary layer, 155, 159, 160, 164
Seismic moment, 32, 36, 38, 48
Shadow method, 304
Slip distribution, 48–50
Slump, 153, 159
Spherical coordinates, 215
SST anomaly, 260, 275, 277, 279, 281
Stationary-phase method, 170
Storm surges, 173, 174

Stratification, 259, 260, 275, 278, 283–293, 297–305
Strike angle, 45, 48
Strike-slip fault, 45
Sudden inundation, 5
Surface gravitational waves, 2

T
Telemetric complex, 235, 236
Teletsunami, 5
Temperature profile, 278, 281, 282, 284–286, 288–290, 301, 302
Tensile fault, 44, 46
Tension of friction, 171, 174
Tide, 234, 238, 239, 243, 247, 249
Time scale, 39
Tokachi-Oki 2003, 124–131
Topographic profile, 243
Topography, 208, 213, 218, 222, 229
T-phase, 22, 23, 100, 112, 125
Tsunami, 1–20, 23, 24
Tsunami catalogue, 5
Tsunami deposit, 241–246
Tsunami earthquake, 33
Tsunami generation, 33–36, 39–41, 43, 56, 61, 68, 70, 71, 80, 82, 84, 100, 102, 124, 125, 132–149, 153–169, 172, 174, 185
Tsunami intensity, 10–15, 18, 19
Tsunami magnitude, 10–15
Tsunami manifestation, 234, 243
Tsunami propagation, 197–215, 220, 221, 223

Tsunami registration, 233, 235, 237
Tsunami run-up, 220, 223–229, 244
Tsunami sensor, 240
Tsunami source, 32, 34–42, 59, 67, 73, 84, 91
Tsunami warning, 15–18, 23
Turbulence, 259, 274, 275, 283, 286, 288, 289, 292, 301, 305
Turbulence energy, 283–291
Turbulent exchange coefficient, 283, 301–303, 305

U
Underwater earthquake, 260, 274, 275, 277–279, 281, 283–293
Underwater volcano, 165–170
Upwelling, 260

V
Velocity potential, 51, 78, 79
Vertical exchange, 259, 260, 274–278, 281, 291, 297, 301, 305
Viscous fluid, 156, 157, 163
Volcanic eruption, 153, 155, 165–171
Volcano, 166–171
Vortex, 42

W
Water compressibility, 102, 122, 124, 131
Wave front, 243, 248–251, 253
Waveguide, 112, 122, 123, 222, 223
Weather phenomena, 260, 276

Printed by Publishers' Graphics LLC
MO20130129.19.19.4